Nutrition and Feeding of Organic Poultry

2nd Edition

Nutrition and Feeding of Organic Poultry

2nd Edition

By

Robert Blair

University of British Columbia, Canada

CABI is a trading name of CAB International

CABI
Nosworthy Way
Wallingford
Oxfordshire OX10 8DE
UK

CABI
745 Atlantic Avenue
8th Floor
Boston, MA 02111
USA

Tel: +44 (0)1491 832111
Fax: +44 (0)1491 833508
E-mail: info@cabi.org
Website: www.cabi.org

Tel: +1 (617)682-9015
E-mail: cabi-nao@cabi.org

A catalogue record for this book is available from the British Library,
London, UK.

Library of Congress Cataloging-in-Publication Data

Names: Blair, Robert, 1933- author.
Title: Nutrition and feeding of organic poultry / by Robert Blair.
Description: 2nd edition. | Boston, MA : CABI, [2018] | Includes
    bibliographical references and index.
Identifiers: LCCN 2018014423 | ISBN 9781786392985 (hardback) | ISBN
    9781786392992 (epdf)
Subjects: LCSH: Poultry--Feeding and feeds. | Organic farming. |
    Feeds--Composition.
Classification: LCC SF494 .B57 2018 | DDC 636.5/0852--dc23
LC record available at https://lccn.loc.gov/2018014423

ISBN-13: 9781786392985

Commissioning editor: Alexandra Lainsbury
Editorial assistant: Tabitha Jay
Production editor: Ali Thompson

Typeset by SPi, Pondicherry, India
Printed and bound in the UK by Bell & Bain Ltd, Glasgow

Contents

Acknowledgements   vii

1. Introduction and Background   1
2. Aims and Principles of Organic Poultry Production   3
3. Elements of Poultry Nutrition   14
4. Approved Ingredients for Organic Diets   50
5. Diets for Organic Poultry Production   174
6. Choosing the Right Breed and Strain   201
7. Integrating Feeding Programmes into Organic Production Systems   219
8. Conclusions and Recommendations for the Future   252

Index   257

This book is enhanced with supplementary resources.
To access the computerized system of on-farm feed formulation please visit:
www.cabi.org/openresources/92985

Acknowledgements

_____

I wish to acknowledge the help and advice received from Alexandra Lainsbury and Sarah Hulbert (formerly) of CAB International in the production of my trilogy of books on the Nutrition and Feeding of Organic Pigs, Poultry and Cattle. Any success that the books have achieved in translating and summarizing the scientific and practical findings on organic production into readable texts are due in part to their efforts.

1

Introduction and Background

In recent years there has been a rapid increase in organic livestock production in many countries. This development is a response to an increased consumer demand for food that is perceived to be fresh, wholesome and flavoursome, free of hormones, antibiotics and harmful chemicals, and without the use of genetically modified (GM) crops. Consumer research indicates that ethical concerns related to standards of animal welfare also play a significant role in the decision to purchase organic food. In addition there is evidence that animal welfare is used by consumers as an indicator of other product attributes, such as safety and impact on human health.

European data show that organic eggs represent 10–20% of total egg sales and there is a willingness of consumers to pay a relatively high price premium for these eggs. Another development showing a change in consumer behaviour is that many supermarkets in North America now sell organic products.

Organic feed is generally more expensive than conventional feed, often resulting in eggs and meat being twice as costly as the conventional products. Therefore while there is an increasing market for organic eggs and meat, they will have to be supplied at a price acceptable to the consumer. This will be a particular challenge for northern regions that have harsher climates and a lower supply of organic feedstuffs than southern, more productive, regions.

A major challenge facing the organic poultry industry at present is a global shortage of organic feedstuffs, exacerbated by the objective in Europe of requiring the feed to be 100% organic by 31 December 2017 and a 110-fold increase in the global production of GM crops since 1996 (ISAAA, 2017). Due to the shortage, this objective could not be achieved, resulting in the EU Commission taking the decision to prolong the feed derogation for organic pigs and poultry that had been due to expire at the end of 2017 (see Chapter 2). At present most countries consider the feed to be organic with a maximum 5–10% of the ingredients being non-organic.

This volume sets out guidance for producers on nutrition and feeding practices that relate to the standards for certification of organic poultry. Details on permitted feed ingredients, with an emphasis on those grown or available locally and on suitable dietary formulations, are included. Although aspects of these topics have been presented at conferences and in trade and scientific publications, no comprehensive text has been published to date.

It is clear that the idealism set out initially in the principles of organic agriculture has had to be tempered by practical considerations. The standards adopted have to aim for a balance between the desire of consumers for organic products and considerations of ethical and ecological integrity, and the practical and financial needs of producers. As a result, synthetic vitamins and pure forms of minerals are allowed in organic poultry feeds, with some restrictions. Some jurisdictions permit the use of certain pure forms of amino acids as feed supplements; therefore this volume will assist producers in formulating diets without and with supplemental amino acids.

The standards and rules laid down to accomplish organic production place several restrictions on diet and feeding. These are detailed in Chapter 2. A main aim of this book is to present advice on how the appropriate diets can be formulated and how feeding programmes can be integrated into an organic production system.

In general, the feed for use in organic poultry production must contain ingredients from three categories only:

1. Agricultural products that have been produced and handled organically, preferably from the farm itself.
2. Non-synthetic substances such as enzymes, probiotics and others considered to be natural ingredients.

3. Synthetic substances that have been approved for use in organic poultry production.

In addition, the diet is intended to ensure quality production of the birds rather than maximizing production, while meeting the nutritional requirements of the stock at various stages of their development. This requirement is extended in some jurisdictions to require that poultry be allowed access to pasture, a requirement based mainly on welfare rather than nutritional considerations since herbage and soil invertebrates do not constitute an important source of nutrients for poultry.

Although the main aim of this volume is to assist nutritionists and organic producers in formulating diets and feeding programmes for organic poultry, the regulatory authorities in several countries may find it of value to address nutritional issues relevant to future revisions of the regulations. It seems clear that the current standards and regulations have been developed mainly by those experienced in crop production and in ecological issues, and that a review of the organic regulations from an animal nutrition perspective would be useful.

Reference

ISAAA (2017) *Global Status of Commercialized Biotech/GM Crops: 2016* (updated May 2017). Brief No. 52. International Service for the Acquisition of Agri-biotech Applications, Ithaca, New York.

2

Aims and Principles of Organic Poultry Production

According to the Codex Alimentarius Commission and the Joint Food and Agriculture Organization of the United Nations (FAO)/World Health Organization (WHO) Food Standards Programme, organic agriculture is:

> 'a holistic production management system which promotes and enhances agroecosystem health, including biodiversity, biological cycles, and soil biological activity ... emphasizes the use of management practices in preference to the use of off-farm inputs as opposed to using synthetic materials. The primary goal is to optimize the health and productivity of interdependent communities of soil life, plants, animals and people ... the systems are based on specific and precise standards of production which aim at achieving optimal agroecosystems which are socially, ecologically and economically sustainable' (Codex Alimentarius Commission, 1999).

Thus organic poultry production differs from conventional production and in many ways is close to the agriculture of Asia. It aims to fully integrate animal and crop production and develop a symbiotic relationship of recyclable and renewable resources within the farm system. Livestock production then becomes one component of a wider, more inclusive organic production system. Organic poultry producers must take into consideration several factors other than the production of livestock. These factors include: (i) the use of organic feedstuffs (including limited use of feed additives); (ii) use of outdoor-based systems; (iii) restrictions on numbers of bought-in stock; (iv) group-housing of breeding stock; and (v) minimizing environmental impact. Organic poultry production also requires certification and verification of the production system. This requires that the organic producer must maintain records sufficient to preserve the identity of all organically managed animals, all inputs and all edible and non-edible organic livestock products produced. The result is that organic food has a very strong brand image in the eye of the consumer and thus should command a higher price in the marketplace than conventionally produced food.

The whole organic process involves four stages: (i) application of organic principles (standards and regulations); (ii) adherence to local organic regulations; (iii) certification by local organic regulators; and (iv) verification by local certifying agencies.

Restrictions on the use of ingredients in organic diets include the following:

- No genetically modified (GM) grain or grain by-products.
- No antibiotics, hormones or drugs. Enzymes are prohibited as feed ingredients used to increase feed conversion efficiency (they may be used under derogation

where necessary for the health and welfare of the animal).

- No animal by-products, except that milk products and some fishmeals are permitted.
- No grain by-products unless produced from certified organic crops.
- No chemically extracted feeds (such as solvent-extracted soybean meal).
- No pure amino acids (AA), either synthetic or from fermentation sources (there are some exceptions to this provision).

Organic Standards

The standards of organic farming are based on the principles of enhancement and utilization of the natural biological cycles in soils, crops and livestock. According to these regulations organic livestock production must maintain or improve the natural resources of the farm system, including soil and water quality. Producers must keep livestock and manage animal waste in such a way that supports instinctive, natural living conditions of the animal, yet does not contribute to contamination of soil or water with excessive nutrients, heavy metals or pathogenic organisms, and optimizes nutrient recycling. Livestock living conditions must accommodate the health and natural behaviour of the animal, providing access to shade, shelter, exercise areas, fresh air and direct sunlight suitable to the animal's stage of production or environmental conditions, while complying with the other organic production regulations. The organic standards require that any livestock or edible livestock product to be sold as organic must be maintained under continuous organic management from birth to market. Feed, including pasture and forage, must be produced organically and health care treatments must fall within the range of accepted organic practices. Organic livestock health and performance are optimized by careful attention to the basic principles of livesoc khusbandry, such as selection of appropriate breeds, appropriate management practices and nutrition, and avoidance of overstocking.

Stress should be minimized at all times. Rather than being aimed at maximizing animal performance, dietary policy should be aimed at minimizing metabolic and physiological disorders; hence the requirement for some forage in their diet. Grazing management should be designed to minimize pasture contamination with parasite larvae. Housing conditions should be such that disease risk is minimized, i.e. ventilation should be adequate, stocking rate should not be excessive and adequate dry bedding should be available.

Nearly all synthetic animal drugs used to control parasites, prevent disease, promote growth or act as feed additives in amounts above those needed for adequate growth and health are prohibited in organic production. Dietary supplements containing animal by-products such as meat meal are also prohibited. No hormones can be used, a requirement which is easy to apply in poultry production since hormone addition to feed has never been practised commercially. When preventive practices and approved veterinary biologics are inadequate to prevent sickness, the producer must administer conventional medications. However, livestock that are treated with prohibited materials must be clearly identified and cannot be sold as organic.

International Standards

The aim of organic standards is to ensure that animals produced and sold as organic are raised and marketed according to defined principles. International standards and state regulations in conjunction with accreditation and certification are therefore very important as guarantees for the consumer.

Currently there is no universal standard for organic food production worldwide. As a result many countries have now established national standards for the production and feeding of organic poultry. They have been derived from those developed originally in Europe by the Standards Committee of the International Federation of Organic Agriculture Movements (IFOAM) and the guidelines for organically produced food developed within the framework of the Codex Alimentarius, a programme created in 1963 by FAO and WHO to develop food standards, guidelines and codes of practice under the Joint FAO/WHO Food Standards Programme. IFOAM Basic Standards were adopted in 1998. Within the Codex, the Organic Guidelines include Organic Livestock production.

The IFOAM standard (IFOAM, 1998) is intended as a worldwide guideline for accredited certifiers to fulfil. IFOAM works closely with certifying bodies around the world to ensure that they operate to the same standards. The main purpose of the Codex is to protect the health of consumers and ensure fair trade practices in the food trade, and also promote coordination of all food standards work undertaken by international governmental and non-governmental organizations (Codex Alimentarius Commission, 1999). The Codex is a worldwide guideline for states and other agencies to develop their own standards and regulations but it does not certify products directly. Thus the standards set out in the Codex and by IFOAM are quite general, outlining principles and criteria that have to be fulfilled. They are less detailed than the regulations dealing specifically with regions such as Europe.

The sections of the Codex regulations relevant to the coverage of this book include the following:

1. The choice of breeds or strains should favour stock that is well adapted to the local conditions and to the husbandry system intended. Vitality and disease resistance are particularly mentioned, and preference should be given to indigenous species.
2. The need for cereals in the finishing phase of meat poultry.
3. The need for roughage, fresh or dried fodder or silage in the daily ration of poultry.
4. Poultry must be reared in open-range conditions and have free access to an open-air run whenever the weather conditions permit. The keeping of poultry in cages is not permitted.
5. Waterfowl must have access to a stream, pond or lake whenever the weather conditions permit.

6. In the case of laying hens, when natural day length is prolonged by artificial light, the competent authority shall prescribe maximum hours respective to species, geographical considerations and general health of the animals.
7. For health reasons buildings should be emptied between each batch of poultry reared and runs left empty to allow the vegetation to grow back.

The general criteria regarding permitted feedstuffs are:

1. Substances that are permitted according to national legislation on animal feeding.
2. Substances that are necessary or essential to maintain animal health, animal welfare and vitality.
3. Substances that contribute to an appropriate diet fulfilling the physiological and behavioural needs of the species concerned; and do not contain genetically engineered/modified organisms and products thereof; and are primarily of plant, mineral or animal origin.

The specific criteria for feedstuffs and nutritional elements state:

1. Feedstuffs of plant origin from non-organic sources can only be used under specified conditions and if they are produced or prepared without the use of chemical solvents or chemical treatment.
2. Feedstuffs of mineral origin, trace elements, vitamins or provitamins can only be used if they are of natural origin. In case of a shortage of these substances, or in exceptional circumstances, chemically well-defined analogical substances may be used.
3. Feedstuffs of animal origin, with the exception of milk and milk products, fish, other marine animals and products derived therefrom, should generally not be used, or as provided by national legislation.
4. Synthetic nitrogen or non-protein nitrogen compounds shall not be used.

Specific criteria for additives and processing aids state:

1. Binders, anti-caking agents, emulsifiers, stabilizers, thickeners, surfactants, coagulants: only natural sources are allowed.

2. Antioxidants: only natural sources are allowed.

3. Preservatives: only natural acids are allowed.

4. Colouring agents (including pigments), flavours and appetite stimulants: only natural sources are allowed.

5. Probiotics, enzymes and microorganisms are allowed.

Although there is no internationally accepted regulation on organic standards, the World Trade Organization and the global trading community are increasingly relying on the Codex and the International Organization of Standardization (ISO) to provide the basis for international organic production standards, as well as certification and accreditation of production systems. Such harmonization will promote world trade in organic produce. The ISO, which was established in 1947, is a worldwide federation of national standards for nearly 130 countries. The most important guide for organic certification is ISO Guide 65:1996, General Requirements for Bodies Operating Product Certification Systems, which establishes basic operating principles for certification bodies. The IFOAM Basic Standards and Criteria are registered with the ISO as international standards.

The International Task Force on Harmonization and Equivalency in Organic Agriculture documented the world situation in 2003 (UNCTAD, 2004), listing 37 countries with fully implemented regulations for organic agriculture and processing. Further developments took place in 2006 when Canada and Paraguay passed organic legislation and other countries elaborated drafts or revised existing legislation (Kilcher et al., 2006). No recent update on the harmonization situation globally appears to be available.

The following sections give a brief description of the legislation in several countries and regions.

Europe

Legislation to govern the production and marketing of food as organic within the European Union (EU) was introduced for plant products in 1993 (Regulation (EEC) No. 2092/91). This Regulation defined organic farming, set out the minimum standards of production and defined how certification procedures must operate. Regulation (EEC) No. 2092/91 was supplemented by various amendments and in 2000 by further legislation (Council Regulation (EC) No. 1804/1999) covering livestock production. In addition to organic production and processing within the EU, the Regulation also covered certification of produce imported from outside the EU.

Regulation (EC) No. 1804/1999 (EC, 1999) allowed the range of products for livestock production to be extended and it harmonized the rules of production, labelling and inspection. It reiterated the principle that livestock must be fed on grass, fodder and feedstuffs produced in accordance with the rules of organic farming. The regulation set out a detailed listing of approved feedstuffs. However, it recognized that under the prevailing circumstances, organic producers might experience difficulty in obtaining sufficient quantities of feedstuffs for organically reared livestock. Accordingly it allowed for authorization to be granted provisionally for the use of limited quantities of non-organically produced feedstuffs where necessary. For poultry the regulations allowed for up to 15% of annual dry matter (DM) from conventional sources until 31 December 2007, 10% from 1 January 2008 until 31 December 2009, and 5% from 1 January 2010 until 31 December 2011. However, the regulations specified that 100% organic diets for poultry would become compulsory in the EU from 1 January 2018, emphasizing the need for the development of sustainable feeding systems based entirely on organic feeds by that time. As noted in Chapter 1, this objective could not be achieved due to the shortage of organic feedstuffs, resulting in the EU Commission taking the decision to prolong the feed derogation for organic pigs and poultry. The revised date for implementation of the requirement that organic poultry and pig feeds consist of 100% organic feedstuffs is now expected to be 2021.

In addition, an important provision of the EU Regulation was to permit the use of trace minerals and vitamins as feed additives to avoid deficiency situations. The approved products are of natural origin or synthetic in the same form as natural products. Other products listed in Annex II, Part D, sections 1.3 (enzymes), 1.4 (microorganisms) and 1.6 (binders, anti-caking agents and coagulants) were also approved for feed use. Roughage, fresh or dried fodder, or silage must be added to the daily ration but the proportion is unspecified. Consideration was given later to the possible approval of pure AA as approved supplements for organic feeds, at the instigation of several Member States. However, approval was not given, on the grounds that the AA approved for commercial feed use were either synthetic or derived from fermentation processes involving GM organisms.

The EC Regulation 2092/91 was repealed and replaced with Regulation 834/2007 in June 2007 (EC, 2007). The regulation set out in more detail the aims and procedures relating to the production of organic livestock (including insects) as in Section 5:

Specific principles applicable to farming
In addition to the overall principles set out in Article 4, organic farming shall be based on the following specific principles:
(a) the maintenance and enhancement of soil life and natural soil fertility, soil stability and soil biodiversity preventing and combating soil compaction and soil erosion, and the nourishing of plants primarily through the soil ecosystem;
(b) the minimization of the use of non-renewable resources and off-farm inputs;
(c) the recycling of wastes and by-products of plant and animal origin as input in plant and livestock production;
(d) taking account of the local or regional ecological balance when taking production decisions;
(e) the maintenance of animal health by encouraging the natural immunological defence of the animal, as well as the selection of appropriate breeds and husbandry practices;
(f) the maintenance of plant health by preventative measures, such as the choice of appropriate species and varieties resistant to pests and diseases, appropriate crop rotations, mechanical and physical methods and the protection of natural enemies of pests;

(g) the practice of site-adapted and land-related livestock production;
(h) the observance of a high level of animal welfare respecting species-specific needs;
(i) the production of products of organic livestock from animals that have been raised on organic holdings since birth or hatching and throughout their life;
(j) the choice of breeds having regard to the capacity of animals to adapt to local conditions, their vitality and their resistance to disease or health problems;
(k) the feeding of livestock with organic feed composed of agricultural ingredients from organic farming and of natural non-agricultural substances;
(l) the application of animal husbandry practices, which enhance the immune system and strengthen the natural defence against diseases, in particular including regular exercise and access to open air areas and pastureland where appropriate;
(m) the exclusion of rearing artificially induced polyploid animals;
(n) the maintenance of the biodiversity of natural aquatic ecosystems, the continuing health of the aquatic environment and the quality of surrounding aquatic and terrestrial ecosystems in aquaculture production;
(o) the feeding of aquatic organisms with feed from sustainable exploitation of fisheries as defined in Article 3 of Council Regulation (EC) No 2371/2002 of 20 December 2002 on the conservation and sustainable exploitation of fisheries resources under the Common Fisheries Policy (13) or with organic feed composed of agricultural ingredients from organic farming and of natural non-agricultural substances.

Under the EU regulations, each member state is required to establish a National Competent Authority to ensure adherence to the law. Between the years 1992 and 1999 the various European governments took quite different approaches to how organic livestock production should be regulated and this difference persists to the present. In addition, within each European country the different certifying bodies also adopted different positions. The end result is a wide variety of standards on organic livestock across Europe. However, every certifying body in Europe must work to standards that at a minimum meet the EU organic legislation (a legal requirement).

North America

USA

The US Department of Agriculture (USDA) National Organic Program (NOP) was introduced in 2002 (NOP, 2000). This is a federal law that requires all organic food products to meet the same standards and be certified under the same certification process. All organic producers and handlers must be certified by accredited organic certification agencies unless exempt or excluded from certification. A major difference between the US and European standards is that organic standards in the USA have been harmonized under the NOP. States, non-profit organizations, for-profit certification groups and others are prohibited from developing alternative organic standards. All organic food products must be certified to the National Organic Standards (NOS). Organic producers must be certified by NOP-accredited certification agencies. All organic producers and handlers must implement an Organic Production and Handling System Plan that describes the practices and procedures that the operation utilizes to comply with the organic practice standards. Both state agencies and private organizations may be NOP-accredited. The NOS establishes the National List, which allows all non-synthetic (natural) materials unless specifically prohibited, and prohibits all synthetic materials unless specifically allowed. In other respects the standards for organic poultry production are similar to European standards.

Canada

Canada issued an official national standard for organic agriculture in 2006 (CGSB, 2006). It was based on a draft of a Canadian Standard for Organic Agriculture which was developed by the Canadian General Standards Board (CGSB, 1999) and recommendations from the Canada Organic Initiative Project (2006). The 1999 draft Standard

provided basic guidelines for organic farming groups and certifying agencies across Canada to develop their own standards. These standards are based on the same set of principles as those in Europe and the USA. The Canadian Food Inspection Agency (CFIA) began enforcing the standards in 2011. A Canadian Organic Office was established to allow the CFIA to provide an oversight to the process of certifying organic farms and products in Canada. The regulations also allow for certified products to carry the official Canada Organic logo on their labels.

Caribbean countries

IFOAM recently set up a regional initiative for Latin America and the Caribbean – El Grupo de America Latina y el Caribe de IFOAM (GALCI) – coordinated from an office in Argentina. Currently, GALCI represents 59 organizations from countries throughout Latin America and the Caribbean, including producers' associations, processors, traders and certification agencies. The purpose and objectives of GALCI include the development of organic agriculture throughout Latin America and the Caribbean.

Mexico

The Government of Mexico introduced a new programme of rules and requirements for organic agriculture certification in 2013, published in its Federal Register (Oficial Diario de la Federación) (GAIN, 2013; SENASICA, 2013). The guidelines are similar to those in the USDA NOP and are equivalent to other internationally accepted guidelines, no doubt to facilitate trade in organic products. One interesting aspect of the Mexican regulations is that they place limits on the stocking rate on land, to ensure that the output of nitrogen in excreta from organic animals does not exceed 500 kg/ha/year.

Latin America

Argentina

In 1992 Argentina was the first country in the Americas to establish standards for the certification of organic products equivalent to those of the EU and validated by IFOAM (GAIN, 2002). Argentinian organic products are admissible in the EU and the USA. Organic livestock and poultry production in Argentina is governed by the Servicio Nacional de Salud (SENASA), a government agency under the Ministry of Agriculture, through Resolution No. 1286/93 and also by the EU Resolution No. 45011. In 1999, the National Law on Organic Production (No. 25127) came into force with the approval of the Senate. This law prohibits marketing of organic products that have not been certified by a SENASA-approved certifying agency. Each organic certification agency must be registered with SENASA.

Brazil

In 1999, the Ministry of Agriculture, Livestock and Food Supply published the Normative Instruction No. 7 (NI7), establishing national standards for the production and handling of organically produced products, including a list of substances approved for and prohibited from use in organic production (GAIN, 2002). The NI7 defines organic standards for production, manufacturing, classification, distribution, packaging, labelling, importation, quality control and certification, of products of both animal and plant origin. The policy also establishes rules for companies wishing to be accredited as certifying agencies, which enforce the NI7 and certify production and operations under the direction of the Orgao Colegiado Nacional (National Council for Organic Production). According to the GAIN (2002) report, about half of the organic production in Brazil is exported, mainly to Europe, Japan and the USA, indicating that the Brazilian standards are compatible with those in the importing countries.

Chile

Chilean national standards came into effect in 1999 under the supervision of the Servicio Agrícola y Ganadero, which is the counterpart of the Plant Protection and Quarantine branch of the US Department of Agriculture. The standards are based on IFOAM standards.

Africa

Several countries in Africa have introduced organic regulations, to ensure the acceptability of products in export markets and to comply with local regulations. In general the regulations have been based on EU regulations relating to organic products.

IFOAM opened an Africa Organic Service Center in Dakar, Senegal, in 2005. A main aim of the Center is to bring together all the different aspects and key people involved in organic agriculture in Africa into a coherent and unified continent-wide movement. Another objective is the inclusion of organic agriculture in national agricultural and poverty reduction strategies.

A major area of organic production is East Africa, which currently leads the continent in production and exports of certified organic products. Cooperation between the Kenya Organic Agriculture Network (KOAN), the Tanzanian Organic Agriculture Movement (TOAM) and the National Organic Agricultural Movement of Uganda (NOGAMU) led to the development in 2007 of the East African organic products standard (EAOPS) (EAS 456:2007).

South Africa and several other countries have introduced national standards for organic agriculture, based on IFOAM recommendations, EU regulations and Codex Alimentarius guidelines.

In keeping with the regulations developed for other countries, such as Mexico, which have climates that allow year-round access of livestock to range land, the organic regulations in Africa generally place limits on the amount of nitrogen that is allowed to be excreted onto the land (e.g. 170 kg N/ha/year).

Australasia

Australia

The Australian National Standard for Organic and Bio-Dynamic Produce (bio-dynamic: an agricultural system that introduces specific additional requirements to an organic system) was first implemented in 1992 as the Australian Export Standard for products labelled organic or bio-dynamic. It was amended in 2005 (edition 3.1). The Standard is issued by the Organic Industry Export Consultative Committee of the Australian Quarantine and Inspection Service and is reviewed periodically, the latest revision (edition 4.1) taking place in 2016 (Australian Organic, 2017). The Standard provides a nationally agreed framework for the organic industry covering production, processing, transportation, labelling and importation. Certifying organizations that have been accredited by the Australian competent authority apply the Standard as a minimum requirement to all products produced by operators certified under the inspection system. This Standard therefore forms the basis of equivalency agreements between approved certifying organizations and importing country requirements. Individual certifying organizations may stipulate additional requirements to those detailed in the Standard.

The Standard states that a developed organic or bio-dynamic farm must operate within a closed input system to the maximum extent possible. External farming inputs must be kept to a minimum and applied only on an 'as needs' basis. The Standard is therefore somewhat more restrictive in terms of the ability of the organic poultry farmer in Australia to improve genotypes. The Standard requires that 'all poultry production shall take place in a pastured range situation, defined as birds being produced under natural conditions, allowing for natural behaviour and social interaction and having access to open range or appropriately fenced and managed area'.

The Standard appears to be similar to European standards in relation to permitted feed ingredients, with feed supplements of agricultural origin having to be of certified organic or bio-dynamic origin. However, a derogation allows that, if this requirement cannot be met, the certifying organization may approve the use of a product that does not comply with the Standard provided that it is free from prohibited substances or contaminants and that

it constitutes no more than 5% of the animals' diet on an annual basis. Permitted feed supplements of non-agricultural origin include minerals, trace elements, vitamins or provitamins only if from natural sources. Treatment of animals for trace mineral and vitamin deficiencies is subject to the same provision of natural origin. AA isolates (pure AA) are not permitted in organic diets.

New Zealand

Revised regulations on organic farming were issued by the New Zealand Food Safety Authority, Ministry of Agriculture and Forestry (NZFSA, 2011). The regulations had previously been issued in draft form in 2000 as an extract from the relevant EU regulation and were subsequently amended to incorporate the US NOS requirements. The regulations set out the minimum requirements for organic production and operators are allowed to adopt higher standards.

The regulations show similarities to European and North American standards; however, some aspects are included. In addressing the issue of climate, the regulations (akin to those in Quebec in the northern hemisphere) allow that the final finishing-poultry production for meat may take place indoors, provided that this indoors period does not exceed one-fifth of the lifetime of the animal. Stocking rates are specified where the spreading of manure from housing on to pasture is undertaken. A detailed list of permitted feed ingredients is included in the regulations: minerals and trace elements used in animal feeding having to be of natural origin or, failing that, synthetic in the same form as natural products. Synthetic vitamins identical to natural vitamins are allowed.

Asia

China

The regulations governing organic animal and poultry production in China are set out in the AgriFood MRL Standard and are summarized below (Pixian Wang, personal communication). The Standard resembles in part the IFOAM standards but contains some unique features, including the following:

8.2 Introduction of Animals and Poultry

8.2.1 When organic animals cannot be introduced, conventional animals can be introduced provided they have been weaned and introduced within 6 weeks of birth.

8.2.2 The number of conventional animals introduced annually is no more than 10% of OFDC (Organic Foods and Development Certification Center) approved adult animals of the same kind. Under certain circumstances, the certifying committee will allow the number of conventional animals introduced annually to be more than 10% but not more than 40%. Introduced animals must go through the corresponding conversion period.

8.2.3 Male breeding animals can be introduced from any source, but can only be raised following approved organic procedures.

8.2.4 All introduced animals must not be contaminated by products of genetic-engineering products, including breeding products, pharmaceuticals, metabolism-regulating agents and biological agents, feeds or additives.

8.3 Feeds

8.3.1 Animals must be raised with organic feed and forage which has been approved by the national organic agency (OFDC) or by an OFDC-certified agency. Of the organic feed and forage, at least 50% must originate from the individual farm or an adjacent farm.

8.3.4 The certification committee allows the farm to purchase regular feed and forage during a shortage of organic feed. However, the regular feed and forage cannot exceed 15% for non-ruminants on a DM basis. Daily maximum intake of conventional feed intake cannot exceed 25% of the total daily feed intake on a DM basis. Exemptions due to severe weather and disasters are permitted. Detailed feed records must be kept and the conventional feed must be OFDC-approved.

8.3.6 The number of animals cannot exceed the stock capacity of the farm.

8.4 Feed Additives

8.4.1 Products listed in Appendix D are allowed to be used as additives.

8.4.2 Natural mineral or trace mineral ores such as magnesium oxide and green sand are allowed. When natural mineral or trace mineral sources cannot be provided, synthesized mineral products can be used if they are approved by OFDC.

8.4.3 Supplemental vitamins shall originate from geminated grains, fish liver oil, or brewing yeast. When natural vitamin sources cannot be provided, synthesized vitamin products can be used if they are approved by OFDC.

8.4.4 Chemicals approved by OFDC in Appendix D are allowed to be used as additives.

8.4.5 Prohibited ingredients include synthesized trace elements and pure AA.

8.5 Complete Feed

8.5.1.1 All the major ingredients in the complete feed must be approved by OFDC or an agency certified by OFDC. The ingredients plus additive minerals and vitamins cannot be less than 95% of the complete feed.

8.5.1.2 Additive minerals and vitamins can be derived from natural or synthesized products, but the complete feed cannot contain prohibited additives or preservatives.

8.5.2 The complete feed must meet the requirements of animals (or poultry) for nutrients and feeding goals. This can be confirmed by either of the following:

- All chemical compositions meet the related national regulations or the related authority regulations.
- Except for water, all other nutrients in the complete feed can meet the requirements of the animals during a different stage (i.e. growth, production or reproduction) if the complete feed is the sole nutrient source. This can be tested by the related national agency using approved procedures.

8.6 Feeding Conditions

8.6.1 The feeding environment (pen, stall) must meet the animal's physiological and behaviour requirements, in terms of space, shelter, bedding, fresh air and natural light.

8.6.2 Where necessary, artificial lighting can be provided to extend the lighting period but cannot exceed 16 hours per day.

8.6.3 All animals must be raised outdoors during at least part of the year.

8.6.4 It is prohibited to feed animals in such a way that they do not have access to soil, or that their natural behaviour or activity is limited or inhibited.

8.6.5 The animals cannot be fed individually, except adult males, sick animals or sows at late gestation stage.

India

The Government of India implemented a National Programme for Organic Production (NPOP) in 2001, the standards for production and accreditation being recognized by Europe and North America as compatible with the IFOAM standards. India is now an important exporter of organic oil seeds and cereal grains.

Japan

The established Japanese Agricultural Standards (JAS) (MAFF, 2001) for organic agricultural production are based on the Codex guidelines for organic agriculture. The Ministry of Agriculture, Forestry and Fisheries issued JAS for organic animal products in 2005 (MAFF, 2005). Since 2001 the JAS have required that organic products sold in Japan conform to the JAS organic labelling standard. Several countries have organic regulations that comply with the JAS guidelines, allowing for the importation of organic products into the Japanese market. Under revised regulations, organic certification bodies are required to be registered (accredited) with MAFF and are now called Registered Certification Organizations.

Republic of Korea

The Republic of Korea introduced an 'Act on the Management and Support for the Promotion of Eco-Friendly Agriculture/Fisheries and Organic Foods' in 2013, to be administered by the Ministry of Agriculture, Food and Rural Affairs (MAFRA). The regulations are compatible with those of the EU, the USA and Canada, allowing trade in organic products between Korea and these countries.

Russia

In 2014 the Russian State Duma approved and signed into effect the National Standard for Organic Products, to become effective in 2015 and be regulated by the Ministry of Agriculture. The Standard and Regulations are based on the EU Council Regulation (EC) No. 834/2007 of June 28, 2007.

Other countries

In most developing countries, there are no markets for certified organic products, but in some countries organic urban markets are developing. Expanding demand for organic foods in developed countries is expected to benefit developing country exports by providing new market opportunities and price premiums, especially for tropical and out-of-season products. Developing country exporters will need to meet the production and certification requirements of those in developed countries.

Impact

These international guidelines, regulations and standards have a strong impact on national standards. It seems clear that increasing convergence or harmonization of these regulations will occur as the markets for organic feedstuffs and poultry products grow and countries seek to export to others.

References

Australian Organic (2017) *Australian Certified Organic Standard. The Requirements for Organic Certification, 11/04/2017, Version 4.* Australian Organic Ltd, Nundah, Queensland.

Minister of Justice. (2009) *Canadian Organic Regulations* pp. 1–20. Organic Product Regulations, 2009. http://laws-lois.justice.gc.ca/PDF/SOR-2009-176.pdf (accessed 21 July 2018).

CGSB (1999) *National Standard for Organic Agriculture.* Canadian General Standards Board, Gatineau, Canada. Available at http://www.pwgsc.gc.ca/cgsb/on_the_net/organic/1999_06_29-e.html (accessed January 2006).

CGSB (2006) *Organic Agriculture: Organic Production Systems General Principles and Management Standards.* CAN/CGSB-32.310-2006. Canadian General Standards Board, Gatineau, Canada. Available at http://www.pwgsc.gc.ca/cgsb/on_the_net/organic/scopes-e.html (accessed September 2006).

Codex Alimentarius Commission (1999) *Proposed Draft Guidelines for the Production, Processing, Labelling and Marketing of Organic Livestock and Livestock Products.* Alinorm 99/22 A, Appendix IV. Codex Alimentarius Commission, Rome.

EC (1999) Council Regulation (EC) No 1804/1999 of 19 July 1999 supplementing Regulation (EEC) No. 2092/91 on organic production of agricultural products and indications referring thereto on agricultural products and foodstuffs to include livestock production. *Official Journal of the European Communities* 2.8.1999, L222, 1–28.

EC (2007) *European Council Regulation on Organic Production and Labelling of Organic Products (Repealing Regulation (EEC) No. 2092/91); Official Journal of the European Communities 189, 20.7.2007, p. 1–23. No. 834/2007 28 June 2007.* No. 834/2007.

GAIN (2002) *Global Agriculture Information Network Report #BR2002.* US Foreign Agricultural Service, US Agricultural Trade Office, Sao Paulo, Brazil.

GAIN (2013) *New Organic Certification and Product Labeling Program in Mexico. Global Agricultural Information Network Report No. MX3313.* US Foreign Agricultural Service, US Agricultural Trade Office, Mexico City.

IFOAM (1998) *IFOAM Basic Standards.* IFOAM General Assembly November 1998. International Federation of Organic Agriculture Movements, Tholey-Theley, Germany.

Kilcher, L., Huber, B. and Schmid, O. (2006) Standards and regulations. In: Willer, H. and Yussefi, M. (eds) *The World of Organic Agriculture. Statistics and Emerging Trends 2006.* International Federation of Organic Agriculture Movements IFOAM, Bonn, Germany and Research Institute of Organic Agriculture FiBL, Frick, Switzerland, pp. 74–83.

MAFF (2001) *The Organic Standard. Japanese Organic Rules and Implementation,* May 2001. Ministry of Agriculture, Forestry and Fisheries, Tokyo. Available at http://www.maff.go.jp/soshiki/syokuhin/hinshitu/organic/eng_yuki_59.pdf (accessed January 2006).

MAFF (2005) *Japanese Agricultural Standard for Organic Livestock Products,* Notification No. 1608, 27 October. Ministry of Agriculture, Forestry and Fisheries, Tokyo. Available at http://www.maff.go.jp/soshiki/syokuhin/hinshitu/e_label/file/SpecificJAS/Organic/JAS_OrganicLivestock.pdf (accessed September 2006).

NOP (2000) *National Standards on Organic Production and Handling, 2000.* United States Department of Agriculture/Agricultural Marketing Service, Washington, DC. Available at http://www.ams.usda.gov/nop/NOP/standards.html (accessed January 2006).

NZFSA (2011) *NZFSA Technical Rules for Organic Production,* Version 7. New Zealand Food Safety Authority, Wellington.

SENASICA (2013) *Mexican Organic Regulations* (in Spanish). Servicio Nacional de Sanidad, Inocuidad y Calidad Agroalimentaria, Mexico City. Available at http://www.senasica.gob.mx/?idnot=1532 (accessed 3 November 2016).

UNCTAD (2004) *Harmonization and Equivalence in Organic Agriculture.* United Nations Conference on Trade and Development, Geneva, Switzerland, 238 pp.

3

Elements of Poultry Nutrition

Like all other animals, poultry require five components in their diet as a source of nutrients: energy, protein, minerals, vitamins and water. A nutrient shortage or imbalance in relation to other nutrients will affect performance adversely. Poultry need a well-balanced and easily digested diet for optimal production of eggs and meat and are very sensitive to dietary quality because they grow quickly and make relatively little use of fibrous, bulky feeds such as lucerne hay or pasture, since they are non-ruminants and do not possess a complicated digestive system that allows efficient digestion of forage-based diets.

Digestion and Absorption of Nutrients

Digestion is the preparation of feed for absorption, i.e. reduction of feed particles in size and solubility by mechanical and chemical means. A summary outline of digestion and absorption in poultry follows. This provides a basic understanding of how the feed is digested and the nutrients absorbed. Readers interested in a more detailed explanation of this topic should consult a recent text on poultry nutrition or physiology.

Birds have a modified gut, in comparison with other non-ruminant species such as pigs or humans (Fig. 3.1). The digestive system can be seen as being relatively simple, probably due to an evolutionary need for a light body weight related to the ability to fly. The mouth is modified into a narrow, pointed beak to facilitate seed-eating, and does not allow for the presence of teeth to permit grinding of the feed into smaller particles for swallowing. Instead, mechanical breakdown of feedstuffs is performed mainly by a grinding action in the gizzard (which is attached to the proventriculus) and contractions of the muscles of the gastrointestinal walls. The function of the proventriculus is analogous to that of the stomach in the pig. Chemical breakdown of the feed particles is achieved by enzymes secreted in digestive juices and by gut microflora. The digestive process reduces feed particles to a size and solubility that allows for absorption of digested nutrients through the gut wall into the portal blood system.

Mouth

Digestion begins here. Saliva produced by the salivary glands moistens the dry feed so that it is easier to swallow. At this point the feed, if accepted, is swallowed whole. The feed then passes quickly to a pouch in the oesophagus, the crop.

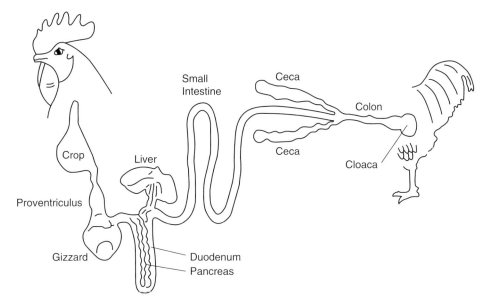

Fig. 3.1. Digestive system of the chicken.

Crop

This is a storage organ from which feed can be metered into the lower oesophagus for passage into the next section of the gut, the proventriculus. There is only minimal amylase activity in the saliva and crop, indicating little digestion of carbohydrates in this organ. There is no digestion of protein in the mouth or crop, either. There is, however, lubrication and further softening of the feed by saliva and by mucus secreted by the crop. The softened feed passes down the oesophagus by a series of muscular contractions (peristalsis) to the next section, the proventriculus.

Proventriculus (stomach)

The proventriculus represents the glandular stomach, where digestive juices are secreted. The juices contain hydrochloric acid (HCl) and the enzyme precursor (zymogen) pepsinogen, which is converted to the active enzyme pepsin in the acidic (pH 2.5) conditions in this organ. This initiates protein digestion, which is continued in the attached gizzard. HCl also serves to dissolve minerals ingested with the feed, such as calcium salts, and it inactivates pathogenic bacteria present in the feed. Mucus is released by the proventriculus to protect the inner wall from acid damage. A grinding action in the gizzard, which is facilitated by the ingestion of grit, continues the process of digestion further by exposing a greater surface area of the feed to chemical breakdown. Partially digested feed in a semi-fluid form known as chyme then moves from the gizzard into the next part of the gut, the small intestine.

There is evidence that gizzard weight can be increased by the presence of whole grains or fibrous material in the diet and higher activities of pancreatic enzymes in the small intestine (Husvéth *et al.*, 2015). Typically the increase in weight of the gizzard and pancreas in growing meat birds is around 25% and 10%, respectively.

Small intestine

The small intestine is a long tube-like structure connecting the gizzard to the large intestine. This is where digestion is completed and absorption of nutrients takes place. Absorption includes various processes that allow the end products of digestion to pass through the membranes of the intestine into the portal bloodstream for distribution throughout the body.

Chyme is mixed with other fluids in the small intestine, the first part of which is known as the duodenum. Duodenal glands produce an alkaline secretion which acts as a lubricant and also protects the duodenal wall against HCl from the gizzard. The pancreas (which is attached to the small intestine) secretes fluid containing bicarbonate and several enzymes (amylase, trypsin, chymotrypsin and lipase) that act on carbohydrates, proteins and fats. The duodenal wall also secretes enzymes, which continue the breakdown process on sugars, protein fragments and fat particles. Bile synthesized by the liver passes into the duodenum via the bile duct. It contains bile salts, which provide an alkaline pH in the small intestine and fulfil an important function in digesting and absorbing fats. The processes comprise emulsification, enhanced by the bile salts, action of pancreatic lipase and formation of mixed micelles which are required for absorption into the intestinal cells.

As a result of these activities the ingested carbohydrates, protein and fats are broken down to small molecules suitable for absorption (monosaccharides, amino acids (AA) and monoglycerides, respectively). In contrast to the situation in the pig, the disaccharide lactose (milk sugar) is only partly utilized by chickens because they lack the enzyme (lactase) necessary for its breakdown. As a result, most milk products are not ideally suited for use in poultry diets.

Muscles in the wall of the small intestine regularly contract and relax, mixing the chyme and moving it towards the large intestine.

Jejunum and ileum

Absorption also takes place in the second section of the small intestine, known as the jejunum, and in the third section, known as the ileum. Digestion and absorption are complete by the time the ingesta have reached the terminal end of the ileum. This area is therefore of interest to researchers studying nutrient bioavailability (relative absorption of a nutrient from the diet) since a comparison of dietary and ileal concentrations of a nutrient provides information on its removal from the gut during digestion and absorption.

Minerals released during digestion dissolve in the digestive fluids and are then absorbed either by specific absorption systems or by passive diffusion.

The processes for the digestion and absorption of fat- and water-soluble vitamins are different, due to their solubility properties. Fat-soluble vitamins and their precursors (A, β-carotene, D, E and K) are digested and absorbed by processes similar to those for dietary fats, mainly in the small intestine. Most water-soluble vitamins require specific enzymes for their conversion from natural forms in feedstuffs into the forms that are ultimately absorbed. Unlike fat-soluble vitamins that are absorbed mostly by passive diffusion, absorption of water-soluble vitamins involves active carrier systems to allow absorption into the portal blood.

Once the nutrients enter the bloodstream, they are transported to various parts of the body for vital body functions. Nutrients are used to maintain essential functions such as breathing, circulation of blood and muscle movement, replacement of worn-out cells (maintenance), growth, reproduction and egg production.

The ingesta, consisting of undigested feed components, intestinal fluids and cellular material from the abraded wall of the intestine, then passes to the next section of the intestine, the large intestine.

Large intestine

The large intestine (lower gut) consists of a colon, which is shorter than in mammals, and a pair of blind caeca attached at the junction with the small intestine. The colon is attached to the cloaca (vent), the common opening for the release of faeces, urine and eggs. Poultry, like other birds, do not excrete liquid urine. Instead they excrete urine as uric acid, which is excreted as a white paste or a dry, white powder. Very little water is required for this process in birds, compared with the excretion of urine in cattle or pigs, and it is related to their ancestry from reptiles. The process also explains the absence of a bladder in poultry.

The contents of the large intestine move slowly and no enzymes are added. Some microbial breakdown of fibre and undigested material occurs in the caeca, but is limited. The extent of breakdown may increase with age of the bird and with habituation to the presence of fibre in the diet. Thus, fibrous feeds, like lucerne, have a relatively low feed value except in ratites such as the ostrich, which are well adapted for the utilization of high-fibre diets.

Remaining nutrients, dissolved in water, are absorbed in the colon. The nutritional significance of certain water-soluble vitamins and proteins synthesized in the large intestine is doubtful because of limited absorption in this part of the gut. The large intestine absorbs much of the water from the intestinal contents into the body, leaving undigested material which is formed into faeces, then mixed with urine and later expelled through the cloaca. Caecal waste is also deposited on the excreta, appearing as a light-brown froth, which should not be confused with diarrhoea.

The entire process of digestion takes about 2.5–25 h in most species of poultry, depending on whether the digestive tract is full, partially full or empty when feed is ingested. Because of the high metabolic rate of the fowl, a more or less continuous supply of feed is required. This is provided for by the crop that acts as a reservoir for the storage of feed prior to its digestion.

Poultry tend to eat meals at about 15 min intervals through the daylight hours and, to some extent, during darkness. They tend to eat larger portions at first light and in the late evening. A meal of normal feed takes about 4 h to pass through the gut in the case of young stock, 8 h in the case of laying hens and 12 h for broody hens. Intact, hard grains take longer to digest than cracked grain.

Feed Intake

Selection of feed is influenced by two types of factors: innate and learned. Although the chicken has relatively few taste buds and does not possess a highly developed sense of smell it is able to discriminate between certain feed sources on the basis of colour, taste or flavour, especially when a choice is available. Discriminating between nutritious and harmful feeds is learned differently in birds than in mammals since chicks are not fed directly by the parents. This learning process is aided in organic production by the presence of the parent birds during the early life of the chick.

Birds appear to rely to a large extent on visual appearance in selecting various feeds; refusal or acceptance of feed on its first introduction being determined by colour and general appearance (El Boushy and van der Poel, 2000). According to the evidence reviewed by these authors, chickens preferred yellow-white maize followed by yellow, orange and finally orange-red maize. Red, red-blue and blue seeds were eaten only when the birds were very hungry. Preference tests showed also that less was eaten of black and green diets. Some of the research indicated that chicks show a preference for diets of the same colour as that fed after hatching. Colour is important also in teaching birds to avoid feeds that produce illness after ingestion.

The review cited above indicates that birds possess a keen sense of taste and can discriminate between feeds on the basis of sweet, salt, sour and bitter. Rancidity and staleness have been shown to reduce intake of feed. However, there appear to be genetic differences in taste discrimination among poultry species. The finding that sucrose in solution appears to be the only sugar for which chickens have a preference may be of use in helping to prevent 'starve-outs' in baby chicks or to help birds during disease outbreaks or periods of stress. Current evidence suggests that most flavours added to poultry feed are ineffective in stimulating intake of feed.

A sense of smell is probably less important in birds than in mammals, birds lacking the behaviour of sniffing.

Other factors identified by El Boushy and van der Poel (2000) as being involved in control of feed intake include temperature, viscosity, osmotic pressure of water, salivary production, nutritive value of feed and toxicity of feed components.

Birds have been shown to possess some degree of 'nutritional wisdom' or 'specific appetites', eating less of diets that are inadequate in nutrient content. Laying stock have the ability to regulate feed intake according to the energy level of the diet; therefore, it is important to adjust the concentration of other nutrients in relation to energy level. Modern broiler stocks appear to have lost the ability to regulate intake according to dietary energy level, requiring breeding stock to be fed rationed amounts. Broilers, on the other hand, appear to have a greater ability than laying stock to select feeds that result in a balanced intake of protein when presented with a variety of feeds (Forbes and Shariatmadari, 1994). Use can be made of this information in planning choice-feeding systems for poultry, as will be outlined in a later chapter.

The findings reviewed by El Boushy and van der Poel (2000) indicated that wheat and sunflower seeds, polished rice, cooked potatoes, potato flakes and fresh fish are very palatable feedstuffs. Oats, rye, rough rice, buckwheat and barley are less palatable, unless ground. Linseed meal appears to be very unpalatable.

Among the physical factors affecting feed intake is particle size. For instance, it has been shown that feed particles are selected by broilers on the basis of size (El Boushy and van der Poel, 2000), intake being greatest with particles of 1.18–2.36 mm. As the birds aged the preference was for particles greater than 2.36 mm. More findings on preferred particle size will be discussed in a later chapter.

Social interaction is another factor influencing intake, chicks being known to eat more in a group situation.

Digestibility

Only a fraction of each nutrient taken into the digestive system is absorbed. This fraction can be measured as the digestibility coefficient, determined through digestibility experiments. Researchers measure both the amount of nutrient present in the feed and the amount of nutrient present in the faeces (not the droppings), or more exactly in the ileum. The difference between the two, commonly expressed as a percentage or in relation to 1 (1 indicating complete digestion), is the proportion of the nutrient digested by the bird. Each feedstuff has its own unique set of digestibility coefficients for all nutrients present. The digestibility of an ingredient or a complete feed can also be measured.

The measurement of digestibility in the bird is more complicated than in the pig, since faeces and urine are excreted together through the cloaca. As a result, it is necessary to separate the faeces and urine, usually by performing a surgical operation on the bird that allows collection of faeces in a colostomy bag.

Digestibility measured in this way is known as 'apparent digestibility', since the faeces and ileal digesta contain substances originating in the fluids and mucin secreted by the gut and associated organs, and also cellular material abraded from the gut wall as the digesta pass. Correction for these endogenous losses allows for the 'true digestibility' to be measured. Generally, the digestibility values listed in feed tables refer to apparent digestibility unless stated otherwise.

Factors affecting digestibility

Some feed ingredients contain components that interfere with digestion. This aspect is dealt with in more detail in Chapter 4.

Digestibility of carbohydrates

Starch is the main energy source in poultry diets and is generally well digested. Complex carbohydrates such as cellulose, which represent much of the fibre in plants, cannot be digested by poultry. There is some microbial hydrolysis of cellulose in the caeca, at least in some avian species, which may contribute to the energy yield from the feed. Other complex carbohydrates that may be present in the feed are hemicelluloses, pentosans and oligosaccharides. They are also difficult to digest and their utilization may be improved by the addition of certain enzymes to the diet. The pentosans and β-glucans found in barley, rye, oats and wheat increase the viscosity of the digesta, consequently interfering with digestion and absorption (NRC, 1994). They also result in sticky droppings, which can lead to foot and leg problems and breast blisters. As a result, it is now a common practice to add the requisite enzymes to conventional poultry diets to achieve breakdown of these components during digestion.

Chitin is the main component of the hard exoskeleton of insects. Domesticated poultry have some ability to digest this component, but studies suggest that the insect skeleton is not an important source of nutrients for poultry (Hossain and Blair, 2007).

Some carbohydrate components in the feed may interfere with digestion. For instance, soybean meal may contain a substantial level of α-galactosaccharide, which has been associated with reduced digestibility of soybean meal-based diets (Araba et al., 1994). Ways of addressing this issue include the use of low-galactosaccharide cultivars of soybean meal and addition of a specific enzyme to the feed.

Cooking improves the digestibility of some feedstuffs such as potato. Steam-pelleting may also improve starch digestibility.

Digestibility of proteins

It is well established that feeding raw soybeans results in growth depression, poor feed utilization, pancreatic enlargement in young chickens, and small egg size in laying hens. These effects are due to antitrypsins in soybeans that reduce digestibility of proteins (Zhang and Parsons, 1993). Antitrypsins inhibit the activities of the proteolytic enzyme trypsin, which results in lower activities of other proteolytic enzymes that require trypsin for activation. Heat treatment of soybeans is effective in deactivating the anti-nutritional compounds.

High levels of tannins in sorghum are associated with reduced dry matter and

protein digestibility and cottonseed meal contains gossypol which, when heated during processing, forms indigestible complexes with the amino acid (AA) lysine (NRC, 1994). The digestibility of protein in lucerne meal may be reduced by its saponin content (Gerendai and Gippert, 1994).

Excess heat applied during feed processing can also result in reduced protein digestibility and utilization, due to reaction of AAs with soluble sugars.

Digestibility of fats

Older birds are better able to digest fats than young birds. For instance, Katongole and March (1980) reported a 20–30% improvement in digestion of tallow for 6- versus 3-week-old broilers and Leghorns. The effect of age appears to be most pronounced for the saturated fats.

Other factors that can influence fat digestibility include the level of fat inclusion in the diet and presence of other dietary components (Wiseman, 1984). Fat composition can influence overall fat digestion because different components can be digested and/or absorbed with varying efficiency.

The addition of fat to the diet can reduce the rate of passage of feed through the gut and influence overall diet digestibility, due to an inhibition of proventricular emptying and intestinal digesta movement. As a result of the decreased rate of passage the digesta spend more time in contact with digestive enzymes, which enhances the extent of digestion of feed components, including non-fat components. This can result in the feed mixture having a higher energy value than can be accounted for from the sum of the energy value of the ingredients, resulting in the 'extra-caloric effect' (NRC, 1994).

Wiseman (1986) reported a reduction in digestibility and in available energy of up to 30% due to oxidation of fat as a result of overheating during processing. A number of naturally occurring fatty acids can also adversely affect overall fat utilization. Two such components are erucic acid present in rapeseed oils and some other Brassica spp., and the cyclopropenoid fatty acids present in cottonseed.

Digestibility of minerals

A high proportion of the phosphorus present in feedstuffs may be in the form of phytate, which is poorly digested by birds because they lack the requisite enzyme in the gut. Consequently, the content of non-phytate phosphorus in feed ingredients is used in formulating poultry diets to ensure the required level of phosphorus, rather than the total phosphorus content. It is now becoming a common practice for a microbial phytase to be added to conventional poultry diets. This achieves a greater release of the bound phosphorus in the gut and a reduced amount to be excreted in the manure and into the environment. Use of microbial phytase may also improve digestion of other nutrients in the diet, associated with breakdown of the phytate complex. Organic producers should take advantage of this knowledge, if supplementation with phytase is permitted by the local organic regulations.

Once fats have been digested, the free fatty acids have the opportunity to react with other nutrients within the digesta. One such possible association is with minerals to form soaps that may or may not be soluble. If insoluble soaps are formed, there is the possibility that both the fatty acid and the mineral will be unavailable to the bird. This appears to be more of a potential problem in young birds fed diets containing saturated fats and high levels of dietary minerals. Soap production seems to be less of a problem with older birds.

Nutrient Requirements

Energy

Energy is produced when the feed is digested in the gut. The energy is then either released as heat or is trapped chemically and absorbed into the body for metabolic purposes. It can be derived from protein, fat or carbohydrate in the diet. In general, cereals and fats provide most of the energy in the diet. Energy in excess of requirement is converted to fat and stored in the body. The provision of energy accounts for the greatest percentage of feed costs.

The total energy (gross energy) of a feedstuff can be measured in a laboratory by burning it under controlled conditions and measuring the energy released in the form of heat. Digestion is never complete under practical situations; therefore, measurement of gross energy does not provide accurate information on the amount of energy useful to the animal. A more precise measurement of energy is digestible energy (DE) which takes into account the energy lost during incomplete digestion and excreted in the faeces. The chemical components of feedstuffs have a large influence on DE values, with increased fat giving higher values and increased fibre and ash giving lower values (Fig. 3.2). Fat provides about 2.25 times the energy provided by carbohydrates or protein.

More accurate measures of useful energy contained in feedstuffs are metabolizable energy (ME), which takes into account energy loss in the urine as well as in the faeces, and net energy (NE), which in addition takes into account the energy lost as heat produced during digestion. Balance experiments can be used to determine ME fairly readily from comparisons of energy in the feed and excreta, the excretion of faeces and urine together in the bird being a convenient feature in this regard. As a result, ME is the most common energy measure used in poultry nutrition in many countries. A more accurate assessment of ME can be obtained by adjusting the ME value for the amount of energy lost or gained to the body in the form of protein nitrogen (N). The ME value corrected to zero N gain or loss is denoted ME_n.

ME obtained by these methods is apparent ME (AME), since all of the energy lost in the excreta is not derived from the feed. Some is derived from endogenous secretions of digestive fluids, sloughed-off intestinal cells and endogenous urinary secretions. True ME (TME) is the term used to describe ME corrected for these losses. TME and TME_n values have been determined for certain feedstuffs by researchers and are used in some countries in the formulation of diets. The endogenous losses

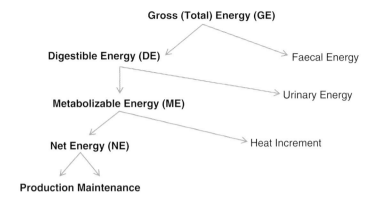

Fig 3.2. Schematic of energy utilization in the bird showing how the various measures of feed energy are derived.

are difficult to measure accurately: one method involves the estimation of losses by withholding the feed for a short period and assuming that the energy contained in the excreta represents endogenous loss (Sibbald, 1982). ME_n values are approximately equivalent to TME_n values for most feedstuffs (NRC, 1994). However, ME_n and TME_n values differ substantially for some ingredients, such as rice bran, wheat middlings and maize distillers' grains plus solubles. Accordingly the NRC (1994) recommended that with these ingredients, ME_n values should not be indiscriminately interchanged with TME_n values for purposes of diet formulation.

Most ME_n values reported for feedstuffs have been determined with young chicks and those for TME_n content have been determined with adult male chickens. Few studies have been carried out to determine either ME_n or TME_n for poultry of different ages and more ME_n and TME_n data are needed for many feed ingredients for chickens, turkeys and other poultry of different ages (NRC, 1994).

Several researchers have developed equations for the estimation of ME based on the chemical composition of the diet (NRC, 1994).

The requirements set out in this volume and taken mainly from the report on the Nutrient Requirements of Poultry (NRC, 1994) are based on ME (AME), expressed as kilocalories (kcal) or megacalories (Mcal)

per kilogram of feed. This energy system is used widely in North America and in many other countries. Energy units used in some countries are based on joules (J), kilojoules (kJ) or megajoules (MJ). A conversion factor can be used to convert calories to joules, i.e. 1 Mcal = 4.184 MJ; 1 MJ = 0.239 Mcal; and 1 MJ = 239 kcal. Therefore, the tables of feedstuff composition in this volume show ME values expressed as MJ or kJ as well as kcal/kg.

Protein and amino acids

The term protein usually refers to crude protein (CP) (measured as N content × 6.25) in requirement tables. Protein is required in the diet as a source of amino acids (AA), which can be regarded as the building blocks for the formation of skin, muscle tissue, feathers, eggs, etc. Body proteins are in a dynamic state with synthesis and degradation occurring continuously; therefore, a constant, adequate intake of dietary AA is required. An inadequate intake of dietary protein (AA) results in a reduction or cessation of growth or productivity and an interference with essential body functions.

There are 22 different AA in the body of the bird, ten of which (arginine, methionine, histidine, phenylalanine, isoleucine, leucine, lysine, threonine, tryptophan and valine) are essential AA (EAA), i.e. cannot be manufactured by the body and must be derived from the diet. Cystine and tyrosine are semi-essential in that they can be synthesized from methionine and phenylalanine, respectively. The others are non-essential AA (NEAA) and can be made by the body.

Methionine is important for feather formation and is generally the first limiting AA. Therefore, it has to be at the correct level in the diet. The level of the first limiting AA in the diet normally determines the use that can be made of the other EAA. If the limiting AA is present at only 50% of requirement then the efficiency of use of the other essential AA will be limited to 50%. This concept explains why a deficiency of individual AA is not accompanied by specific deficiency signs: a deficiency of any EAA results in a generalized protein deficiency. The primary sign is usually a reduction in feed intake that is accompanied by increased feed wastage, impaired growth and production and general unthriftiness. Excess AA are not stored in the body but are excreted in the urine as N compounds.

Although a protein requirement per se is no longer appropriate in requirement tables, stating a dietary requirement for both protein and EAA is a convenient way to ensure that all AAs needed physiologically are provided correctly in the diet (NRC, 1994).

In most poultry diets, a portion of each AA that is present is not biologically available to the animal. This is because most proteins are not fully digested and the AAs are not fully absorbed. The AA in some proteins such as egg or milk are almost fully bioavailable, whereas those in other proteins such as certain plant seeds are less bioavailable. It is therefore more accurate to express AA requirements in terms of bioavailable (or digestible) AA.

Protein and AA requirements vary according to the age and stage of development. Growing meat birds have high AA requirements to meet the needs for rapid growth and tissue deposition. Mature cockerels have lower AA requirements than laying hens, even though their body size is greater and feed consumption is similar. Body size, growth rate and egg production of poultry are determined by the genetics of the bird in question. AA requirements, therefore, also differ among types, breeds and strains of poultry.

Dietary requirements for AA and protein are usually stated as proportions of the diet. However, the level of feed consumption has to be taken into account to ensure that the total intake of protein and AA is appropriate. The protein and AA requirements derived by the NRC (1994) relate to poultry kept in moderate temperatures (18–24°C). Ambient temperatures outside of this range cause an inverse response in feed consumption; i.e. the lower the temperature, the greater is the feed intake and vice versa (NRC, 1994). Consequently, the dietary levels of protein and AA to meet the requirements should be increased in warmer environments and decreased in cooler environments, in accordance with expected differences in feed intake. These adjustments are designed to help ensure the required daily intake of AA.

For optimal performance the diet must provide adequate amounts of EAA, adequate energy and adequate amounts of other essential nutrients. The CP requirement values outlined by the NRC (1994) assume a maize/soy diet, of high digestibility. It is advisable to adjust the dietary target values when diets based on feedstuffs of lower digestibility are formulated. The bioavailability of EAA in a wide range of feedstuffs has been measured by researchers. The primary method has been to measure the proportion of a dietary AA that has disappeared from the gut when digesta reach the terminal ileum, using surgically altered birds. Interpretation of the data is, however, somewhat complicated. The values determined by this method are more correctly termed 'ileal digestibilities' rather than bioavailabilities, because AA are sometimes absorbed in a form that cannot be fully used in metabolism. Furthermore, unless a correction is made for endogenous AA losses, the values are 'apparent' rather than 'true'.

The estimates of requirement are based on the assumption that the profile of dietary bioavailable EAA should remain relatively constant during all growth stages, and that a slightly different profile is more appropriate for egg production. The desirable profile has been called ideal protein (IP). The CP need is minimized as the dietary EAA pattern approaches that of IP. The nearer the EAA composition of the diet is to IP, the more efficiently the diet is utilized and the lower the level of N excretion. Energy is also used most efficiently at this point; thus, both protein and energy utilization are maximized.

Van Cauwenberghe and Burnham (2001) and Firman and Boling (1998) reviewed various estimates of ideal proportions of AAs in broiler, layer and turkey diets based on digestible AA and lysine as the first limiting AA. These estimates are shown in Tables 3.1–3.3.

Cereal grains, such as maize, barley, wheat and sorghum, are the main ingredients of poultry diets and usually provide 30–60% of the total AA requirements. Other sources of protein such as soybean meal and canola meal must be provided to ensure adequate amounts and a proper balance of essential AA. The protein levels necessary to provide adequate intakes of essential AA will depend on the feedstuffs used. Feedstuffs that contain 'high-quality' proteins (i.e. with an AA pattern similar to the bird's needs) or mixtures of feedstuffs in which the AA pattern of one complements the pattern in another, will meet the essential AA requirements at lower dietary protein levels than feedstuffs with a less desirable AA pattern. This is important if one of the goals is to minimize N excretion.

The profile of AA in a feedstuff is a main determinant of its value as a protein source. If the profile is close to that of IP (as in fish or meat), it is considered a high-quality protein. Correct formulation of the diet ensures that the dietary AA (preferably on a bioavailable basis) are as close to IP as possible and with minimal excesses of EAA.

Estimated AA requirements are shown in the Tables 3.7–3.16 at the end of this chapter, based on the concept of IP (NRC, 1994). Factors that affect the level of feed intake have an influence on requirements, a reduction in expected feed intake requiring the concentration of dietary AA to be increased. Correspondingly, the concentration of AAs may be reduced when feed intake is increased.

Minerals

Minerals perform important functions in the animal body and are essential for proper growth and reproduction. In addition to being constituents of bone and eggs they take part in other essential processes. A lack of minerals in the diet can result in deficiency signs, including reduced or low feed intake, reduced rate of growth, leg problems, abnormal feather development,

Table 3.1. Estimated ideal dietary AA pattern for broilers, relative to lysine at 100 (from Van Cauwenberghe and Burnham, 2001).

Amino acid	NRC, 1994	Baker and Han, 1994	Lippens et al., 1997[a]	Gruber, 1999[a]	Mack et al., 1999
Lysine	100	100	100	100	100
Arginine	114	105	125	108	ND
Isoleucine	73	67	70	63	71
Methionine	46	36	ND	37	ND
Methionine + cystine	82	72	70	70	75
Threonine	73	70	66	66	63
Tryptophan	18	16	ND	14	19
Valine	82	77	ND	81	81

[a]Cited in Van Cauwenberghe and Burnham (2001).
ND = not determined.

Table 3.2. Estimated ideal dietary AA pattern for layers, relative to lysine at 100 (from Van Cauwenberghe and Burnham, 2001).

Amino acid	NRC, 1994	CVB, 1996	ISA, 1996/97[a]	MN, 1998[a]
Lysine	100	100	100	100
Arginine	101	ND	ND	130
Isoleucine	94	74	82	86
Methionine	43	45	51	49
Methionine + cystine	84	84	88	81
Threonine	68	64	70	73
Tryptophan	23	18	22	20
Valine	101	81	93	102

[a]Cited in Van Cauwenberghe and Burnham (2001).
ND = not determined.

Table 3.3. Estimated ideal dietary AA pattern for starting hen turkeys, relative to lysine at 100 (from Firman and Boling, 1998).

Amino acid	
Lysine	100
Arginine	105
Histidine	36
Isoleucine	69
Leucine	124
Methionine + cystine	59
Phenylalanine + tyrosine	105
Threonine	55
Tryptophan	16
Valine	76

Table 3.4. Minerals required by poultry.

Macrominerals	Trace minerals
Calcium (Ca)[a]	Cobalt (Co)
Chlorine (Cl)[a]	Copper (Cu)[b]
Magnesium (Mg)	Iodine (I)[c]
Phosphorus (P)[a]	Iron (Fe)[b]
Potassium (K)	Manganese (Mn)[b]
Sodium (Na)[a]	Selenium (Se)[b]
Sulfur (S)	Zinc (Zn)[b]

[a]Included as dietary ingredients.
[b]Included in a premix.
[c]Included as iodized salt.

goitre, unthriftiness, breeding and reproductive problems, and increased mortality.

Poultry need at least 14 mineral elements (Table 3.4) and it is possible that other minerals may also be essential in the body. Under natural conditions it is likely that poultry can obtain part of their mineral requirements by ingesting pasture and pecking in the soil. However, these sources cannot be guaranteed to provide all of the requirements consistently. Therefore, poultry diets must be supplemented with minerals.

Minerals required in large amounts are known as macrominerals. These include calcium, phosphorus, sulfur, sodium, chloride, potassium and magnesium. Minerals required in small amounts are called microminerals or trace minerals. These include iron, zinc, copper, manganese, iodine and selenium. Cobalt is also required, but it does not need to be supplied as a trace mineral, because it is a part of the vitamin B_{12} molecule. In practical diets, copper and iron are often present at sufficient levels without supplementation. Trace elements function as part of larger organic molecules. Iron is a part of haemoglobin and cytochromes, and iodine is a part of the hormone thyroxine. Copper, manganese, selenium and zinc function as essential accessory factors to enzymes. The requirements for certain trace minerals are often met by concentrations present in conventional feed ingredients. Soils vary in their content of trace minerals and plants vary in their uptake of minerals. Consequently, feedstuffs grown in certain geographical areas may be marginal or deficient in specific elements. Thus, poultry diets usually require supplementation to ensure an adequate intake of trace minerals. Mineral salts used as feed supplements are not usually pure compounds but contain variable amounts of other minerals.

Of the essential mineral elements, those likely to be deficient in poultry diets are calcium, phosphorus, sodium, copper, iodine, manganese, selenium and zinc. Deficiencies of the other essential mineral elements are less common and the feeds used probably contain them in sufficient quantities. There are some indications that magnesium supplementation may be beneficial in certain situations.

Required minerals can be categorized as follows:

Calcium and phosphorus

Calcium and phosphorus are essential for the formation and maintenance of the skeleton. Together they make up over 70% of the mineral content of the avian body, mainly combined with each other. These values indicate the importance of calcium and phosphorus in the diet. An inadequate supply of either one in the diet will limit the utilization of the other. These two minerals are discussed together because there is a close relationship between them. Most of the calcium in the diet of the growing bird is used for bone formation, whereas in the mature laying bird most of the dietary calcium is used for eggshell formation. Another function of calcium is in blood-clotting. An excess of dietary calcium interferes with the availability of other minerals, such as phosphorus, magnesium, manganese and zinc. A ratio of approximately 2:1 calcium to non-phytate phosphorus (by weight) is appropriate for most poultry diets, with the exception of diets for laying hens. A much higher level of calcium is needed for eggshell formation, and a ratio as high as 12:1 calcium to non-phytate phosphorus (by weight) is more appropriate for layers. Phosphorus, in addition to its function in bone formation, is also required in the utilization of energy and in structural components of cells.

A deficiency of calcium is more likely than a deficiency of phosphorus. Cereal grains, which constitute most of the avian diet, are quite low in calcium, though generally the calcium present in cereal grains and most feedstuffs is of higher availability than that of phosphorus. Legumes and pasture provide some calcium.

The phosphorus content of cereal grains and grain by-products is higher, though about one-half or more is in the form of organically bound phytate, which is poorly digested by poultry. Only about 10% of the phytate phosphorus in maize and wheat is digested by poultry (NRC, 1994). The phosphorus in animal products and phosphorus supplements is generally considered to be well utilized. The phosphorus in oilseed meals also has a low bioavailability. In contrast, the phosphorus in protein sources of animal origin is largely inorganic (meaning in this context not containing carbon; organic compounds are those containing carbon), and most animal protein sources (including milk and meat products) have a high phosphorus bioavailability. The phosphorus in dehydrated lucerne meal is highly available. Steam-pelleting has been shown to improve the bioavailability of phytate phosphorus in some studies but not in others. The phosphorus in inorganic phosphorus supplements also varies in bioavailability. As a result, the requirements are now set out in terms of available phosphorus or non-phytate phosphorus. An adequate amount of vitamin D is also necessary for proper metabolism of calcium and phosphorus, but a very high level of vitamin D can mobilize excessive amounts of calcium and phosphorus from bones.

Less is known about the availability of calcium in feedstuffs, but the level of calcium is generally so low that the bioavailability is of little consequence. The calcium in common supplementary sources such as ground limestone, oyster shell and dicalcium phosphate is highly available. Blair *et al.* (1965) showed that the availability of calcium for the chick was higher in dicalcium phosphate than in ground limestone.

Signs of calcium or phosphorus deficiency are similar to those of vitamin D deficiency (NRC, 1994). They include depressed growth and poor bone mineralization, resulting in rickets in young birds and osteomalacia in older birds. Calcium is removed from the bones to meet the demands of egg production when the layer diet contains insufficient calcium. Deficient chicks and poults have soft, rubbery bones that fracture readily. An egg contains about 2 g of calcium in the shell; therefore, the calcium need of the laying hen is high. A deficiency results in soft-shelled eggs and reduced egg production. A weakness termed 'layer fatigue' has also been linked to calcium deficiency (as well as phosphorus or vitamin D deficiency), though it is usually reported in caged birds.

Excess calcium not only decreases the utilization of phosphorus but also increases the requirement for zinc in the presence of phytate and may result in zinc deficiency. Excess calcium also increases the requirement for vitamin K.

Sodium, potassium and chloride

Sodium, potassium and chloride are the primary dietary ions that influence the electrolytic balance and acid–base status, and the proper dietary balance of sodium, potassium and chloride is necessary for growth, bone development, eggshell quality and AA utilization. Potassium is the third most abundant mineral in the body after calcium and phosphorus and is the most abundant mineral in muscle tissue. It is involved in electrolyte balance and neuromuscular function. The content of potassium in poultry diets is usually adequate. Chloride is present in gastric juice and chlorine is part of the HCl molecule which assists in the breakdown of feed in the proventriculus. Sodium is essential for nerve membrane stimulation and ionic transport across cell membranes. Signs of sodium, potassium or chloride deficiency include reduced appetite, poor growth, dehydration and increased mortality.

Poultry can tolerate high dietary levels of sodium chloride, provided that they have access to ample non-saline drinking water.

Magnesium

Magnesium is a cofactor in several enzyme systems and is a constituent of bone. The magnesium present in poultry diets is usually adequate. Signs of magnesium deficiency include lethargy, panting, gasping and convulsions followed by death.

Sulfur

Sulfur is an essential element but is present in the diet in adequate amounts, making supplementation unnecessary.

Trace minerals

Six trace minerals have been shown to be needed as supplements in poultry diets: iron, copper, zinc, manganese, iodine and selenium. Subclinical trace mineral deficiencies probably occur more frequently than are recognized by poultry producers. Some soils are naturally deficient in trace minerals. In addition, crops and plants vary in their uptake of minerals. Consequently, feedstuffs grown in certain geographical areas may be marginal or deficient in specific elements. Certain areas in North America experience a high rainfall, which results in leaching of the soil and selenium deficiency. As a result, selenium deficiencies have been observed in livestock in Asia when fed US-produced maize and soybean meal but not when fed locally grown feed. Feed suppliers are usually aware of deficient (and adequate) levels of the trace minerals present in feedstuffs and will provide trace-mineral mixes formulated appropriately.

Several studies have shown that omitting trace minerals from poultry diets depresses productivity and tissue mineral concentrations. Patel et al. (1997) found that removal of supplemental trace minerals and vitamins from the diet during the period 35–42 days post hatching decreased daily weight gain in three different broiler strains. In addition, removal of supplemental riboflavin from the finisher diet 7 days prior to slaughter resulted in a 43% decrease in the content of riboflavin in breast muscle. Shelton and Southern (2006) reported that omission of a trace mineral premix from broiler diets had no effect on productivity during the early stage of growth but had progressively deleterious effects on productivity with increasing age of the birds. In addition, removal of trace minerals had a negative effect on bone strength and on tissue trace mineral concentrations. A study conducted in Turkey by İnal *et al.* (2001) with laying hens showed that omission of a trace mineral and vitamin supplement resulted in reduced egg production, feed intake, egg size and zinc content of eggs. These findings are of importance to organic producers, in view of their relevance to production efficiency and product quality.

Cobalt

Cobalt is a component of the vitamin B_{12} molecule but a deficiency of cobalt has not been demonstrated in poultry fed a diet adequate in vitamin B_{12}. Therefore, supplementation with this element is not normally necessary. Diets containing no ingredients of animal origin (which contain vitamin B_{12}) contain no vitamin B_{12}. Therefore, poultry fed on all-plant diets may require dietary cobalt, unless the diet is supplemented with vitamin B_{12}. In practice, many feed manufacturers use a cobalt-iodized salt for all species since cobalt is needed in ruminant diets. This avoids the need to stock separate salt types for ruminant and non-ruminant diets and the inclusion of cobalt provides some insurance in case the poultry diet is lacking sufficient vitamin B_{12}.

Copper

Copper is required for the activity of enzymes associated with iron metabolism, elastin and collagen formation, melanin production and the integrity of the central nervous system. It is required with iron for normal red blood cell formation. Copper is also required for bone formation, brain cell and spinal cord structure, the immune response and feather development and pigmentation. A deficiency of copper leads to poor iron mobilization, abnormal blood formation and decreased synthesis of elastin, myelin and collagen. Leg weakness, various types and degrees of leg crookedness and incoordination of muscular action also result. Tibial dyschondroplasia is an example of a leg disorder in poultry that can be caused by a copper deficiency. Poor collagen and/or elastin formation can also lead to cardiovascular lesions and aortic rupture, particularly in turkeys.

Iodine

It has been known for over 100 years that iodine is required for the proper functioning of the thyroid gland and that an iodine deficiency causes goitre. As a result, iodized salt is now used to prevent this disease in animals and humans. Iodine metabolism is greatly influenced by selenium nutrition, thus influencing basal metabolic rate and several physiological processes. Some dietary factors are goitrogenic. Cruciferous plants contain potential goitrogens of the thiouracil type, while brassicas and white clover contain cyanogenetic glycosides that are goitrogenic (Underwood and Suttle, 1999). Canola meal has resulted from the selection of rapeseed that is low in glucosinolate, a common goitrogen. There are also goitrogenic substances in other feeds such as carrots, linseed, cassava, sweet potatoes, lima beans, millet, groundnuts, cottonseed and soybeans which impair hormone release from the thyroid gland. Goitre can then occur even though the iodine level in the diet may appear to be adequate.

A high calcium level in drinking water is also known to reduce iodine absorption and result in goitre, particularly if the dietary iodine level is borderline. Signs of iodine deficiency include an enlargement of the thyroid gland (which might not be noticed because of the feathers on the neck), poor growth and reduced hatchability of the eggs. At necropsy, the thyroid is enlarged and haemorrhagic.

Most feedstuffs contain only low levels of iodine. The exception is seaweed, which can contain 4000–6000 mg iodine/kg.

Iron

Most of the iron in the body is in the form of haemoglobin in red blood cells and myoglobin in muscle. The remainder is in the liver, spleen and other tissues. Haemoglobin is essential for the proper functioning of every organ and tissue of the body. Iron has a rapid turnover rate in the chicken; therefore, it must be provided in a highly available form in the diet on a daily basis. Iron deficiency can result in microcytic, hypochromic anaemia in poultry. Any internal infection such as coccidiosis can also interfere with iron absorption and lead to a deficiency.

Soil contains iron and may provide sufficient for poultry raised outdoors on pasture. It is important, however, that the soil be free of disease organisms and parasites.

Manganese

Manganese is essential for the synthesis of chondroitin sulfate, a mucopolysaccharide that is an important component of bone cartilage. Manganese is also required to activate enzymes involved in the synthesis of polysaccharides and glycoproteins and it is a key component of pyruvate carboxylase, which is a critical enzyme in carbohydrate metabolism. Lipid metabolism is also dependent on manganese. A deficiency of manganese in poultry results in perosis, bone shortening (chondrodystrophy) and retarded down formation in the embryos, bowing of the legs and poor eggshell quality in laying hens. Decreased growth rate and feed efficiency also occur with a manganese deficiency.

Selenium

Selenium is an important component of glutathione peroxidase, an enzyme that destroys peroxides before they can damage body tissues. Vitamin E is also effective as an antioxidant. Therefore, both selenium and vitamin E prevent peroxide damage to body cells. This aids the body's defence mechanisms against stress. Most feeds contain compounds that can form peroxides. Unsaturated fatty acids are a good example. Rancidity in feeds causes formation of peroxides that destroy nutrients. Vitamin E, for example, is easily destroyed by rancidity. Selenium spares vitamin E by its antioxidant effect. Selenium and vitamin E are interrelated in their biological functions. Both are needed by birds and both have metabolic roles in the body in addition to their antioxidant effect. In some instances, vitamin E will substitute in varying degrees for selenium, or vice versa. However, there are deficiency symptoms

that respond only to selenium or vitamin E. Although selenium cannot replace vitamin E in nutrition, it reduces the amount of vitamin E required and delays the onset of E deficiency signs. Selenium plays an important role in increasing the immune response, together with vitamin E. Sudden death is a common finding with selenium deficiency. Other selenoproteins in poultry play an important role in the prevention of exudative diathesis (a severe oedema produced by a marked increase in capillary permeability due to cell damage) and in maintaining normal pancreatic function and fertility.

Gross necropsy lesions of a selenium deficiency are identical to those of a vitamin E deficiency (NRC, 1994) and include exudative diathesis and myopathy of the gizzard. Paleness and dystrophy of the skeletal muscles (white muscle disease) are also common. The incidence and degree of selenium deficiency may be increased by environmental stress. Selenium is generally included in trace mineral premixes. Common sources for supplementation of poultry diets are sodium selenite and sodium selenate. Selenium yeast is also used in conventional diets.

Excess dietary selenium has to be avoided because of its potential toxicity at high levels in the diet and the feed regulations in several countries are designed to prevent this occurrence.

Zinc

Zinc is widely distributed throughout the body and is present in many enzyme systems involved in metabolism. It is required for normal protein synthesis and metabolism and is also a component of insulin so that it functions in carbohydrate metabolism. Zinc plays an important role in poultry, particularly for layers, as a component of a number of enzymes such as carbonic anhydrase, which is essential for eggshell formation in the shell gland. Other important zinc enzymes in the bird include carboxypeptidases and DNA polymerases. These enzymes play important roles in the immune response, in skin and wound healing and in hormone production. Classic signs of a zinc deficiency in poultry include a suppressed immune system, poor feathering and dermatitis of the feet, low hatchability and poor shell quality. Zinc absorption is reduced with diets high in calcium or phytate. The zinc in soybean meal, cottonseed meal, sesame meal and other plant protein supplements has low availability, due to the presence of phytate in the feedstuffs which combines with zinc to form zinc phytate.

Vitamins

Vitamins are organic (carbon-containing) compounds required for normal growth and the maintenance of animal life. The absence of a given vitamin from the diet, or its impaired absorption or utilization, results in a specific deficiency disease or syndrome.

A commonly accepted definition of a vitamin is an organic compound that meets the following criteria:

1. It is a component of natural food or feed but is distinct from carbohydrate, fat, protein and water.
2. It is present in feedstuffs in minute quantities.
3. It is essential for development of normal tissue and for health, growth and maintenance.
4. When absent from the diet, or not properly absorbed or utilized, it results in a specific deficiency disease or syndrome.
5. It cannot be synthesized by the animal and therefore must be obtained from the diet.

There are exceptions to the above. Most or all vitamins can be synthesized chemically. Vitamin D can be synthesized in the skin of animals by exposure to ultraviolet irradiation, and nicotinic acid (niacin) can be synthesized in the body from the amino acid tryptophan.

Although vitamins are required in small amounts, they serve essential functions in maintaining normal growth and reproduction. Few vitamins can be synthesized by the bird in sufficient amounts to meet its needs. Some are found in adequate amounts in the feedstuffs commonly used in poultry diets; others must be supplemented. Although the total amount of a vitamin may appear to be adequate, some vitamins are present in bound or unavailable forms in feedstuffs. Supplementation is then essential.

Classification of vitamins

Vitamins are either fat-soluble or water-soluble and are commonly classified in this way (Table 3.5). Vitamin A was the first vitamin discovered and is fat-soluble. Others were later discovered in this group: vitamins D, E and K. Being fat-soluble these vitamins are absorbed into the body with dietary fat, by similar processes. Their absorption is influenced by the same factors influencing fat absorption. Fat-soluble vitamins can be stored in appreciable quantities in the animal body. When they are excreted from the body, they appear in the droppings (excreta).

The first water-soluble vitamin discovered was called vitamin B to distinguish it from vitamin A. Later other B vitamins were discovered and given names such as vitamin B_1, B_2, etc. Now the specific chemical names are used. In distinction to the fat-soluble vitamins, the water-soluble vitamins are not absorbed with fats and they are not stored in appreciable quantities in the body (with the possible exception of B_{12} and thiamin). Excesses of these vitamins are excreted rapidly in urine, requiring a constant dietary supply.

Poultry require 14 vitamins (Table 3.6), but not all have to be provided in the diet. Scott *et al.* (1982) presented good descriptions of the effects of vitamin deficiencies in poultry.

Poultry do not require vitamin C in their diet, because their body tissues can synthesize this vitamin. The other vitamins must be provided in the diet in proper amounts for poultry to grow and reproduce. The egg normally contains sufficient vitamins to supply the needs of the developing embryo. For this reason, eggs are one of the best animal sources of vitamins in the human diet.

Table 3.5. Summary of characteristics of fat-soluble and water-soluble vitamins.

	Fat-soluble	Water-soluble
Chemical composition	C, H, O only	C, H, O + N, S and Co
Occurrence in feeds	Provitamins or precursors may be present	No precursors known (except tryptophan can be converted to niacin)
Function	Specific roles in structural units Exist as several similar compounds	Energy transfer; all are required in all cells, as coenzymes One exact compound
Absorption	Absorbed with fats	Simple diffusion
Storage in body	Substantial; primarily in liver, adipose tissue; not found in all tissues	Little or no storage (except vitamin B_{12} and possibly thiamin)
Excretion	Faecal (exclusively)	Urinary (mainly); bacterial products may appear in faeces
Importance in diet	All animals	Non-ruminants only (generally)
Grouping	A, D, E, K	B complex, C, choline

Table 3.6. Vitamins required by poultry.

Fat-soluble	Water-soluble
Vitamin A[a]	Biotin[a]
Vitamin D[a]$_3$	Choline[a]
Vitamin E[a]	Folacin[a]
Vitamin K[a]	Niacin[a]
	Pantothenic acid[a]
	Riboflavin[a]
	Thiamin
	Pyridoxine
	B_{12} (cobalamin)[a]
	Vitamin C (ascorbic acid)

[a]Supply requirement in dietary supplement.

Fat-soluble vitamins

Vitamin A or a precursor must be provided in the diet. This vitamin occurs in various forms (vitamers): retinol (alcohol), retinal (aldehyde), retinoic acid and vitamin A palmitate (ester). Requirements for vitamin A are usually expressed in international units (IU) per kilogram of diet. The international standards for vitamin A activity are as follows: 1 IU of vitamin A = vitamin A activity of 0.3 µg crystalline vitamin A alcohol (retinol), 0.344 µg vitamin A acetate, or 0.55 µg vitamin A palmitate. One IU of vitamin A activity is equivalent to the activity of 0.6 µg of β-carotene; alternatively, 1 mg β-carotene = 1667 IU vitamin A (for poultry).

Vitamin A has essential roles in vision, bone and muscle growth, reproduction and maintenance of healthy epithelial tissue. Naturally occurring precursors of vitamin A are found in some seeds, leafy green vegetables and forages such as lucerne. The common form of the precursor is β-carotene, which can be converted into vitamin A in the intestinal wall. Carotene is present in considerable quantities in pasture, lucerne hay or meal, and yellow maize. Carotene and vitamin A are rapidly destroyed by exposure to air, light and rancidity, especially at high temperature. Since it is difficult to assess the amount of vitamin A present in the feed, diets should be supplemented with this vitamin.

Deficiency symptoms in poultry include muscular incoordination, uric acid deposits in the ureters and kidneys and general unthriftiness. Hens receiving insufficient vitamin A produce fewer eggs and the eggs frequently do not hatch. Other deficiency signs in poultry include reduced feed intake, susceptibility to respiratory and other infections and, ultimately, death.

Vitamin D is needed by birds for absorption and deposition of calcium. The effects of a deficiency are particularly severe in the young bird. Chicks receiving a diet lacking or low in vitamin D soon develop rickets similar to that resulting from a deficiency of calcium or phosphorus. Growing bones fail to calcify normally and the birds are retarded in growth, unthrifty and often unable to walk. Hens fed diets deficient in vitamin D lay eggs with progressively thinner shells until production ceases. Embryo development is incomplete, probably because the embryo cannot absorb calcium from the eggshell.

Like other fat-soluble vitamins, vitamin D is absorbed in the gut with other lipids. The two major natural sources of vitamin D are cholecalciferol (vitamin D_3, the animal form) and ergocalciferol (vitamin D_2, the plant form). Poultry can only utilize the D_3 form effectively, whereas pigs and other livestock can use both. Most feedstuffs, except for sun-cured hays, are low in this vitamin; therefore, supplementation becomes necessary, especially during winter. Vitamin D can be synthesized in the body by the action of sunlight on a precursor (7-dehydrocholesterol) in the skin, which in summer can provide all of the requirement for vitamin D in poultry housed outdoors. Radiation in the ultraviolet band (UVB) (290–315 nm) portion of the solar spectrum acts on 7-dehydrocholesterol in the skin to produce previtamin D_3, which is then converted in the body to the active forms of the vitamin. Latitude and season affect both the quantity and quality of solar radiation reaching the earth's surface, especially in the UVB region of the spectrum. Studies (Webb *et al.*, 1988) have shown that 7-dehydrocholesterol in human skin exposed to sunlight on cloudless days in Boston (42.2° N) from November to February produced no previtamin D_3. In Edmonton (52° N), this ineffective winter period extended from October to March. Further south (34° N and 18° N), sunlight effectively photo-converted 7-dehydrocholesterol to previtamin D_3 in the middle of winter. Presumably a similar situation prevails in the southern hemisphere. These results demonstrate the dramatic influence of changes in solar UVB radiation on vitamin D_3 synthesis in skin and indicate the effect of latitude on the length of the 'vitamin D winter' during which dietary supplementation of the vitamin is necessary for poultry housed outdoors. Organic poultry producers need to be aware of these findings. (Humans should also take note of

these findings since it is now known that over half the population of senior citizens in Germany are periodically deficient in vitamin D.)

Without supplementation there is a seasonal fluctuation in body stores of the vitamin in poultry housed outdoors, requiring dietary supplementation during winter. Once this deficiency was recognized, dietary supplementation with vitamin D became common practice.

The potency of vitamin D sources is measured in IU or ICU (International Chick Units), 1 IU of vitamin D being defined as equivalent to the activity of 0.025 µg crystalline D_3.

Vitamin E is required for normal growth and reproduction. The most important natural source is α-tocopherol found in plant oils and seeds. The ester form (e.g. vitamin E acetate) can be synthesized and is used for feed supplementation. One IU of vitamin E is defined as being equivalent to the activity of 1 mg DL-α-tocopherol acetate. The nutritional role of vitamin E is closely interrelated with that of selenium and is involved mainly in the protection of lipid membranes, such as cell walls, from oxidative damage. Although these signs are similar to those of selenium deficiency, it is not possible to substitute selenium completely for vitamin E. Both nutrients are required in the diet.

In growing chicks, a deficiency can result in: (i) encephalomalacia or 'crazy chick disease'; (ii) exudative diathesis, an oedema caused by excessive capillary permeability; or (iii) muscular dystrophy. Encephalomalacia occurs when the diet contains unsaturated fats that are susceptible to rancidity. Some antioxidants, in addition to vitamin E, are also effective against encephalomalacia. Exudative diathesis is prevented by dietary selenium; and muscular dystrophy is a complex disease influenced by vitamin E, selenium and the AA methionine and cystine. Poor hatchability of fertile eggs can occur when diets of breeding hens are deficient in vitamin E. To prevent possible vitamin E deficiency, diets for growing poultry and breeding hens are usually supplemented with a source of vitamin E and possibly a suitable antioxidant.

Vitamin K occurs naturally in various forms: phylloquinone (K_1) in plants and menaquinone (K_2), which is synthesized in the gut by microbes. Vitamin K is involved in the synthesis of prothrombin in the liver, a blood-clotting factor, hence its name as the coagulation or anti-haemorrhagic vitamin. Chicks or poults fed a diet deficient in this vitamin are likely to develop haemorrhages following a bruise or injury to any part of the body and may bleed to death. Mature fowls are not so easily affected, but when breeding hens are fed diets deficient in vitamin K, the chicks have low reserves of the vitamin and are therefore susceptible to severe bleeding because of greatly prolonged blood-clotting time. When needed, vitamin K is usually added to diets for growing chicks and breeding hens as a synthetic water-soluble form of the vitamin.

Water-soluble (B) vitamins

Eight B vitamins are important in poultry nutrition. In general they participate in biochemical reactions as enzyme cofactors that mostly affect the transfer of energy.

Biotin plays a role in the synthesis of lipids and in glucose metabolism. Poultry diets in regions using wheat as the main cereal source (e.g. Western Canada, Australia, Scandinavia) commonly require supplementation with this vitamin. Good sources of this vitamin include groundnut meal, safflower meal, yeasts, lucerne meal, canola meal, fishmeal and soybean meal. A deficiency of biotin in the diet of young chicks results in skin lesions similar to those observed in pantothenic acid deficiency. The feet become rough and calloused and later crack open and become haemorrhagic. Eventually lesions appear at the corners of the mouth and the eyelids may become granular. Biotin deficiency has also been observed in turkeys, requiring supplementation. Chicks or poults fed raw eggs develop biotin deficiency because biotin is inactivated by avidin, one of the proteins in egg white. Cooked egg white does not produce this effect. Biotin is also involved in the prevention of perosis and is essential for good hatchability of eggs. The amount needed for good health and egg production in mature hens is quite low.

Choline is not a vitamin in the strict sense, but is generally included in the water-soluble group. It is a structural component of cells and is involved in nerve impulses. Together with methionine it serves as an important source of methyl groups, which are necessary in metabolism. Poultry synthesize this vitamin but the process is often inefficient in young birds, making supplementation advisable for broilers and turkeys. Adult birds are able to synthesize enough choline. Good dietary sources include fish solubles, fishmeal, soybean meal and distillers' solubles. Along with manganese, folic acid, nicotinic acid and biotin, choline is necessary for the prevention of perosis (slipped tendon) in young chicks and poults. A lack of choline also results in retarded growth and poor feed utilization.

Cobalamin (vitamin B_{12}) is closely related to folic acid in its metabolism. All plants, fruits, vegetables and grains are devoid of this vitamin. Microorganisms produce all of the cobalamin found in nature. Any occurring in plant material is the result of microbial contamination. Therefore, poultry diets containing no animal products require supplementation. Adequacy of vitamin B_{12} is most critical for growing chicks, poults and breeding hens. Deficiency signs include slow growth, perosis in young stock, decreased efficiency of feed utilization, increased mortality and reduced hatchability of eggs.

Folacin (folic acid) is involved in metabolism and in the biosynthesis of purines and pyrimidines. It is a very stable vitamin but does not occur naturally in feedstuffs. Instead it occurs in reduced forms as polyglutamates, which are readily oxidized. These forms are converted to folic acid in the body. Diets commonly contain sufficient folacin but this is not assured. Folacin is therefore usually included in the vitamin supplement added to poultry diets to ensure adequacy. A deficiency in young chicks or poults results in retarded growth, poor feathering and perosis. Coloured plumage may lack normal pigmentation, and a characteristic anaemia is also present. Cervical paralysis is an additional symptom in deficient turkeys.

Niacin (nicotinic acid) is a constituent of two coenzymes (NAD and NADP), important in metabolism. It is often deficient in diets because feed grains (particularly maize) contain niacin in a form mostly unavailable to poultry. Legumes are good sources, also yeast, wheat bran and middlings, fermentation by-products and some grasses. This vitamin can be synthesized by the bird from the AA tryptophan, but the efficiency of conversion is low. A deficiency of the vitamin in young chicks results primarily in poor growth, enlargement of the hock joint and perosis. Turkeys are particularly susceptible to hock disorders. Other signs of deficiency are a dark inflammation of the tongue and mouth cavity, loss of appetite and poor feathering. Affected chicks become nervous and irritable. With reduced feed consumption, growth is greatly retarded. A synthetic form of nicotinic acid is generally used for dietary supplementation.

Pantothenic acid is a component of coenzyme A (CoA). Diets are often deficient in this vitamin since cereal grains and plant proteins are a poor source of this vitamin. Good sources include brewer's yeast, lucerne and fermentation by-products. Young chicks and poults fed a diet deficient in pantothenic acid grow slowly and have ragged feathering. Scabby lesions appear at the corners of the mouth, on the edges of the eyelid and around the vent. In severe cases they also occur on the feet. A deficiency in breeding flocks results in reduced hatchability and the chicks that do hatch frequently show high early mortality. Calcium pantothenate is commonly used for dietary supplementation.

Pyridoxine is a component of several enzyme systems involved in N metabolism. In general, diets provide an adequate amount, in the free form or combined with phosphate. Some feedstuffs such as linseed and certain varieties of beans may contain pyridoxine antagonists. Pyridoxine is one of the vitamins that suffers during feed processing, 70–90% of the content in wheat being lost during milling (Nesheim, 1974). A severe deficiency results in jerky movements and aimless running about, followed by convulsions, exhaustion and death.

In mature fowls there is a loss of appetite followed by rapid loss of weight and death. Reduced egg production and poor hatchability can also be observed.

Riboflavin, a water-soluble vitamin, is the one most likely to be deficient in poultry diets, since cereal grains and plant proteins are poor sources of riboflavin. Therefore, all poultry diets need to be supplemented with this vitamin. Milk products have been used traditionally in poultry diets as good sources of riboflavin. Other good sources are green forages and fermentation by-products. Riboflavin is required as a component of two important coenzymes (FAD and FMN) and chickens and turkeys receiving diets deficient in this vitamin grow poorly and often develop a lameness called curled-toe paralysis. Breeding hens need supplements of riboflavin in the diet, otherwise their eggs will not hatch properly. Diets are usually supplemented with a synthetic source of this vitamin.

Thiamin is important as a component of the coenzyme thiamin pyrophosphate (TPP) (cocarboxylase). Good sources are lucerne, grains and yeast. A deficiency is less frequently encountered than deficiencies of other vitamins, since thiamin occurs in abundance in whole grains, which make up the major part of poultry diets. A diet deficient in thiamin results in nervous disorders in both young and old birds, and eventual paralysis of the peripheral nerves (polyneuritis).

Ascorbic acid (vitamin C) is a water-soluble vitamin but is not part of the B group. It is a metabolic requirement for all species but is a dietary requirement only for those that lack the enzyme for its synthesis (primates, guinea pigs, certain birds, fish). Therefore, it is not required in poultry diets. It is involved in the formation and maintenance of intercellular tissues having collagen or related substances as basal constituents.

Response to signs of vitamin deficiency

Vitamin deficiency signs are rarely specific. Thus, if a deficiency of A, D or E is suspected, it is advisable to check with a nutritionist or veterinarian and administer all three by supplementing the feed or the drinking water (using a water-miscible form). If a B vitamin deficiency is suspected, it is advisable to check with a nutritionist or veterinarian and administer a B vitamin complex by supplementing the feed, or preferably the drinking water, since these vitamins are water-soluble and poultry do not eat well when deficient in B vitamins. The prevailing organic standards may permit injection of vitamins to correct deficiencies, but this should be checked with the certifying agency.

Water

Water is also a required nutrient, the requirement being about two to three times the weight of feed eaten. The most important consideration with poultry is to ensure that there is an adequate supply of fresh, uncontaminated water available at all times. Water should always be available *ad libitum*, from drinkers designed for poultry. Water quality is important. The guidelines are based on total dissolved solids (TDS) of up to 5000 mg/kg and pH between 6 and 8 being generally acceptable.

Birds are also very sensitive to the temperature of the drinking water, preferring cold water over water that is above the ambient temperature. This can affect intake of feed.

Feed Analysis

A feed ingredient or diet can be analysed chemically to provide information on the contents of the components discussed above. Generally this does not provide information on the amount of the nutrient biologically available to the animal.

Proximate (approximate) analysis is a scheme developed originally in 1865 by Henneberg and Stohmann of the Weende Experiment Station in Germany to analyse the main components. It is often referred to as the Weende System and has been refined over time. The system consists of

determinations of water (moisture), ash, crude fat (ether extract), crude protein (CP) and crude fibre (CF). It attempts to separate carbohydrates into two broad classifications: CF (indigestible carbohydrate) and N-free extract (NFE, or digestible carbohydrate). NFE is measured by difference rather than by direct analysis.

The information gained is as follows:

1. Moisture (water). This can be regarded as a component that dilutes the content of nutrients and its measurement provides more accurate information on actual contents of nutrients.

2. Dry matter. This is the amount of dry material present after the moisture (water) content has been deducted.

3. Ash. This provides information on mineral content. Further analyses can provide information on sand contamination and on specific minerals present, such as calcium.

4. Organic matter. This is the amount of protein and carbohydrate material present after ash has been deducted from dry matter.

5. Protein. This is determined as nitrogen content × 6.25. It is a measure of protein present, based on the assumption that the average nitrogen content is 16 g N/100 g protein. Some of the nitrogen in most feeds is present as non-protein nitrogen (NPN) and, therefore, the value calculated by multiplying nitrogen content by 6.25 is referred to as crude rather than true protein. True protein is made up of amino acids, which can be measured using specialized techniques.

6. Non-nitrogenous material:

6.1. Fibre: measured as crude fibre. Part of this fraction is digestible; therefore more exact methods of fibre analysis have been developed, especially for use in cattle and sheep feeding since these species eat mainly forages. The aim is to separate the forage fibre into two fractions: (i) plant cell contents, a highly digestible fraction consisting of sugars, starches, soluble protein, pectin and lipids (fats); and (ii) plant cell wall constituents, a fraction of variable digestibility. The soluble fraction is termed neutral-detergent solubles (NDS) (cell contents) and the fibrous residue is called neutral-detergent fibre (NDF) (cell wall constituents). A second method is the acid-detergent fibre (ADF) analysis, which breaks down NDF into a soluble fraction containing primarily hemicellulose and some insoluble protein and an insoluble fraction containing cellulose, lignin and bound nitrogen. Lignin has been shown to be a major factor influencing the digestibility of forages.

It is important to note that CF is still the fibre component used by the NRC (1994) and is the component required by Feed Regulatory authorities to be stated on the feed tag of purchased feed, at least in North America.

6.2. Nitrogen-free extract: the digestible carbohydrates, i.e. starch and sugars. This is a measure of energy and is calculated as 100 − (% crude protein + % crude fat + % crude fibre + % ash + % moisture).

7. Fat. This is measured as crude fat (sometimes called oil or ether extract since ether is used in the extraction process). More detailed analyses can be done to measure individual fatty acids.

Vitamins are not measured directly in the Proximate Analysis system, but can be measured in the fat- and water-soluble extracts by appropriate methods.

Rapid methods based on techniques such as near-infrared reflectance spectroscopy (NIRS) have been introduced to replace chemical methods for routine feed analysis, but bioavailability of nutrients is based mainly on the results of animal studies.

Publications on Nutrient Requirements

Nutrient requirements in North America are based on the recommendations of the National Research Council – National Academy of Sciences, Washington, DC. The recommendations cover pigs, poultry, dairy cattle, horses, laboratory animals, etc. and are published as a series of books. The recommendations for each species are usually updated about every 10 years, the current Nutrient Requirements of Poultry being the 9th revised edition (1994). A committee was set up in 2017 to prepare the 10th edition. The information is used widely by the feed industry in North America and in many other regions.

No comparable recommendations exist in other countries. UK nutrient requirement standards were prepared in the past by national committees (e.g. ARC, 1975) but they have not been updated recently. Australian feeding standards were published in 1987 (SCA, 1987) but have not been revised. The most recent French publication on requirements is the Institut National de la Recherche Agronomique publication (INRA, 1984), which covers pigs, poultry and rabbits. One of the limitations of published estimates of requirements is their applicability generally. For instance, a main issue influencing nutrient requirements for energy and AAs in the growing bird is the capacity for the genotype in question to deposit lean tissue as the bird grows to maturity or develops reproductive capacity. Responses to higher dietary concentrations of AAs will be positive only in birds with a genetic potential to deposit lean tissue rather than fat, or to produce a large number of eggs. As a result, it is difficult to establish nutritional standards for AAs that can be applied generally to all genotypes. For this reason, the conventional broiler and layer industries in Europe, Asia, Australia and North America commonly use nutrient requirement models based on requirement data but tailored to specific strains and genotypes of poultry. These models require accurate information on input/output data and are beyond the scope of the average organic producer. There is currently no set of nutritional standards designed specifically for organic poultry. These standards have to be derived from existing standards for commercial poultry.

One of the criticisms of the NRC publications is that some of the data are old and out of date because the research in question was carried out some time ago. Also, the time lag in the derivation of new research findings, its peer review and publication in scientific journals and its incorporation into the NRC recommendations makes the information less applicable to superior genotypes. However, this criticism is of less importance to organic producers. Many organic producers use traditional breeds and strains of poultry that have not been subjected to the selection pressure imposed on leading genotypes used in conventional production. Consequently, they should find the NRC publications a useful guide to nutrient requirements. Furthermore, it could be argued that, of the various requirement estimates available, the ARC (1975) estimates are the most applicable to organic production because of the genotypes used in deriving them. The data are, however, incomplete. It is debatable whether requirement tables such as those produced by NRC and ARC are applicable in developing countries. For instance, Preston and Leng (1987) argued that in developing countries the objective should be to optimize the use of available resources and minimize the use of imported ingredients. Under these conditions it is very difficult to apply NRC or ARC requirements economically and optimal production is, as a result, less than maximal.

Blair *et al.* (1983) reviewed the existing international nutrition standards for poultry and the British Society of Animal Science (BSAP) conducted a similar review in 2002. The BSAP review provided a good assessment of the factors that need to be taken into account in setting standards based on estimated requirements and of information currently lacking in the database on nutrient requirements for the various species of poultry and on nutrient availability in feed ingredients.

This volume takes the view that the NRC recommendations are of primary interest to organic poultry producers worldwide. Accordingly the nutrient requirements set out in Tables 3.7–3.16 below (from NRC, 1994) are suggested as the basis for the establishment of nutritional standards applicable in average flocks of organic poultry, the birds being drawn from traditional breeds and strains. On the other hand, organic producers using modern hybrids may find the requirement values recommended by the breeding company for that particular genotype to be more useful than the NRC values.

Derivation of Standards

Standards can be derived from the above data for application by producers and by the feed industry. Application of the standards is aimed at providing a balanced diet, the features of which can be outlined as follows:

- The AME (or TME) level is correct for the class of fowl in question.
- The CP is in correct proportion to AME (or TME).
- The essential AA requirements have been met and the balance of AAs is appropriate.
- Sufficient minerals have been added to meet the requirements for:
  i. macrominerals;
  ii. trace minerals.
- Sufficient vitamins have been added to meet requirements.
- The diet contains no dangerous excesses of nutrients or deleterious compounds.

In addition, it is desirable to ensure that suitable ingredients have been selected and that they have been mixed to produce a uniform diet. This aspect is outlined in Chapter 5.

Table 3.7. NRC (1994) estimated nutrient requirements of growing Leghorn-type chickens, amount/kg diet (90% moisture basis)[a].

Stage	White-egg layers				Brown-egg layers			
	0–6 weeks	6–12 weeks	12–18 weeks	18 weeks – 1st egg	0–6 weeks	6–12 weeks	12–18 weeks	18 weeks – 1st egg
Final body weight (g)	450	980	1375	1475	500	1100	1500	1600
Typical AME (kcal)	2850	2850	2900	2900	2800	2800	2850	2850
Crude protein (g)	180	160	150	170	170	150	140	160
Amino acids (g)								
Arginine	10.0	8.3	6.7	7.5	9.4	7.8	6.2	7.2
Glycine + serine	7.0	5.8	4.7	5.3	6.6	5.4	4.4	5.0
Histidine	2.6	2.2	1.7	2.0	2.5	2.1	1.6	1.8
Isoleucine	6.0	5.0	4.0	4.5	5.7	4.7	3.7	4.2
Leucine	11.0	8.5	7.0	8.0	10.0	8.0	6.5	7.5
Lysine	8.5	6.0	4.5	5.2	8.0	5.6	4.2	4.9
Methionine	3.0	2.5	2.0	2.2	2.8	2.3	1.9	2.1
Methionine + cystine	6.2	5.2	4.2	4.7	5.9	4.9	3.9	4.4
Phenylalanine	5.4	4.5	3.6	4.0	5.1	4.2	3.4	3.8
Phenylalanine + tyrosine	10.0	8.3	6.7	7.5	9.4	7.8	6.3	7.0
Threonine	6.8	5.7	3.7	4.7	6.4	5.3	3.5	4.4
Tryptophan	1.7	1.4	1.1	1.2	1.6	1.3	1.0	1.1
Valine	6.2	5.2	4.1	4.6	5.9	4.9	3.8	4.3
Minerals (g/kg)								
Calcium	9.0	8.0	8.0	20.0	9.0	8.0	8.0	18
Phosphorus (non-phytate)	4.0	3.5	3.0	3.2	4.0	3.5	3.0	3.5
Chloride	1.5	1.2	1.2	1.5	1.2	1.1	1.1	1.1
Magnesium	0.6	0.5	0.4	0.4	0.57	0.47	0.37	0.37
Potassium	2.5	2.5	2.5	2.5	2.5	2.5	2.5	2.5
Sodium	1.5	1.5	1.5	1.5	1.5	1.5	1.5	1.5
Trace minerals (mg)								
Copper	5.0	4.0	4.0	4.0	5.0	4.0	4.0	4.0
Iodine	0.35	0.35	0.35	0.35	0.33	0.33	0.33	0.33
Iron	80.0	60.0	60.0	60.0	75.0	56.0	56.0	56.0
Manganese	60.0	30.0	30.0	30.0	56.0	28.0	28.0	28.0
Selenium	0.15	0.1	0.1	0.1	0.14	0.1	0.1	0.1
Zinc	40.0	35.0	35.0	35.0	38.0	33.0	33.0	33.0

Continued

Table 3.7. Continued.

Stage	White-egg layers				Brown-egg layers			
	0–6 weeks	6–12 weeks	12–18 weeks	18 weeks – 1st egg	0–6 weeks	6–12 weeks	12–18 weeks	18 weeks – 1st egg
Vitamins (IU)								
Vitamin A	1500	1500	1500	1500	1420	1420	1420	1420
Vitamin D₃	200	200	200	300	190	190	190	280
Vitamin E	10.0	5.0	5.0	5.0	9.5	4.7	4.7	4.7
Vitamins (mg)								
Biotin	0.15	0.1	0.1	0.1	0.14	0.09	0.09	0.09
Choline	1300	900	500	500	1225	850	470	470
Folacin	0.55	0.25	0.25	0.25	0.52	0.23	0.23	0.23
Niacin	27.0	11.0	11.0	11.0	26.0	10.3	10.3	10.3
Pantothenic acid	10.0	10.0	10.0	10.0	9.4	9.4	9.4	9.4
Pyridoxine	3.0	3.0	3.0	3.0	2.8	2.8	2.8	2.8
Riboflavin	3.6	1.8	1.8	2.2	3.4	1.7	1.7	1.7
Thiamin	1.0	1.0	0.8	0.8	1.0	1.0	0.8	0.8
Vitamin K	0.5	0.5	0.5	0.5	0.47	0.47	0.47	0.47
Vitamins (µg)								
Cobalamin (vitamin B₁₂)	9.0	3.0	3.0	4.0	9.0	3.0	3.0	3.0
Linoleic acid (g)	10.0	10.0	10.0	10.0	10.0	10.0	10.0	10.0

[a]Based on a maize/soy diet. Some values in the above table were stated as being tentative.

Table 3.8. NRC (1994) estimated nutrient requirements of Leghorn-type laying hens, amounts/kg diet (90% moisture basis) and amounts per day[a].

	Amounts/kg diet at different feed intakes: white-egg layers			Amounts per day		
	80	100	120	White-egg breeders	White-egg layers	Brown-egg layers
Feed intake (g/day)	80	100	120	100	100	110
Crude protein (g)	188	150	125	15.0	15.0	16.5
Amino acids (g)						
Arginine	8.8	7.0	5.8	0.7	0.7	0.77
Histidine	2.1	1.7	1.4	0.17	0.17	0.19
Isoleucine	8.1	6.5	5.4	0.65	0.65	0.72
Leucine	10.3	8.2	6.8	0.82	0.82	0.9
Lysine	8.6	6.9	5.8	0.69	0.69	0.76
Methionine	3.8	3.0	2.5	0.3	0.3	0.33
Methionine + cystine	7.3	5.8	4.8	0.58	0.58	0.65
Phenylalanine	5.9	4.7	3.9	0.47	0.47	0.52
Phenylalanine + tyrosine	10.4	8.3	6.9	0.83	0.83	0.91
Threonine	5.9	4.7	3.9	0.47	0.47	0.52
Tryptophan	2.0	1.6	1.3	0.16	0.16	0.18
Valine	8.8	7.0	5.8	0.7	0.7	0.77
Minerals (g)						
Calcium	40.6	32.5	27.1	3.25	3.25	3.6
Phosphorus (non-phytate)	3.1	2.5	2.1	0.25	0.25	0.28
Chloride	1.6	1.3	1.1	0.13	0.13	0.15
Magnesium	0.63	0.5	0.42	0.05	0.05	0.06
Potassium	1.9	1.5	1.3	0.15	0.15	0.17
Sodium	1.9	1.5	1.3	0.15	0.15	0.17
Trace minerals (mg)						
Copper	ND	ND	ND	ND	ND	ND
Iodine	0.044	0.035	0.029	0.01	0.004	0.004
Iron	56	45	38	6.0	4.5	5.0
Manganese	25	20	17	2.0	2.0	2.2
Selenium	0.08	0.06	0.05	0.006	0.006	0.006
Zinc	44	35	29	4.5	3.5	3.9
Vitamins (IU)						
Vitamin A	3750	3000	2500	300	300	330
Vitamin D₃	375	300	250	30	30	33
Vitamin E	6	5	4	1.0	0.5	0.55

Continued

Table 3.8. Continued.

	Amounts/kg diet at different feed intakes: white-egg layers			Amounts per day		
				White-egg breeders	White-egg layers	Brown-egg layers
Vitamins (mg)						
Biotin	0.13	0.1	0.08	0.01	0.01	0.011
Choline	1310	1050	875	105	105	115
Folacin	0.31	0.25	0.21	0.035	0.025	0.028
Niacin	12.5	10.0	8.3	1.0	1.0	1.1
Pantothenic acid	2.5	2.0	1.7	0.7	0.2	0.22
Pyridoxine	3.1	2.5	2.1	0.45	0.25	0.28
Riboflavin	3.1	2.5	2.1	0.36	0.25	0.28
Thiamin	0.38	0.7	0.6	0.07	0.07	0.08
Vitamin K	0.6	0.5	0.4	0.1	0.05	0.06
Vitamins (µg)						
Cobalamin (vitamin B_{12})	4.0	4.0	4.0	8.0	0.4	0.4
Linoleic acid (g)	12.5	10.0	8.3	1.0	1.0	1.1

[a]Based on a maize/soy diet. Some values were stated as being tentative.
ND = not determined.

Table 3.9. Estimates of ME required per laying hen per day in relation to body weight and rate of egg production (from NRC, 1994).

Body weight (kg)		Rate of egg production (%)						
		0	50	60	70	80	90
1.0	ME (kcal)	130	192	205	217	229	242
1.5	ME (kcal)	177	239	251	264	276	289
2.0	ME (kcal)	218	280	292	305	317	330
2.5	ME (kcal)	259	321	333	346	358	371
3.0	ME (kcal)	296	358	370	383	395	408

Table 3.10. NRC (1994) estimated nutrient requirements of broiler chickens, amounts/kg diet (900 g/kg DM basis)[a].

	Starting 0–3 weeks	Growing 3–6 weeks	Finishing
AME (kcal)[a]	3200	3200	3200
Crude protein (g)	230	200	180
Amino acids (g)			
Arginine	12.5	11.0	10.0
Glycine + serine	12.50	11.4	9.7
Histidine	3.5	3.2	2.7
Isoleucine	8.0	7.3	6.2
Leucine	12.0	10.9	9.3
Lysine	11.0	10.0	8.5
Methionine	5.0	3.8	3.2
Methionine + cystine	9.0	7.2	6.0
Phenylalanine	7.2	6.5	5.6
Phenylalanine + tyrosine	13.4	12.2	10.4
Threonine	8.0	7.4	6.8
Tryptophan	2.0	1.8	1.6
Valine	9.0	8.2	7.0
Minerals (g)			
Calcium	10.0	9.0	8.0
Phosphorus (non-phytate)	4.5	3.5	3.0
Chloride	2.0	1.5	1.2
Magnesium	0.6	0.6	0.6
Potassium	3.0	3.0	3.0
Sodium	2.0	1.5	1.2
Trace minerals (mg)			
Copper	8.0	8.0	8.0
Iodine	0.35	0.35	0.35
Iron	80	80	80
Manganese	60	60	60
Selenium	0.15	0.15	0.15
Zinc	40	40	40
Vitamins (IU)			
Vitamin A	1500	1500	1500
Vitamin D$_3$	200	200	200
Vitamin E	10	10	10
Vitamins (mg)			
Biotin	0.15	0.15	0.12
Choline	1300	1000	750
Folacin	0.55	0.55	0.5
Niacin	35	30	25
Pantothenic acid	10.0	10.0	10.0
Pyridoxine	3.5	3.5	3.0
Riboflavin	3.6	3.6	3.0
Thiamin	1.8	1.8	1.8
Vitamin K	0.5	0.5	0.5
Vitamins (µg)			
Cobalamin (vitamin B$_{12}$)	10.0	10.0	7.0
Linoleic acid (g)	10.0	10.0	10.0

[a]Typical ME level used in conventional diets. Some values were stated as being tentative.

Table 3.11. NRC (1994) estimated nutrient requirements of male (M) and female (F) turkeys, amounts/kg diet (90% moisture basis)[a].

	Amounts/kg diet							
	Growing turkeys						Breeding turkeys	
	0–4 M / 0–4 F	4–8 M / 4–8 F	8–12 M / 8–11 F	12–16 M / 11–14 F	16–20 M / 14–17 F	20–24 M / 17–20 F	Holding	Laying
AME (kcal)	2800	2900	3000	3100	3200	3300	2900	2900
Crude protein (g)	280	260	220	190	165	140	120	140
Amino acids (g)								
Arginine	16.0	14.0	11.0	9.0	7.5	6.0	5.0	6.0
Glycine + serine	10.0	9.0	8.0	7.0	6.0	5.0	4.0	5.0
Histidine	5.8	5.0	4.0	3.0	2.5	2.0	2.0	3.0
Isoleucine	11.0	10.0	8.0	6.0	5.0	4.5	4.0	5.0
Leucine	19.0	17.5	15.0	12.5	10.0	8.0	5.0	5.0
Lysine	16.0	15.0	13.0	10.0	8.0	6.5	5.0	6.0
Methionine	5.5	4.5	4.0	3.5	2.5	2.5	2.0	2.0
Methionine + cystine	10.5	9.5	8.0	6.5	5.5	4.5	4.0	4.0
Phenylalanine	10.0	9.0	8.0	7.0	6.0	5.0	4.0	5.5
Phenylalanine + tyrosine	18.0	16.0	12.0	10.0	9.0	9.0	8.0	10.0
Threonine	10.0	9.5	8.0	7.5	6.0	5.0	4.0	4.5
Tryptophan	2.6	2.4	2.0	1.8	1.5	1.3	1.0	1.3
Valine	12.0	11.0	9.0	8.0	7.0	6.0	5.0	5.8
Minerals (g)								
Calcium	12.0	10.0	8.5	7.5	6.5	5.5	5.0	22.5
Phosphorus (non-phytate)	6.0	5.0	4.2	3.8	3.2	2.8	2.5	3.5
Chloride	1.5	1.4	1.4	1.2	1.2	1.2	1.2	1.2
Magnesium	0.5	0.5	0.5	0.5	0.5	0.5	0.5	0.5
Potassium	7.0	6.0	5.0	5.0	4.0	4.0	4.0	6.0
Sodium	1.7	1.5	1.2	1.2	1.2	1.2	1.2	1.2
Trace minerals (mg)								
Copper	8.0	8.0	6.0	6.0	6.0	6.0	6.0	8.0
Iodine	0.4	0.4	0.4	0.4	0.4	0.4	0.4	0.4
Iron	80	60	60	60	50	50	50	60
Manganese	60	60	60	60	60	60	60	60
Selenium	0.2	0.2	0.2	0.2	0.2	0.2	0.2	0.2
Zinc	70	65	50	40	40	40	40	65

Continued

Table 3.11. Continued.

| | Amounts/kg diet | | | | | | | |
| | Growing turkeys | | | | | | Breeding turkeys | |
	0–4 M / 0–4 F	4–8 M / 4–8 F	8–12 M / 8–11 F	12–16 M / 11–14 F	16–20 M / 14–17 F	20–24 M / 17–20 F	Holding	Laying
Vitamins (IU)								
Vitamin A	5000	5000	5000	5000	5000	5000	5000	5000
Vitamin D$_3$	1100	1100	1100	1100	1100	1100	1100	1100
Vitamin E	12	12	10	10	10	10	10	25
Vitamins (mg)								
Biotin	0.25	0.2	0.125	0.125	0.1	0.1	0.1	0.2
Choline	1600	1400	1100	1100	950	800	800	1000
Folacin	1.0	1.0	0.8	0.8	0.7	0.7	0.7	1.0
Niacin	60	60	50	50	40	40	40	40
Pantothenic acid	10.0	9.0	9.0	9.0	9.0	9.0	9.0	16.0
Pyridoxine	4.5	4.5	3.5	3.5	3.0	3.0	3.0	4.0
Riboflavin	4.0	3.6	3.0	3.0	2.5	2.5	2.5	4.0
Thiamin	2.0	2.0	2.0	2.0	2.0	2.0	2.0	2.0
Vitamin K	1.75	1.5	1.0	0.75	0.75	0.5	0.5	1.0
Vitamins (µg)								
Cobalamin (vitamin B$_{12}$)	3.0	3.0	3.0	3.0	3.0	3.0	3.0	3.0
Linoleic acid (g)	10.0	10.0	8.0	8.0	8.0	8.0	8.0	11.0

[a]Based on a maize/soy diet. Some values were stated as being tentative.

Table 3.12. NRC (1994) estimated nutrient requirements of geese, amounts/kg diet (90% moisture basis)[a].

	0–4 weeks	After 4 weeks	Breeding
AME (kcal)	2900	3000	2900
Crude protein (g)	200	150	150
Amino acids (g)			
Lysine	10.0	8.5	6.0
Methionine + cystine	6.0	5.0	5.0
Minerals (g)			
Calcium	6.5	6.0	22.5
Phosphorus (non-phytate)	3.0	3.0	3.0
Vitamins (IU)			
Vitamin A	1500	1500	4000
Vitamin D$_3$	200	200	200
Vitamins (mg)			
Choline	1500	1000	ND
Niacin	65	35	20
Pantothenic acid	15.0	10.0	10.0
Riboflavin	3.8	2.5	4.0

[a]Based on a maize/soy diet. Some values were stated as being tentative.
ND = not determined.

Table 3.13. NRC (1994) estimated nutrient requirements of ducks (White Pekin), amounts/kg diet (90% moisture basis)[a].

	0–2 weeks	2–7 weeks	Breeding
AME (kcal)	2900	3000	2900
Crude protein (g)	220	160	150
Amino acids (g)			
Arginine	11.0	10.0	ND
Isoleucine	6.3	4.6	3.8
Leucine	12.6	9.1	7.6
Lysine	9.0	6.5	6.0
Methionine	4.0	3.0	2.7
Methionine + cystine	7.0	5.5	5.0
Tryptophan	2.3	1.7	1.4
Valine	7.8	5.6	4.7
Minerals (g)			
Calcium	6.5	6.0	27.5
Phosphorus (non-phytate)	4.0	3.0	ND
Chloride	1.2	1.2	1.2
Magnesium	0.5	0.5	0.5
Sodium	1.5	1.5	1.5

Continued

Table 3.13. Continued.

	0–2 weeks	2–7 weeks	Breeding
Trace minerals (mg)			
Manganese	50	ND	ND
Selenium	0.2	ND	ND
Zinc	60	ND	ND
Vitamins (IU)			
Vitamin A	2500	2500	4000
Vitamin D$_3$	400	400	900
Vitamin E	10.0	10.0	10.0
Vitamins (mg)			
Niacin	55	55	55
Pantothenic acid	11.0	11.0	11.0
Pyridoxine	2.5	2.5	3.0
Riboflavin	4.0	4.0	4.0
Vitamin K	0.5	0.5	0.5

[a]Based on a maize/soy diet. Some values were stated as being tentative.
ND = not determined.

Table 3.14. NRC (1994) estimated nutrient requirements of ring-necked pheasants, amounts/kg diet (90% moisture basis).

	0–4 weeks	5–8 weeks	9–17 weeks	Breeding
AME (kcal)	2800	2800	2700	2800
Crude protein (g)	280	240	180	150
Amino acids (g)				
Glycine + serine	18.0	15.5	10.0	5.0
Lysine	15.0	14.0	8.0	6.8
Methionine	5.0	4.7	3.0	3.0
Methionine + cystine	10.0	9.3	6.0	6.0
Minerals (g)				
Calcium	10.0	8.5	5.3	25.0
Phosphorus (non-phytate)	5.5	5.0	4.5	4.0
Chloride	1.1	1.1	1.1	1.1
Sodium	1.5	1.5	1.5	1.5
Trace minerals (mg)				
Manganese	70	70	60	60
Zinc	60	60	60	60
Vitamins (mg)				
Choline	1430	1300	1000	1000
Niacin	70	70	40	30
Pantothenic acid	10.0	10.0	10.0	16.0
Riboflavin	3.4	3.4	3.0	4.0
Linoleic acid (g)	10.0	10.0	10.0	10.0

Table 3.15. NRC (1994) estimated nutrient requirements of Japanese quail (*Coturnix*), amounts/kg diet (90% moisture basis).

	Starting and growing	Breeding
AME (kcal)	2900	2900
Crude protein (g)	240	200
Amino acids (g)		
Arginine	12.5	12.6
Glycine + serine	11.5	11.7
Histidine	3.6	4.2
Isoleucine	9.8	9.0
Leucine	16.9	14.2
Lysine	13.0	10.0
Methionine	5.0	4.5
Methionine + cystine	7.5	7.0
Phenyalanine	9.6	7.8
Phenyalanine + tyrosine	18.0	14.0
Threonine	10.2	7.4
Tryptophan	2.2	1.9
Valine	9.5	9.2
Minerals (g)		
Calcium	8.0	25.0
Phosphorus (non-phytate)	3.0	3.5
Chloride	1.4	1.4
Magnesium	0.3	0.5
Potassium	4.0	4.0
Sodium	1.5	1.5
Trace minerals (mg)		
Copper	5.0	5.0
Iodine	0.3	0.3
Iron	120	60
Manganese	60	60
Selenium	0.2	0.2
Zinc	25	50
Vitamins (IU)		
Vitamin A	1650	3300
Vitamin D$_3$	750	900
Vitamin E	12	25
Vitamins (mg)		
Biotin	0.3	0.15
Choline	2000	1500
Folacin	1.0	1.0
Niacin	40	20
Pantothenic acid	10	15
Pyridoxine	3.0	3.0
Riboflavin	4.0	4.0
Thiamin	2.0	2.0
Vitamin K	1.0	1.0
Vitamins (μg)		
Cobalamin (vitamin B$_{12}$)	3.0	3.0
Linoleic acid (g)	10.0	10.0

Table 3.16. NRC (1994) estimated nutrient requirements of Bobwhite quail, amounts/kg diet (90% moisture basis).

	0–6 weeks	After 6 weeks	Breeding
AME_n (kcal)	2800	2800	2800
Crude protein (g)	260	200	240
Amino acids (g)			
Methionine + cystine	10.0	7.5	9.0
Minerals (g)			
Calcium	6.5	6.5	24.0
Phosphorus (non-phytate)	4.5	3.0	7.0
Chloride	1.1	1.1	1.1
Sodium	1.5	1.5	1.5
Trace minerals (mg)			
Iodine	0.3	0.3	0.3
Vitamins (mg)			
Choline	1500	1500	1000
Niacin	30	30	20
Pantothenic acid	12.0	9.0	15.0
Riboflavin	3.8	3.0	4.0
Linoleic acid (g)	10.0	10.0	10.0

References

Araba, M., Gos, J., Kerr, I. and Dyer, D. (1994) Identity preserved varieties: high oil corn and low stachyose soyabean. *Proceedings of the Arkansas Nutrition Conference*. Fayetteville, Arkansas, pp. 135–142.

ARC (1975) *The Nutrient Requirements of Farm Livestock, No. 1 Poultry*. Agricultural Research Council, London.

Baker, D.H. and Han, Y. (1994) Ideal amino acid profile for chicks during the first three weeks post-hatching. *Poultry Science* 73, 1441–1447.

Blair, R., English, P.R. and Michie, W. (1965) Effect of calcium source on calcium retention in the young chick. *British Poultry Science* 6, 355–356.

Blair, R., Daghir, N.J., Peter, V. and Taylor, T.G. (1983) International nutrition standards for poultry. *Nutrition Abstracts and Reviews – Series B* 53, 669–713.

CVB (1996) Amino acid requirement of laying hens and broiler chicks. In: Schutte, J.B. (ed.) *Report No. 18*. Dutch Bureau of Livestock Feeding, Lelystad, The Netherlands.

El Boushy, A.R.Y. and van der Poel, A.F.B. (2000) Palatability and feed intake regulations. In: *Handbook of Poultry Feed from Waste: Processing and Use*, 2nd edn. Kluwer Academic Publishers, Dordrecht, The Netherlands, pp. 348–397.

Firman, J.D. and Boling, S.D. (1998) Ideal protein in Turkeys. *Poultry Science* 77, 105–110.

Forbes, J.M. and Shariatmadari, F. (1994) Diet selection for protein by poultry. *World's Poultry Science Journal* 50, 7–24.

Gerendai, D. and Gippert, T. (1994) The effect of saponin content of alfalfa meal on the digestibility of nutrients and on the production traits of Tetra-SL layers. *Proceedings of the 9th European Poultry Conference*, Glasgow, UK, pp. 503–504.

Hossain, S.M. and Blair, R. (2007) Chitin utilization by broilers and its effect on body composition and blood metabolites. *British Poultry Science* 48, 33–38.

Husvéth, F., Pál, L., Galamb, E., Ács, K.C., Bustyaházai, L. *et al.* (2015) Effects of whole wheat incorporated into pelleted diets on the growth performance and intestinal function of broiler chickens. *Animal Feed Science and Technology* 210, 144–151.

İnal, F., Coskun, B., Gülsen, N. and Kurtoğlu, V. (2001) The effects of withdrawal of vitamin and trace mineral supplements from layer diets on egg yield and trace mineral composition. *British Poultry Science* 42, 77–80.

INRA (1984) *L'alimentation des Animaux Monogastriques: Porc, Lapin, Volailles.* Institut National de la Recherche Agronomique, Paris, France.

Katongole, J.B.D. and March, B.E. (1980) Fat utilization in relation to intestinal fatty acid binding protein and bile salts in chicks of different ages and different genetic sources. *Poultry Science* 59, 819–827.

Mack, S., Bercovici, D., De Groote, G., Leclercq, B., Lippens, M. *et al.* (1999) Ideal amino acid profile and dietary amino acid specification for broiler chickens of 20–40 days of age. *British Poultry Science* 40, 257–265.

Nesheim, R.O. (1974) Nutrient changes in food processing: a current review. Federation Proceedings 33, 2267–2269.

NRC (1994) *Nutrient Requirements of Poultry*, 9th revised edn. National Academy of Sciences, National Academy Press, Washington, DC.

Patel, K.P., Edwards, H.M. and Baker, D.H. (1997) Removal of vitamin and trace mineral supplements from broiler finisher diets. *Journal of Applied Poultry Research* 6, 191–198.

Preston, T.R. and Leng, R.A. (1987) *Matching Ruminant Production Systems with Available Resources in the Tropics and Subtropics.* Penambull Books, Armidale, New South Wales, Australia.

SCA (1987) *Feeding Standards for Australian Livestock – Poultry.* Standing Committee on Agriculture. CSIRO Editorial and Publishing Unit, East Melbourne, Victoria, Australia.

Scott, M.L., Nesheim, M.C. and Young, R.J. (1982) *Nutrition of the Chicken*, 3rd edn. M.L Scott & Associates, Ithaca, New York.

Shelton, J.L. and Southern, L.L. (2006) Effects of phytase addition with or without a trace mineral premix on growth performance, bone response variables, and tissue mineral concentrations in commercial broilers. *Journal of Applied Poultry Research* 15, 94–102.

Sibbald, I.R. (1982) Measurement of bioavailable energy in poultry feedingstuffs: a review. *Canadian Journal of Animal Science* 62, 983–1048.

Underwood, E.J. and Suttle, N. (1999) *The Mineral Nutrition of Livestock*, 3rd edn. CAB International, Wallingford, UK.

Van Cauwenberghe, S. and Burnham, D. (2001) New developments in amino acid and protein nutrition of poultry, as related to optimal performance and reduced nitrogen excretion. *Proceedings of the 13th European Symposium on Poultry Nutrition*, October 2001, Blankenberge, Belgium, pp. 1–12.

Webb, A.R., Kline, L. and Holick, M.F. (1988) Influence of season and latitude on the cutaneous synthesis of vitamin D3: exposure to winter sunlight in Boston and Edmonton will not promote vitamin D3 synthesis in human skin. *Journal of Clinical Endocrinology and Metabolism* 67, 373–378.

Wiseman, J. (1984) Assessment of the digestibility and metabolizable energy of fats for non-ruminants. In: Wiseman, J. (ed.) *Fats in Animal Nutrition.* Butterworths, London, pp. 227–297.

Wiseman, J. (1986) Anti-nutritional factors associated with dietary fats and oils. In: Haresign, W. and Cole, D.J.A. (eds) *Recent Advances in Animal Nutrition.* Butterworths, London, pp. 47–75.

Zhang, Y. and Parsons, C.M. (1993) Effect of extrusion and expelling on the nutritional quality of conventional and kunitz trypsin inhibitor-free soybeans. *Poultry Science* 72, 2299–2308.

4

Approved Ingredients for Organic Diets

As with pigs and other livestock, the standards of organic poultry farming are based on the principles of enhancement and utilization of the natural biological cycles in soils, crops and livestock. Accordingly, organic poultry production should maintain or improve the natural resources of the farm system, including soil and water quality. Another aim is to maximize the use of farm-grown feed ingredients in poultry and livestock production.

Feed, including pasture and forage, must be produced organically and health care treatments must fall within the range of accepted organic practices. Organic poultry health and performance should be optimized by application of the basic principles of husbandry, such as selection of appropriate breeds and strains, appropriate management practices and nutrition, and avoidance of overstocking. Rather than being designed to maximize performance, the feeding programmes should be designed to minimize metabolic and physiological disorders, hence the requirement for some forage in the diet. Grazing management should be designed to minimize pasture contamination with parasitic larvae. Housing conditions should be such that disease risk is minimized.

Nearly all synthetic animal drugs used to control parasites, prevent disease, promote growth or act as feed additives in amounts above those needed for adequate growth and health are prohibited in organic poultry production. Dietary supplements containing animal by-products such as meat meal are also prohibited. No hormones can be used, a requirement that is easy to apply in organic poultry production since hormone addition to feed has never been a commercial practice and since diethyl stilboestrol (DES) which was used in implantable form in poultry many years ago was banned in 1959.

Permitted feedstuffs for organic poultry production are detailed below. These feed ingredients are considered to be necessary or essential in maintaining bird health, welfare and vitality; they contribute to an appropriate diet fulfilling the physiological and behavioural needs of the species concerned; and they do not contain genetically modified organisms and products thereof.

The specific criteria for feedstuffs and nutritional elements state that:

- feedstuffs of plant origin from non-organic sources can only be used under specified conditions and provided they have been produced or prepared without the use of chemical solvents or chemical treatment;
- feedstuffs of mineral origin, trace elements, vitamins and provitamins can only be used if they are of natural origin. In the case of a shortage of these

substances, or in exceptional circum-stances, chemically well-defined equivalent substances may be used;

- feedstuffs of animal origin, with the exception of milk and milk products, fish, other marine animals and products derived therefrom should generally not be used, or as permitted by national legislation; and
- synthetic nitrogen or non-protein nitrogen compounds shall not be used.

Specific criteria for additives and processing aids state that:

- binders, anti-caking agents, emulsifiers, stabilizers, thickeners, surfactants, coagulants: only natural sources are allowed;
- antioxidants: only natural sources are allowed;
- preservatives: only natural acids are allowed;
- colouring agents (including pigments), flavours and appetite stimulants: only natural sources are allowed; and
- probiotics, enzymes and certain microorganisms are allowed.

The regulations include the requirement for roughage and fresh/dried fodder or silage in the daily ration. All feed ingredients used must be certified as being produced, handled and processed in accordance with the standards specified by the certifying body.

The nutritional characteristics, average composition and (where possible) the recommended inclusion rates of the feedstuffs that are considered most likely to be used in organic poultry diets are set out below.

Due to a lack of data on feedstuffs that have been grown organically, the nutritional data refer mainly to feedstuffs that have been grown conventionally. Eventually a database of organic feedstuffs composition should be developed. The available evidence suggests that organic feed grains and protein crops are slightly lower in nutrient content than their conventionally-grown counterparts (Grashorn and Ritteser, 2016). According to the latter researchers, the contribution of methionine from organic feedstuffs is therefore lower in poultry diets,

an important observation since methionine is generally the first-limiting amino acid in poultry diets. These authors made the important observation that although the protein content of organic cereal grains and protein crops was lower in organic sources, the content of methionine in the major feedstuffs in relation to protein content was similar. Also, the digestibility of amino acids was similar in organic and conventional feedstuffs.

New Zealand is one of the few countries to include a list of approved feed ingredients in its organic regulations (Table 4.1). This is a very useful feature of its regulations. In addition, the regulations stipulate that the feeds must meet the Agricultural Compounds and Veterinary Medicines (ACVM) Act and regulations, and the Hazardous Substances and New Organisms (HSNO) Act, or are exempt, thus providing additional assurance to the consumer. This list appears to be based on the EU list (Table 4.1), possibly because of export requirements.

The EU list is somewhat similar to that for New Zealand, but it details non-organic feedstuffs that can be used in limited quantities in organic feeds for poultry. It may be inferred from the EU list that organic sources of the named ingredients are acceptable.

Most countries follow the EU system and do not publish an approved list, stating that all feedstuffs used must meet organic guidelines. An example is the USA, where the regulations also state that all feed, feed additives and feed supplements must comply with Food and Drug Administration (FDA) regulations.

Canada has a much less detailed Permitted Substances List (CAN/CGSB-32.311-2006) than the New Zealand or EU lists, stating that: '… energy feeds and forage concentrates (grains) and roughages (hay, silage, fodder, straw) shall be obtained from organic sources and may include silage preservation products (e.g. bacterial or enzymatic additives derived from bacteria, fungi and plants and food by-products [e.g. molasses and whey]). Note that if weather conditions are unfavourable to fermentation, lactic, propionic and formic acid may be used'.

Table 4.1. Comparison of approved organic feedstuffs in New Zealand and approved non-organic feedstuffs in the EU.

Group	New Zealand-approved list (only those named in each category) MAF Standard OP3, Appendix Two, 2006 (NZFSA, 2011)	EU-approved list of non-organic feedstuffs (up to defined limits) Council Regulation EC No 834/2007, 2007 (EU, 2007)
1. Feed materials from plant origin	1.1 Cereals, grains, their products and by-products: oats as grains, flakes, middlings, hulls and bran; barley as grains, protein and middlings; rice germ expeller; millet as grains; rye as grains and middlings; sorghum as grains; wheat as grains, middlings, bran, gluten feed, gluten and germ; spelt as grains; triticale as grains; maize as grains, bran, middlings, germ expeller and gluten; malt culms; brewer's grains. (Rice as grain, rice broken, rice bran, rye feed, rye bran and tapioca were delisted in 2004.)	1.1 Cereals, grains, their products and by-products: oats as grains, flakes, middlings, hulls and bran; barley as grains, protein and middlings; rice as grains, rice broken, bran and germ expeller; millet as grains; rye as grains, middlings, feed and bran; sorghum as grains; wheat as grains, middlings, bran, gluten feed, gluten and germ; spelt as grains; triticale as grains; maize as grains, bran, middlings, germ expeller and gluten; malt culms; brewer's grains
	1.2 Oilseeds, oil fruits, their products and by-products: rapeseed, expeller and hulls; soybean as bean, toasted, expeller and hulls; sunflower seed as seed and expeller; cotton as seed and seed expeller; linseed as seed and expeller; sesame seed as expeller; palm kernels as expeller; pumpkin seed as expeller; olives, olive pulp; vegetable oils (from physical extraction). (Turnip rapeseed expeller was delisted in 2004.)	1.2 Oilseeds, oil fruits, their products and by-products: rapeseed, expeller and hulls; soybean as bean, toasted, expeller and hulls; sunflower seed as seed and expeller; cotton as seed and seed expeller; linseed as seed and expeller; sesame seed as seed and expeller; palm kernels as expeller; turnip rapeseed as expeller and hulls; pumpkin seed as expeller; olive pulp (from physical extraction of olives)
	1.3 Legume seeds, their products and by-products: chickpeas as seeds, middlings and bran; ervil as seeds, middlings and bran; chickling vetch as seeds submitted to heat treatment, middlings and bran; peas as seeds, middlings and bran; broad beans as seeds, middlings and bran; horse beans as seeds, middlings and bran; vetches as seeds, middlings and bran; lupin as seeds, middlings and bran	1.3 Legume seeds, their products and by-products: Chickpeas as seeds; ervil as seeds; chickling vetch as seeds submitted to an appropriate heat treatment; peas as seeds, middlings and bran; broad beans as seeds, middlings and bran; horse beans as seeds; vetches as seeds; lupin as seeds
	1.4 Tuber roots, their products and by-products: sugarbeet pulp, potato, sweet potato as tuber, potato pulp (by-product of the extraction of potato starch), potato starch, potato protein and manioc (cassava)	1.4 Tuber roots, their products and by-products: Sugarbeet pulp, dried beet, potato, sweet potato as tuber, manioc as roots, potato pulp (by-product of the extraction of potato starch), potato starch, potato protein and tapioca
	1.5 Other seeds and fruits, their products and by-products: carob, carob pods and meals thereof, pumpkins, citrus pulp, apples, quinces, pears, peaches, figs, grapes and pulps thereof; chestnuts, walnut expeller, hazelnut expeller; cocoa husks and expeller; acorns	1.5 Other seeds and fruits, their products and by-products: carob pods, citrus pulp, apple pomace, tomato pulp and grape pulp

Continued

Table 4.1. Continued.

Group	New Zealand-approved list (only those named in each category) MAF Standard OP3, Appendix Two, 2006 (NZFSA, 2011)	EU-approved list of non-organic feedstuffs (up to defined limits) Council Regulation EC No 834/2007, 2007 (EU, 2007)
	1.6 Forages and roughages: lucerne (alfalfa), lucerne meal, clover, clover meal, grass (obtained from forage plants), grass meal, hay, silage, straw of cereals and root vegetables for foraging	1.6 Forages and roughages: lucerne, lucerne meal, clover, clover meal, grass (obtained from forage plants), grass meal, hay, silage, straw of cereals and root vegetables for foraging
	1.7 Other plants, their products and by-products: molasses, seaweed meal (obtained by drying and crushing seaweed and washed to reduce iodine content), powders and extracts of plants, plant protein extracts (solely provided to young animals), spices and herbs	1.7 Other plants, their products and by-products: molasses as a binding agent in compound feedingstuffs, seaweed meal (obtained by drying and crushing seaweed and washed to reduce iodine content), powders and extracts of plants, plant protein extracts (solely provided to young animals), spices and herbs
2. Feed materials of animal origin	2.1 Milk and milk products: raw milk, milk powder, skimmed milk, skimmed milk powder, buttermilk, buttermilk powder, whey, whey powder, whey powder low in sugar, whey protein powder (extracted by physical treatment), casein powder, lactose powder, curd and sour milk	2.1 Milk and milk products: raw milks (as defined in Article 2 of Directive 92/46/EEC), milk powder, skimmed milk, skimmed milk powder, buttermilk, buttermilk powder, whey, whey powder, whey powder low in sugar, whey protein powder (extracted by physical treatment), casein powder and lactose powder
	2.2 Fish, other marine animals, their products and by-products: fish, fish oil and cod-liver oil not refined; fish molluscan or crustacean autolysates, hydrolysate and proteolysates obtained by an enzyme action, whether or not in soluble form, solely provided to young animals; fishmeal	2.2 Fish, other marine animals, their products and by-products: fish, fish oil and cod-liver oil not refined; fish molluscan or crustacean autolysates, hydrolysate and proteolysates obtained by an enzyme action, whether or not in soluble form, solely provided to young animals; fishmeal
3. Feed materials of mineral origin	3.1 Sodium products: unrefined sea salt, coarse rock salt, sodium sulfate, sodium carbonate, sodium bicarbonate, sodium chloride	3.1 Sodium products: unrefined sea salt, coarse rock salt, sodium sulfate, sodium carbonate, sodium bicarbonate, sodium chloride
	3.2 Calcium products: lithotamnion and maerl shells of aquatic animals (including cuttlefish bones), calcium carbonate, calcium lactate, calcium gluconate	3.2 Calcium products: lithotamnion and maerl shells of aquatic animals (including cuttlefish bones), calcium carbonate, calcium lactate, calcium gluconate
	3.3 Phosphorus products: bone dicalcium phosphate precipitate, defluorinated dicalcium phosphate, defluorinated monocalcium phosphate	3.3 Phosphorus products: bone dicalcium phosphate precipitate, defluorinated dicalcium phosphate, defluorinated monocalcium phosphate
	3.4 Magnesium products: magnesium sulfate, magnesium chloride, magnesium carbonate, magnesium oxide (anhydrous magnesia)	3.4 Magnesium products: anhydrous magnesia, magnesium sulfate, magnesium chloride, magnesium carbonate
	3.5 Sulfur products: sodium sulfate	3.5 Sulfur products: sodium sulfate

Continued

Table 4.1. Continued.

Group	New Zealand-approved list (only those named in each category) MAF Standard OP3, Appendix Two, 2006 (NZFSA, 2011)	EU-approved list of non-organic feedstuffs (up to defined limits) Council Regulation EC No 834/2007, 2007 (EU, 2007)
Feed additives	Trace elements	Trace elements
	E1 Iron products: ferrous carbonate, sulfate monohydrate and/or heptahydrate, ferric oxide	E1 Iron products: ferrous carbonate, ferrous sulfate monohydrate, ferric oxide
	E2 Iodine products: calcium iodate, anhydrous calcium iodate, hexahydrate, sodium iodide	E2 Iodine products: calcium iodate, anhydrous calcium iodate hexahydrate, potassium iodide
	E3 Cobalt products: cobaltous sulfate monohydrate and/or heptahydrate, basic cobaltous carbonate monohydrate	E3 Cobalt products: cobaltous sulfate monohydrate and/or heptahydrate, basic cobaltous carbonate monohydrate
	E4 Copper products: copper oxide, basic copper carbonate monohydrate, copper sulfate pentahydrate	E4 Copper products: copper oxide, basic copper carbonate monohydrate, copper sulfate pentahydrate
	E5 Manganese products: manganous carbonate, manganous oxide, manganous sulfate monohydrate and/or tetrahydrate	E5 Manganese products: manganous carbonate, manganous oxide, manganic oxide. Manganous sulfate, monohydrate and/or tetrahydrate
	E6 Zinc products: zinc carbonate, zinc oxide, zinc sulfate monohydrate and/or heptahydrate	E6 Zinc products: zinc carbonate, zinc oxide, zinc sulfate monohydrate and/or heptahydrate
	E7 Molybdenum products: ammonium molybdate, sodium molybdate	E7 Molybdenum products: ammonium molybdate, sodium molybdate
	E8 Selenium products: sodium selenate, sodium selenite	E8 Selenium products: sodium selenate, sodium selenite
Vitamins and provitamins	Vitamins approved for use under New Zealand Legislation: preferably derived from ingredients occurring naturally in feeds, or synthetic vitamins identical to natural vitamins only for non-ruminant animals. When the organic feed or organic pork product is to be exported to the USA, the vitamins and trace minerals used have to be FDA-approved	Vitamins, provitamins and chemically well-defined substances having a similar effect. Vitamins authorized under Directive 70/524/EEC: preferably derived from raw materials occurring naturally in feedstuffs, or synthetic vitamins identical to natural vitamins (only for non-ruminant animals)
Enzymes	Enzymes approved for use under New Zealand Legislation	Enzymes authorized under Directive 70/524/EEC
Microorganisms	Microorganisms approved for use under New Zealand Legislation	Microorganisms authorized under Directive 70/524/EEC
Preservatives	E 236 Formic acid	E 236 Formic acid only for silage
	E 260 Acetic acid	E 260 Acetic acid only for silage
	E 270 Lactic acid	E 270 Lactic acid only for silage
	E 280 Propionic acid	E 280 Propionic acid only for silage
Binders, anti-caking agents and coagulants	E 551b Colloidal silica	E 551b Colloidal silica
	E 551c Kieselgur	E 551c Kieselgur
	E 558 Bentonite	E 553 Sepiolite
	E 559 Kaolinitic clays	E 558 Bentonite

Continued

Table 4.1. Continued.

Group	New Zealand-approved list (only those named in each category) MAF Standard OP3, Appendix Two, 2006 (NZFSA, 2011)	EU-approved list of non-organic feedstuffs (up to defined limits) Council Regulation EC No 834/2007, 2007 (EU, 2007)
	E 561 Vermiculite	E 559 Kaolinitic clays
	E 562 Sepiolite	E 561 Vermiculite
	E 599 Perlite	E 599 Perlite
Antioxidant substances	E 306 Tocopherol-rich extracts of natural origin	E 306 Tocopherol-rich extracts of natural origin
Certain products used in animal nutrition	Brewer's yeasts	Brewer's yeast
Processing aids for silage	Sea salt, coarse rock salt, whey, sugar, sugarbeet pulp, cereal flour and molasses	Sea salt, coarse rock salt, enzymes, yeasts, whey, sugar, sugarbeet pulp, cereal flour, molasses and lactic, acetic, formic and propionic bacteria
Pure amino acids	None	None

Note: Maize as bran is listed twice in the New Zealand and earlier EU regulations; probably a typographical error – corrected above.

Based on the information above, which is drawn from both the northern hemisphere (EU) and southern hemisphere (New Zealand), Table 4.1 can be suggested as a potential list of the feedstuffs available for organic poultry production in many countries. Not all of the feed ingredients in Table 4.1 are suitable for inclusion in poultry diets, since the lists include those more suited for ruminant feeding. In addition, some of the ingredients are not usually available in sufficient quantity.

One of the questions raised by the publication of lists of approved feed ingredients in organic regulations is how new ingredients are added. An example is lentils, which can be grown organically (mainly for the human market) and are available in some countries for poultry feeding. Therefore, the sections below contain feed ingredients that are not included in Table 4.1 but meet the criteria for inclusion on organic diets. Another interesting question is whether fish products such as fishmeal are organic in the conventional sense. Fortunately they are accepted as such since they are valuable sources of amino acids (AA), particularly when pure AA are not allowed. However, they have to be from sustainable sources and any antioxidant added to prevent spoilage must be from the approved list.

The status of other products such as potato protein could be questioned as being organic in the conventional sense, as they are industrial by-products. Again, their inclusion in approved lists of organic feed ingredients is fortunate since they too are valuable sources of AA. Their designation as organic may therefore be based on expediency rather than organic principles.

Although the New Zealand and EU regulations state that pure amino acids such as lysine and methionine are not allowed in organic feed, the regulations in several regions allow a temporary usage of these feed additives to improve the quality of the protein. For instance, Canada has banned the use of AA from non-synthetic sources only but has granted an exception for the use of synthetic DL-methionine, DL-methionine-hydroxy hydroxyl analogue and DL-methionine-hydroxy analogue calcium in poultry diets, with this exception to be re-evaluated at the next revision of the Canadian standard. Currently the Canadian provinces of Quebec and British Columbia allow some pure amino acids in organic diets, a distinction being made between those of fermented origin (approved, e.g. lysine) and those of synthetic origin (prohibited, e.g. methionine).

Another example of a new feedstuff that might be acceptable in organic feeds is insect larval meal, which is being produced in several countries from substrates such as household food waste. Some certifying agencies take the view that the meal produced is a permissible ingredient (and a good source of protein) in organic feed, while others take the view that the insect meal is unacceptable as an organic feed ingredient since the substrate used cannot be verified as organic. Producers planning to use this type of product should, therefore, obtain approval from a local certifying agency before doing so.

Approved lists are also open to interpretation. An example is calcium carbonate, an approved organic source of calcium (Ca). Is ground limestone, a natural and common source of calcium carbonate and prepared from mined calcareous rock, approved as 'calcium carbonate'? It is a well established ingredient in conventional poultry diets and one assumes that it is acceptable in organic diets. In cases such as this the producer should verify with the certifying agency that this interpretation is correct. This example adds weight to suggestions that it would be very helpful if lists of approved feedstuffs could be very specific.

Professor Lorin Harris, Director of the International Feedstuffs Institute at Utah State University, devised an International Feed Vocabulary to overcome confusion in naming feeds. The system is now used universally. In this system feed names are constructed by combining components within six facets: (i) origin, including scientific name (genus, species, variety), common name (genus, species, variety) and chemical formula where appropriate; (ii) part fed to animals as affected by process(es); (iii) process(es) and treatment(s) to which the origin of part eaten was subjected prior to being fed to the animal; (iv) stage of maturity and development (applicable to forages and animals); (v) cutting (primarily applicable to forages); (vi) grade (official grades and guarantees, etc.). In addition, feeds are separated into eight classes: (i) dry forages and roughages; (ii) pasture, range plants or forages fed green; (iii) silages; (iv) energy feeds; (v) protein supplements; (vi) mineral supplements; (vii) vitamin supplements; and (viii) additives. Each class represents a special characteristic peculiar to a given group of feed products. A six-digit International Feed Number (IFN) is assigned to each feed and is used in the accompanying tables of feedstuffs composition. The first digit of this number denotes the class of feed. The remaining digits are assigned consecutively but never duplicated. The reference number is used in computer programs to identify the feed for use in calculating diets, for summarization of the data, for printing feed composition tables and for retrieving online data on a specific feed. Since some feedstuffs are known under several common names internationally, the IFN is given for each feed in the sections below, which set out the nutritional characteristics and recommended inclusion rates of feedstuffs that are considered most likely to be used in organic poultry diets.

Cereal Grains and By-products

The primary sources of energy in poultry diets are cereal grains. In addition to these grains, the processing of cereals for the human market yields by-products that are important as feed ingredients. Most of the cereals suitable for use in organic poultry production belong to the grass family (*Poaceae*). Their seeds (grains) are high in carbohydrate and they are generally palatable and well digested. Nutrient composition can be quite variable, depending on differences in crop variety, fertilizer practices and growing, harvesting and storage conditions (Svihus and Gullord, 2002) (Table 4.2).

Variability may be higher in organic grains than in conventional grains, because of the fertilizer practices in organic grain production, but the data are inadequate at present. Cereal by-products tend to be more variable than the grains, therefore their use in poultry diets may have to be limited to achieve consistency of the formulations.

Table 4.2. Variability in the nutritional value of wheat, barley and oats grown in Norway (total of 60 batches) (Svihus and Gullord, 2002).

	Barley			Oats			Wheat		
	Min.	Max.	Mean	Min.	Max.	Mean	Min.	Max.	Mean
Crude protein (g/kg)	96	115	107	74	132	100	109	154	130
Starch (g/kg)	587	641	614	468	545	509	614	712	665
Crude fibre (g/kg)	38	64	49	92	123	108	20	26	25
Specific weight (kg/hl)	58.4	73.2	67.8	57.3	62.5	59.7	77	83.1	79.4

The fibre in grains is contained mainly in the hull (husk) and can be variable, depending on the growing and harvesting conditions. This can affect the starch content of the seed and, as a consequence, the energy value (Table 4.2). The hull is quite resistant to digestion and also has a lowering effect on the digestibility of nutrients.

On a dry basis, the crude protein (CP) content ranges from about 100 g/kg to 160 g/kg and is often variable. Generally the crude protein content is slightly lower than in conventionally-grown cereals (Jacob, 2007; Blair, 2012; Kyntäjä et al., 2014) and according to Grela and Semeniuk (2008) organic cereals also contain a higher content of crude fibre and a lower level of energy than conventionally grown grains. The report of Kyntäjä et al. (2014) presented detailed data on the nutritional value of a wide range of organic feedstuffs but without comparable data on their conventional counterparts.

The protein in grains is low in important AA (lysine, methionine, threonine and tryptophan), relative to the birds' requirement. Grains also tend to be low in vitamins and minerals. Therefore, cereal grain-based diets must be supplemented with other ingredients to meet AA and micronutrient requirements. The Grela and Semeniuk (2008) study suggested a slightly lower content of minerals in organic cereals, but the content appeared to be affected mainly by factors such as soil type, fertilization practice and plant species. Yellow maize is the only cereal grain to contain vitamin A, owing to the presence of provitamins (mainly β-carotene). All the grains are deficient in vitamins D and K. The ether extract (oil) in cereal grains is contained in the germ and varies from less than 20 g/kg (dry basis) in wheat to over 50 g/kg in oats. It is high in oleic and linoleic acids, which are unstable after the seed is ground. As a result, rancidity can develop quickly and result in reduced palatability of the feed or in feed refusal.

The cereals in general are good sources of vitamin E and may supply all of the requirements for this vitamin, provided that the grain is used quickly after processing to prevent the development of rancidity and off-flavour. The oil of wheat germ is one of the best-known natural sources of vitamin E, but is unstable.

Of the principal B vitamins, the cereals are good sources of thiamin, but they are low in riboflavin. Maize, oats and rye are much lower in niacin than are barley and wheat, with only about one-third of the niacin being available. Maize is also low in pantothenic acid and all the grains are deficient in vitamin B_{12}. All the cereal grains, especially maize, are deficient in Ca. They contain much higher levels of phosphorus (P) but much of the P is bound as phytate, which is largely unavailable to poultry. Also, phytate affects the availability of Ca and other minerals. Plant breeders are aware of the phytate issue and are developing new cultivars of cereals with reduced phytate content. A type of barley with a 75% reduction in phytate P was introduced in Canada in 2006. Cereals generally supply enough magnesium, but insufficient levels of sodium and possibly potassium. None of the cereals contains high levels of trace minerals.

Thus, feed grains meet only part of the requirement for dietary nutrients. Other feed components are needed to balance the diet completely. Combining the grains and other ingredients into a final dietary mixture to meet the birds' nutritional needs requires information about the nutrient content of each feedstuff and its suitability as a feed ingredient.

Maize, wheat, oats, barley and sorghum are the principal cereals, the whole grains of which are used for feed. Of these, maize is the only grain in layer diets that will result in a yellow-pigmented yolk. Producers using other grains need to ensure supplementation of the diet with alternative sources of egg-pigmenting agents if they wish to market eggs with yellow-coloured yolks. Generally maize, sorghum and wheat are highest in energy value for poultry. Barley, oats and rye are lower in energy. Some rye is used in poultry feeding. Although it is similar to wheat in composition, it is less palatable than other grains and may contain ergot, a toxic fungus. Triticale, a hybrid of rye and wheat, is also used for poultry feeding in some countries. There do not appear to be any genetically modified (GM) varieties of wheat, sorghum, barley or oats being grown, unlike the situation with maize. In the USA, for instance, substantial quantities of GM maize varieties developed with insect and herbicide resistance are grown. Such bioengineered varieties are obviously unsuitable for organic poultry production.

Average composition values of commonly used feeds are presented in the tables at the end of this chapter (Appendix 4.1: Tables 4A.1 to 4A.45) and can be used as a guide in formulating diets for poultry. However, it is recommended that, where possible, chemical analysis of the grain or feed product be conducted prior to feeding, to determine more exactly its nutrient composition and quality. Analyses for moisture, protein and kernel weight are generally adequate for grains.

Several by-products of grain milling and processing are valuable ingredients for poultry diets. The grain seed consists of an outer hull or bran fraction, covering the endosperm fraction which consists mainly of starch and some protein. At the base of the seed is the germ which contains most of the fat (oil), fat-soluble vitamins and minerals. Processing of grains for the human market usually involves removal of the starch, leaving the other fractions as animal feed. The composition of these by-products varies according to the process used. Grain screenings (cleanings) are used in conventional animal feeding. These contain broken and damaged grains, weed seeds, dust, etc. and probably do not meet the quality standards for organic poultry diets.

In general, grain should be ground to a fine, uniform consistency for mixing into diets to be pelleted or crumbled. A coarser grind can be used for mash diets. In this case the particle size should be similar to that of the other ingredients, to prevent the birds from over- or under-consuming the grain relative to the other dietary constituents. The more dominant birds in the flock may pick out the larger particles, leaving the smaller particles for less dominant birds. This can result in uneven flock performance.

One aspect of grain feeding that is of interest to organic poultry producers is the inclusion of whole (unground) grain in the diet. Feeding systems employing whole grain are closer to the natural poultry feeding situation than the feeding of diets containing ground grain. In addition, the bird is well equipped to digest whole grains since its gut possesses a grinding organ (the gizzard) attached to the proventriculus. Blair *et al.* (1973) found that egg production and metabolizable energy (ME) intake were similar when light hybrid layers were fed a conventional mash diet or a diet based on whole grain plus concentrate pellets in the correct proportion. Another aspect of feed particle size that is of relevance to organic production is that a coarser grind or whole grain may assist in improving gizzard function and disease resistance. This aspect will be discussed further in Chapter 7.

Barley and by-products

Barley (*Hordeum* spp.) (Fig. 4.1) is grown more widely throughout the world than any other cereal. It is grown in regions of North America, Europe and Australia that are less well adapted to maize, typically where the growing season is relatively short and climatic conditions cool and dry. It is the major feed grain grown in Canada, mainly in the prairies. Barley is also a good rotation crop with wheat, tends to be higher yielding, matures earlier and is more resistant to drought and salinity problems. Barley is classified as six-row or two-row, depending

Fig. 4.1. Barley is an excellent cool-season grain crop for organic poultry feeding.

on the physical arrangement of kernels on the plant. Two-row varieties are adapted to drier climates and six-row cultivars to the wetter areas.

Traditionally the higher grades of barley have been used for malting and the lower grades for livestock feed. Barley can be an excellent grain source for poultry diets (Jacob and Pescatore, 2012; Anon., 2013). It has been used for some time as the principal grain for poultry feeding in the western areas of North America, the UK and many countries of Europe because of its better adaptation to climate. It is also used in place of yellow maize in regions that demand a white-skinned bird with a firmer layer of subcutaneous fat than is found with maize-based diets.

Nutritional features

Barley is considered a medium-energy grain, containing more fibre and being lower in ME than maize. It is an important feedstuff for poultry (Jeroch and Dänicke, 1995). The proportion of hull to kernel is variable, resulting in a variable ME value for poultry. The nutritive value and suitability of this grain as a feedstuff for growing poultry are influenced also by varying concentration (commonly 40–150 g/kg) of a non-starch polysaccharide (NSP), β-glucan. Water-soluble NSPs increase the viscosity of the intestinal contents, resulting in a reduction in bird performance. They also cause wet and sticky droppings, which may result in foot and leg problems and breast blisters. The protein content is higher than in maize and can range from about 90 g/kg to 160 g/kg. According to Kyntäjä et al. (2014) barley grown organically in Europe has a CP content of 112 g/kg (DM basis), whereas the value normally reported by other researchers for conventionally grown barley is around 125 g/kg. A large study in Poland found that the average CP value was 108.7 g/kg in organic barley compared with 112.8 g/kg in conventional barley (Grela and Semeniuk, 2008). The corresponding values for crude fibre (CF) were 42.1 g/kg and 39.3 g/kg. The AA profile of the protein is better than in maize and is closer to that of

oats or wheat; also the bioavailability of the essential amino acids (EAA) is high. Barley contains more P than other common cereal grains.

Poultry diets

Barley is more suitable in diets for laying stock than in diets for meat birds because of its higher fibre content and lower ME value relative to maize or wheat. Layer diets can contain a high proportion of barley without negative effects on egg production, though feed efficiency is poorer than with higher-energy diets. When formulating barley-based layer diets the low content of linoleic acid should be taken into account, otherwise egg size may be reduced. The negative effects of NSPs can be reduced or eliminated by the addition of β-glucanase of microbial origin to the diet. Supplementation with enzyme allows barley to be suitable for inclusion in broiler diets (Classen et al., 1988a; Choct et al., 1995). Research studies have shown that supplementation of barley-based diets with enzyme can result in broiler performance comparable to that attained with maize (Marquardt et al., 1994). Barley should be ground to a fine, uniform consistency for mixing into diets to be pelleted or crumbled. A coarser grind can be used for mash diets.

Hull-less barley

Nutritional features

Hull-less barley varieties have been developed, in which the hull separates during threshing. These varieties contain more protein and less fibre than conventional barley and theoretically should be superior in nutritive value to conventional barley. However, Ravindran et al. (2007) found that the ME (N-corrected basis) was similar in hull-less and hulled barley. The chemical composition of six Brazilian hull-less barley cultivars was studied by Helm and de Francisco (2004) and reported as follows. The highest constituents were starch (575–631 g/kg), crude protein (125–159 g/kg) and total dietary fibre (TDF) (124–174 g/kg), the starch and crude protein contents being in agreement with those previously reported for Swedish (Elfverson et al., 1999) and Canadian (Li et al., 2001) varieties. The other reported values (g/kg) were ash content 15.1–22.7, ether extract 29.1–40.0, starch 574.6–631.4, insoluble dietary fibre 80.7–121.6, soluble dietary fibre 43.0–64.5 and β-glucan 37.0–57.7.

Poultry diets

Classen et al. (1988b) in two experiments fed 20-week-old White Leghorn hens for 40 weeks on wheat-based diets with 0, 200, 400, 600 or 800 g wheat/kg replaced by hull-less barley (experiment 1) and 357 or 714 g wheat/kg replaced by hull-less or conventional barley (experiment 2). Hens that were fed diets with hull-less barley at 714 g/kg were heavier and produced larger eggs than hens fed diets containing a similar level of conventional barley. The results indicated that hull-less barley was at least equivalent to wheat and was better than conventional barley as a cereal grain for laying hens.

Research findings with growing birds have been less conclusive, related probably to the level of β-glucan in hull-less barley which has more significant effects in young birds than in adult birds. For instance, Bekta et al. (2006) found in a broiler study in Poland that a diet based on hull-less barley gave poorer growth performance than a diet based on wheat, even when the barley diet was supplemented with β-glucanase enzyme. In addition, hull-less barley trades at a premium over regular barley and is commonly used in other markets such as human or pig feeding. Consequently, some nutritionists regard hull-less barley as a substitute for conventional barley on the assumption that they are similar in nutritive value.

There is evidence that hull-less barley is improved in nutritive value for young broilers by supplementation with β-glucanase (Salih et al., 1991). Ravindran et al. (2007) found that β-glucanase supplementation improved the ME of several barley cultivars, but the magnitude of response was markedly greater in waxy genotypes than in normal starch genotypes. These data suggest that starch characteristics and type of β-glucan influence the energy value of barley for broiler chickens and that these characteristics may be equally or more important than fibre contents in determining the feeding value of barley for poultry. Producers wishing to use hull-less barley in organic diets should therefore obtain information on the particular cultivar available.

Brewer's dried grains

Brewer's dried grains (often referred to as spent grains) are the extracted dried residue of barley malt alone or in mixture with other cereal grain or grain products resulting from the manufacture of wort or beer. This by-product consists largely of structural carbohydrates (cellulose, hemicellulose) together with the protein remaining after barley has been malted and mashed to release sugars for brewing (Westendorf and Wohlt, 2002). Other grains may be included with the barley.

Nutritional features

Because of the removal of sugars and starches, the spent grains are higher in CP and lower in energy than the original grain. The CP, oil and CF contents of brewer's grains are approximately twice as high as in the original grain. According to data from Westendorf and Wohlt (2002) the CP ranges from 210 to 290 g/kg on a dry matter (DM) basis (US data). Some recent data cited by these authors indicated a higher average CP content of 290–330 g/kg (DM basis). Westendorf and Wohlt (2002) speculated that the increase might be due to improved varieties of barley, maize and rice used for brewing, different brewing methods or changes in the recovery or pooling of wastes generated during the brewing process.

Other by-products of the brewing process are malt sprouts, brewer's condensed solubles (produced from the mechanical dewatering of brewer's grains) and brewer's yeast. Most of the brewer's grains are marketed in the wet form as dairy cattle feed (Westendorf and Wohlt, 2002). However, some dried product may be available economically as a fed ingredient for poultry diets. North American definitions of the main by-products of brewing that are suitable for poultry feeding are as follows (AAFCO, 2005).

Brewer's dried grains are the dried extracted residue of barley malt alone or in mixture with other cereal grain or grain products resulting from the manufacture of wort or beer, and may contain pulverized dried spent hops in an amount not to exceed 3% evenly distributed. IFN 5-00-516 Barley brewer's grains dehydrated.

Malt sprouts are obtained from malted barley by the removal of the rootlets and sprouts which may include some of the malt hulls, other parts of malt and foreign material unavoidably present. The traded product must contain not less than 24% CP. The term malt sprouts, when applied to a corresponding portion of other malted cereals, must be in a qualified form: i.e. 'Rye Malt Sprouts', 'Wheat Malt Sprouts', etc. Malt sprouts are also known as malt culms in some countries. IFN 5-00-545 Barley malt sprouts dehydrated; IFN 5-04-048 Rye malt sprouts dehydrated; IFN 5-29-796 Wheat malt sprouts dehydrated.

Malt cleanings are obtained from the cleaning of malted barley or from the re-cleaning of malt that does not meet the minimum CP standard of malt sprouts. They must be designated and sold according to the CP content. IFN 5-00-544 Barley malt cleanings dehydrated.

Brewer's wet grains are the extracted residue resulting from the manufacture of wort from barley malt alone or in mixture with other cereal grains or grain products. The guaranteed analysis should include the maximum moisture content. IFN 5-00-517 Barley brewer's grains wet.

Brewer's condensed solubles are obtained by condensing liquids resulting as by-products from manufacturing beer or wort. The traded product must contain not less than 20% total solids and 70% carbohydrates on a DM basis, and the guaranteed analysis must include maximum moisture content. IFN 5-12-239 Barley brewer's solubles condensed.

Brewer's dried yeast is the dried, non-fermentative, non-extracted yeast of the botanical classification *Saccharomyces* resulting as a by-product from the brewing of beer and ale. The traded product must contain not less than 35% CP. It must be labelled according to its CP content. IFN 7-05-527 Yeast brewer's dehydrated.

Poultry diets

Brewer's grains are commonly fed to farm livestock, usually cattle or pigs. Most of the brewer's grains used in poultry diets are in the dry form.

Draganov (1986) reviewed findings on the utilization of brewer's grains in poultry and livestock diets and concluded that it could be used as a partial substitute for wheat bran, soybean meal and sunflower seed meal in poultry diets, e.g. at levels up to 200 g/kg in broiler diets. Subsequent research, however, indicated that dried brewer's grains caused a reduction in feed intake when included in broiler diets at more than 100 g/kg (Onifade and Babatunde, 1998). Results with laying hens indicated a reduction in body weight, rate of lay or in egg size when the diet contained more than 50 g/kg (Eldred *et al.*, 1975; Ochetim and Solomona, 1994). Shim *et al.* (1989) concluded that wet brewer's grains were not a suitable feed ingredient for ducks up to 4 weeks old, but could be used at 100–200 g/kg in the finisher diet. It is difficult to generalize on these findings, since the grains used in the brewing process and processing conditions differ in different countries. This by-product is probably best utilized in diets for breeding flocks or in diets for laying hens in the late lay period, particularly in regions such as Africa where the economics of by-products may favour the use of dried brewers' grains for poultry feeding.

Buckwheat (*Fagopyrum* spp.)

Buckwheat is a member of the *Polygonaceae* family, not a true cereal. It is most commonly grown for human consumption, and therefore may only be available in limited quantities for poultry feeding.

Nutritional features

The protein quality of buckwheat is considered to be the highest of the grains, being high in lysine. However, buckwheat is relatively low in ME compared to other grains (11.0 MJ/kg; Farrell, 1978) due to its high fibre and low oil contents. Another significant factor limiting the use of buckwheat in poultry diets is the presence of an antinutritional factor (ANF), fagopyrin, which has been reported to cause skin lesions, vesication of the bare parts of the head, incoordination and intense itching when poultry consuming buckwheat were exposed to sunlight, especially for a prolonged period (Cheeke and Shull, 1985). As a result, buckwheat should be used with caution in diets for organic poultry.

Poultry diets

Farrell (1978) evaluated buckwheat as a feed for rats, pigs, chickens and laying hens. Growth studies with rats and chicks showed that buckwheat in single-grain diets was superior to other cereal grains. The most favourable response to buckwheat was achieved in pig growth studies, but Farrell (1978) concluded that buckwheat as the sole grain source was not suitable for the species examined and resulted in some feed refusals. Removal of the fibrous hulls from buckwheat by sifting did not increase growth rate, but feed conversion was improved. Although chemical analysis indicated that the AA were adequate to maintain egg production, this was not confirmed in laying trials even when additions of lysine and methionine were made to the diet. Growth rate on an all-buckwheat diet was not different from that on an all-wheat (120 g CP/kg) diet, but addition of lysine improved growth rate and feed conversion. Addition of meat meal or sunflower meal to buckwheat at 100g/kg gave similar growth and feed conversion to those with a commercial grower diet.

Buckwheat may have more potential as a poultry feedstuff in highland regions such as North-east India. Gupta *et al.* (2002) reported that three buckwheat crops could be obtained per year from the same fields and that buckwheat could be incorporated successfully in chicken grower diets at levels up to 300 g/kg, although DM and Ca utilization were reduced at levels above 150 g/kg.

These findings indicate that buckwheat can replace part of the grains in poultry diets. They also suggest that it is not sufficient to evaluate dietary ingredients in a laboratory setting when the results are intended to apply to birds raised organically. The environmental conditions also need to be considered.

Maize (*Zea mays*)

This cereal is also known as Indian corn in the Americas. It can be grown in more countries than any other grain crop because of its versatility. It is the most important feed grain in the USA because of its palatability, high energy value, and high yields of digestible nutrients per unit of land. Consequently, it is used as a yardstick in comparing other feed grains for poultry feeding. The plains of the USA provide some of the best growing conditions for maize, making it the world's top maize producer. Other major maize-producing countries are China, Brazil, the EU, Mexico and Argentina.

Nutritional features

Maize is high in carbohydrate, most of which is highly digestible starch, and is low in fibre. It has a relatively high oil content, and thus a high ME value. Other grains, except wheat, have a lower ME value than maize. Maize oil has a high proportion of unsaturated fatty acids and is an excellent source of linoleic acid. The use of yellow maize grain should be restricted in poultry diets if it results in carcass fat that is too soft or too yellow for the market in question. White maize can be used to avoid the fat coloration. Yellow and white maize are comparable in energy, protein and minerals, but yellow maize has more carotene than white maize. Yellow maize is a valuable component of layer diets when yellow-pigmented yolks are desired. Yellow maize also contains the pigment cryptoxanthin, which can be converted into vitamin A in the animal body.

Protein concentration in maize is normally about 85 g/kg but the protein is not well balanced in AA content, with lysine, threonine, isoleucine and tryptophan being limiting. Varieties such as Opaque-2 and Floury-2 have improved AA profiles but do not appear to yield as well as conventional varieties. As a result, they are not grown extensively. Producers wishing to use such varieties should check their acceptability with the organic certifying agency.

Maize is very low in Ca (about 0.3 g/kg). It contains a higher level of P (2.5–3.0 g/kg) but much of the P is bound in phytate form so that it is poorly available to poultry. As a result, a high proportion of P passes through the gut and is excreted in the droppings. The diet may be supplemented with phytase enzyme to improve P utilization. Another approach would be to use one of the newer low phytic acid maize varieties that have about 35% of their P bound in phytate acid compared with 70% for conventional maize. These varieties allow the P to be more effectively utilized by the bird and less excreted in the droppings. As with other improved varieties of maize, producers wishing to use such varieties should check their acceptability with the organic certifying agency. Maize is low in potassium and sodium, as well as in trace minerals.

Maize is low in vitamins but contains useful amounts of biotin and carotenoids. Niacin is present in a bound form and together with a low level of tryptophan, a niacin precursor, leads to this vitamin being particularly limiting in maize-based diets unless supplemented.

The quality of maize is excellent when harvested and stored under appropriate conditions, including proper drying to moisture levels of 100–120 g/kg. Fungal toxins (zearalenone, aflatoxin and ochratoxin) can develop in grain that is harvested damp or allowed to become damp during storage. These toxins can cause adverse effects in poultry. Varieties differ markedly in storage characteristics, due to husk coating and endosperm type.

Poultry diets

Maize is suitable for feeding to all classes of poultry. One of the benefits of using yellow maize in poultry diets is that birds are attracted to the yellow colour of the grain. It should be ground medium to medium-fine for use in mash diets and fine for inclusion in pelleted diets. The grain should be mixed into feed immediately after grinding since it is likely to become rancid during storage.

Maize by-products

Grain-processing plants offer by-products that are suitable for livestock use, if acceptable by organic certification agencies. In producing ground maize for the human market the hull and germ are removed, leaving hominy feed, consisting of bran, germ and some endosperm. It resembles the original grain in composition, but has higher contents of fibre, protein and oil.

Hominy feed

Hominy feed is an excellent feed for poultry, similar to whole maize in ME value because of the higher oil content. It is a good source of linoleic acid. One of the benefits of using maize by-products such as this is that the grain used is of very high quality since the main product is intended for the human market. This helps to ensure that the maize is free from mycotoxin contamination and insect and rodent infestation. Leeson et al. (1988) showed that high-fat hominy produced by a dry-milling process was comparable to maize in broiler and turkey poult diets.

Maize gluten feed

Maize (corn) gluten feed (CGF) is a maize by-product that is available in some regions as a wet or dry product. The dry product is traded internationally. Because its method of production can vary, it tends to be of variable feed value depending on the exact process in use. Therefore, it should be purchased on the basis of a guaranteed analysis. CGF is a by-product of starch production. In the North American process the maize is soaked ('steeped') in water and sulfur dioxide, for 30–35 h at 35–47°C, to soften it for the initial milling step. During this process, some nutrients dissolve in the water (steep liquor). When the steeping is complete, this liquor is drawn off and concentrated to produce 'solubles'. During the wet-milling process, the germ is separated from the kernel and may be further processed to remove the oil. In a secondary wet-milling process the remaining portion of the kernel is separated into the hull (bran), gluten and starch fractions. The bran fraction is then mixed with steep liquor and the germ fraction, and marketed as CGF, wet or dehydrated. The gluten fraction may be marketed as maize gluten meal or may be added back to the CGF. The dried CGF is made into pellets to facilitate handling. It analyses typically at 210 g protein/kg, 25 g fat/kg and 80 g CF/kg. Wet CGF (450 g/kg DM) is similarly combined but not dried. It is a perishable product that has to be used within 6–10 days and must be stored in an anaerobic environment.

Maize gluten meal

Maize gluten meal is not a very suitable feed for poultry since it is of relatively low palatability and being very deficient in lysine has an imbalanced AA profile. However, it can be incorporated in broiler and layer diets as a source of natural yellow pigment since it is rich in xanthophyll pigments (up to 300 mg/kg). Maize gluten meal may no longer be available or economic for feed use as it is used extensively as a natural weed-killer in horticulture.

Distiller's dried grains

This maize by-product can be used in poultry feeding and, like maize gluten meal, it is a good source of yellow pigment. It is derived from ethanol production (as a fuel source or as liquor). This North American by-product is exported to several regions, including Europe. In the production process dry-milling is used, followed by cooking and fermentation of the starch fraction with yeast, to produce ethanol. Removal of the starch leaves the nutrients in the remaining residue at about three times the content in the original grain. Evaporation of the remaining liquid produces solubles that are usually added back to the residue to produce distiller's grains plus solubles. Usually this product is dehydrated

and marketed as dry distiller's grains plus solubles (DDGS). One of the benefits of this product is the contribution of nutrients provided by yeast.

DDGS typically analyses at 270 g protein/kg, 110 g fat/kg and 90 g CF/kg. Batal and Dale (2006) reported that true metabolizable energy (TME$_n$) ranged from 2490 to 3190 kcal/kg (86% DM basis) with a mean of 2820 kcal/kg. Variation was noted among samples, presumably reflecting differences in the original maize composition, fermentation and disposition of solubles (Cromwell et al., 1993; Batal and Dale, 2006). The reported ranges were: CP 234–287 g/kg, fat 29–128 g/kg, neutral detergent fibre (NDF) 288–403 g/kg, acid detergent fibre (ADF) 103–181 g/kg and ash 34–73 g/kg. Lysine concentration ranged from 4.3 to 8.9 g/kg. Most samples surveyed were golden in colour, and true AA digestibility values were relatively consistent among these samples. The average total concentration (g/kg) and digestibility coefficients of several most limiting AA for the eight DDGS samples were as follows: lysine 7.1 (70); methionine 5.4 (87); cystine 5.6 (74); threonine 9.6 (75); valine, 13.3 (80), isoleucine 9.7 (83); and arginine 10.9 (84). In general, DDGS samples that were more yellow and lighter in colour had higher total and digestible AA levels. The variation in TME$_n$ and AA digestibility observed among samples strongly indicated that confirmatory analyses should be conducted prior to using samples from a new supplier.

POULTRY DIETS. Lumpkins et al. (2004) evaluated DDGS as a feed ingredient for broilers. The product was included in one experiment at 0 or 150 g/kg in diets of high (220 g CP/kg and 3050 kcal ME$_n$/kg) or low (200 g CP/kg and 3000 kcal ME$_n$/kg) density. Weight gain and feed efficiency (gain/feed ratio) were better with the high-density diets than with the low-density diets, but within the two density levels there was no difference in growth performance with 0 or 150 g DDGS/kg. In another experiment the diets contained 0, 60, 120 or 180 g DDGS/kg. There was no significant difference in performance or carcass yield throughout the 42-day experiment except for a depression in body weight gain and feed conversion efficiency when chicks were fed diets with 180 g DDGS/kg in the starter period. These studies indicated that DDGS was an acceptable feed ingredient for broiler diets and could be used effectively at 60 g/kg in the starter diet and 120–150 g/kg in the grower and finisher diets.

The same authors showed that DDGS was an acceptable feed ingredient in layer diets (Lumpkins *et al.*, 2004). The hens were fed a commercial or low-density diet formulated to contain 0 or 150 g DDGS/kg for 25–43 weeks of age. No differences were observed in the production characteristics of hens fed diets containing 0 or 150 g DDGS/kg. However, there was a significant reduction in hen-day egg production through 35 weeks of age when hens were fed the low-density diet with 150 g DDGS/kg. A maximal inclusion level of 100–120 g DDGS/kg was suggested for diets with a conventional level of energy and a lower level for low-density diets. Research with turkey hens indicated that DDGS could be included at a level of 100 g/kg in growing and finishing diets (Roberson, 2003).

Światkiewicz and Koreleski (2006) studied the effects of dietary inclusion of up to 200 g DDGS/kg on egg production in laying hens. They found that in the first stage of the laying cycle (26–43 weeks of age) levels of DDGS did not significantly affect rate of lay, egg weight, feed intake or feed conversion. In the second phase (44–68 weeks), there were no differences in egg production parameters in the groups fed diets with up to 150 g/kg. The diet with 200 g DDGS/kg gave a lower rate of lay and a reduced egg weight, which was alleviated partly by supplementation with NSP-hydrolysing enzymes. DDGS level did not affect albumen height, Haugh units, eggshell thickness, shell density, breaking strength or sensory properties of boiled eggs. However, egg yolk colour score significantly increased with increasing DDGS levels in the diets, reflecting the high content of xanthophylls in this ingredient.

Oats (*Avena sativa*)

Oats are grown in cooler, wetter regions. Before 1910 the area seeded to oats often exceeded the area for wheat in Canada, in order to feed horses. Today the world's leading oat producers are Russia, the EU, Canada, the USA and Australia. Oats can be grown on many organic farms and utilized on-farm (Anon., 2013).

Nutritional features

The chemical composition can vary widely, depending on variety, climate and fertilizer practices. Threshed oats remain enclosed in the hulls, leaving the chaffs (glumes) on the straw. Oats are higher in hull, CF and ash contents and are lower in starch than maize, grain sorghum or wheat. The proportion of hulls can range from 200 to 450 g/kg. They thus have a much lower energy value than other main cereals. The nutritional value is inversely related to the hull content, which can be approximated from the thousand-kernel weight. Oats vary in protein content from about 110 to 170 g/kg, with an AA profile similar to wheat, being limiting in lysine, methionine and threonine. The protein content is affected mainly by the proportion of hull, since it is present almost entirely in the kernel. A large study in Poland found that the average CP value was 113.1 g/kg in organic oats compared with 115.7 g/kg in conventional oats (Grela and Semeniuk, 2008). The corresponding values for CF were 147.2 and 134.5 g/kg. Jacob (2007) reported a higher content of CP in organic oats: 123.1 g/kg versus 110 g/kg in conventional oats. Oats have a higher oil content than maize, but this does not compensate for the high fibre content. As with other small grains oats contain β-glucans and may have to be supplemented with an appropriate enzyme supplement to avoid their adverse effects.

Poultry diets

Because of their high fibre and low energy content, oats are best used in pullet developer diets and in diets for breeder flocks. Newer varieties of oats with a lower hull content may be more suitable in diets for growing meat birds.

A comparison of feeding maize, oats and barley in relation to the growth of White Leghorn chickens, gastrointestinal weights of males and sexual maturity of hens was reported by Ernst *et al.* (1994). Five diets were used, each with approximately 225 g CP/kg and ME 3000 kcal/kg. The control diet contained maize only, and the remaining test diets had 200 or 400 g maize/kg replaced with oats or barley. All diets were fed to growing and laying hens, half being fed grit up to 15 weeks of age and half not. Average body weight gains and feed/gain ratios were not affected by diet at any age. At 12 weeks of age, cockerels fed the oat diet had significantly heavier digestive tracts and gizzard weights than those fed on the maize-only diet. Grit also increased gizzard weight at this age. Hens fed on the 200 g/kg oat diet reached sexual maturity first, 8 days ahead of those fed on maize. These findings, particularly those on digestive tract and gizzard development, are of interest to organic producers in relation to the production of birds with access to forage and with maximum natural resistance to diseases such as coccidiosis that can be of significant importance in organic production. Research from Australia suggested that birds with a better-developed gizzard are better able to withstand a coccidiosis challenge (Cumming, 1992).

Aimonen and Näsi (1991) concluded that good-quality oats can substitute for barley in the diet of laying hens without any negative effects on egg production. They also found that enzyme supplementation increased the average ME of the oat diet from 11.8 to 12.1 MJ/kg (DM basis) and improved feed conversion efficiency by 3%.

The inclusion of oats may have to be avoided during hot weather because of a higher heat increment during digestion, which can result in reduced feed intake and growth rate.

Naked (hull-less) oats (*Avena nuda*)

Naked oat varieties have been developed, though yields are sometimes low. These varieties lose their hull during harvesting and consequently are much lower in fibre than hulled oats. Like conventional oats they contain β-glucans. Producers intending to use naked oats should ensure that they are not from GM varieties.

Naked oats have higher protein and fat contents than conventional oats, resulting in an ME value similar to that of maize. The balance of AA is also better than in conventional oats or maize. Thus, naked oats have the potential of being an excellent feedstuff for poultry. As a result, there is currently great interest in this grain crop, particularly in Europe.

Maurice *et al.* (1985) evaluated naked oats as a potential feedstuff for poultry feeding and reported that this grain was superior to maize in AA profile and mineral content. Total lipid was 68.5 g/kg with a linoleic acid content of 309 g/kg, which is higher than that of most cereal grains, and contributed to a relatively high ME value of 13.31 MJ/kg. The naked oats contained 4 g total P/kg and a phytic acid concentration of 10.7 g/kg.

Poultry diets

Early results in Canada showed a reduction in the growth of starter broiler chickens when fed diets containing some cultivars of naked oats. In addition, there was a problem of sticky droppings. In contrast, the growth of older broilers was satisfactory (Burrows, 2004). Subsequent experiments showed that the growth depression in the starter broiler chickens resulted from decreased nutrient availability and feed intake caused by the presence of β-glucan in the oat grain. Attempts to improve the naked oat by steam and pelleting processes were unsuccessful, but supplementation with β-glucanase was effective and allowed high levels of naked oats to be included in the diet. One interesting finding from these studies was that broiler meat showed a moderate increase in fat stability as the level of oats in the diet increased, no doubt due to the type of lipid contained in the oats. Subsequent research in the USA suggested that naked oats could be included at 400 g/kg in the diet of broilers with no adverse effect (Maurice *et al.*, 1985) and research in Australia suggested that naked oats could comprise the sole grain source in broiler diets, provided that the diets were pelleted and supplemented with enzyme preparations (Farrell *et al.*, 1991). The cultivars of naked oats used in the Australian studies contained 31–118 g total lipid/kg and 98–181 g CP/kg.

Canadian research with laying birds showed that naked oats could be included at dietary levels up to 600 g/kg to replace maize, soybean meal and fat (Burrows, 2004). The total weight of eggs produced was equal to that of a maize–soybean diet when naked oats were included at up to 600 g/kg and was reduced by only 4% at 800 g naked oats/kg, when no soybean meal was used. Hsun and Maurice (1992) obtained similar results. Feed intake, egg weight and egg production were not affected by including up to 660 g naked oats/kg in the diet, but feed intake, egg production, egg weight and eggshell strength were decreased with the naked oats at a dietary level of 880 g/kg. Sokól et al. (2004) reported egg production of over 90% in layers fed a diet based on maize or on naked oats as a complete replacement for the maize. Supplementation of the naked oats diet with β-glucanase had no influence on productivity parameters or the fatty acid profile of egg yolks. Good results have also been reported with turkeys and quail.

One finding with layers fed diets containing high levels of feedstuffs, such as naked oats in place of yellow maize, is a reduction in yolk colour. Therefore, organic producers selling eggs to markets that demand intensely coloured yolks should consider supplementing diets based on barley, oats or wheat with a natural source of pigment such as lucerne meal, maize gluten meal, grass meal or marigold meal (Karadas et al., 2006).

Rice (*Oryza sativa*)

As shown in Table 4.1, this grain is not included universally in lists of permitted ingredients, therefore its acceptability should be checked with the local certifying agency.

The polished white rice that is an important staple food for a large section of the human population is obtained by milling paddy rice. Rice is first dehulled to yield the following fractions (approximate proportions in parentheses): brown rice (80%) and hulls (20%). The brown rice is then milled to yield rice bran (10%), white rice (60%) and polishings + broken rice (10%). The main by-product – rice bran – is a mixture of hulls, germ and bran that is suitable for poultry feeding. The polishings and broken grains are usually added to the bran. The percentage of by-products depends on milling rate, type of rice and other factors.

Rough rice (paddy or padi rice) can be used in poultry diets after grinding but is generally not available. Rice that does not meet the quality standards for humans after processing is a good feed ingredient for poultry diets, provided that it is not moulded and contaminated with toxic fungi.

Rice by-products

Rice bran is the most important rice by-product. It is suitable as a grain substitute, equivalent to wheat in feeding value if of good quality. It is a good source of water-soluble vitamins and is palatable to poultry when used fresh. One of the problems with rice bran is the high oil content (140–180 g/kg), which is very unsaturated and unstable. At high ambient temperature and in the presence of moisture the oil breaks down to glycerol and free fatty acids, due to the presence of a lipolytic enzyme that becomes active when the bran is separated from the rice. The result is an unpleasant taste and odour and reduced palatability of the product. Peroxidation of the oil was attributed to the incidence of vitamin E-responsive conditions reported in Australia (Farrell and Hutton, 1990). A ratio of vitamin E to linoleic acid of 0.6 mg/g has been recommended as a precautionary measure (Farrell and Hutton, 1990). Stabilization of the oil in other ways (e.g. by use of an antioxidant) should also be considered, if acceptable by the certifying agency. Another aspect of the oil is its high content of unsaturated fatty acids can lead to soft fat in the carcass. The oil can be extracted from the bran to avoid the problems noted above. De-oiled rice bran can be used at higher levels than regular rice bran. Apart from extraction of the oil the rancidity process can be delayed by heating or drying immediately after milling to reduce the moisture content to below 40 g/kg. Heating to 100°C for 4 or 5 min with live steam is sufficient to retard the increase in free fatty acids. The rice bran can also be heated dry if spread out on trays at 200°C for 10 min.

Poultry diets

Rice bran at up to 200 g/kg has been successfully in diets for growing and laying poultry in several countries, with rice bran replacing part of the grain component (Ravindran and Blair, 1991; Farrell, 1994). Ducks appear to be able to tolerate higher dietary levels. Rice bran at up to 300 g/kg can be included successfully in broiler diets, following supplementation with xylanase, phytase and lipase enzymes (Mulyantini *et al.*, 2005).

Rice bran oil appears to give similar results to those obtained with tallow when added as an energy source to broiler diets at levels up to 40 g/kg (Purushothaman *et al.*, 2005).

Rye (*Secale cereale*)

Rye has an energy value intermediate to that of wheat and barley, and the protein content is similar to that of barley and oats. A large study in Poland found that the average CP value was 82.7 g/kg in organic rye compared with 86.8 g/kg in conventional rye (Grela and Semeniuk, 2008). The corresponding values for CF were 21.3 and 21.4 g/kg DM. The nutritional value of rye is reduced by the presence of several ANFs such as β-glucans and arabinoxylans which are known to cause increased viscosity of the intestinal contents and reduced digestibility, and other undesirable effects such as an increased incidence of dirty eggs. These effects are more pronounced in hot and dry environments, which accentuate the rate of cereal ripening prior to harvest (Campbell and Campbell, 1989), as occurs in Spain and other Mediterranean countries. Rye may also contain ergot, which is a toxic fungus that reduces poultry health and performance. As a result it is not a favoured ingredient in poultry diets in several countries.

Sorghum (*Sorghum vulgare*), milo

Sorghum, commonly called grain sorghum or milo, is the fifth most important cereal crop grown in the world. Much of it is used in the human market. As a continent, Africa is the largest producer of sorghum. Other leading producers include the USA, India, Mexico, Australia and Argentina. Sorghum is one of the most drought-tolerant cereal crops and is more suited than maize to harsh weather conditions such as high temperature and less consistent moisture.

Nutritional features

Grain sorghum is generally higher in CP than maize and is similar in ME content. However, one disadvantage of grain sorghum is that it can be more variable in composition because of growing conditions. CP content usually averages around 89 g/kg, but can vary from 70 to 130 g/kg. Therefore, a protein analysis prior to formulation of diets is recommended. The hybrid yellow-endosperm varieties are more palatable to poultry than the darker brown sorghums, which possess a higher tannin content to deter wild birds from damaging the crop. High-tannin cultivars can be used successfully after storage under high-moisture conditions (reconstitution) for a period of 10 days to inactivate the tannins. Reconstitution has been shown to result in an increased protein digestibility of 6–16% units and an increase in dietary ME of 0.1–0.3 kcal/g (Mitaru *et al.*, 1983, 1985). True ileal protein digestibility was 45.5% and 66.7% for two untreated high-tannin sorghum cultivars and 89.9% for a low-tannin cultivar. Values for AA digestibility were 43.1–73.7% and 84.9–93.0%, respectively. For the reconstituted high-tannin sorghums, protein digestibility was 77.4% and 84.5%, and AA digestibility was 73.5–90.9%. Reconstitution of low-tannin sorghum was without effect. These findings suggest a way for poultry producers with only high-tannin sorghum available for poultry feeding to use this grain in poultry diets. Lysine is the most limiting AA in sorghum protein, followed by tryptophan and threonine.

A high proportion of the sorghum grown in the USA is used for ethanol production, yielding by-products such as distiller's dried grains with solubles (sorghum-DDGS) for animal feeding.

Poultry diets

Research results show that low-tannin grain sorghum cultivars can be successfully used as the main or only grain source in poultry diets (Adamu *et al.*, 2006; Etuk *et al.*, 2012) as a substitute for maize. Mitaru *et al.* (1983, 1985) showed that reconstitution of high-tannin sorghum resulted in a significant improvement in utilization of energy and protein by broilers and in growth performance. Starting broilers fed diets based on three cultivars of high-tannin-reconstituted sorghum had higher daily body weight gains (23–83 g) and better feed efficiencies than broilers fed on diets containing untreated sorghum. Research results showed that low-tannin grain sorghum cultivars can be successfully used as the main or only grain source in poultry diets (Adamu *et al.*, 2006). Proper grinding of grain sorghum is important because of the hard seedcoat.

Spelt (*Triticum aestivum* var. *spelta*)

Spelt is a subspecies of wheat grown widely in Central Europe. Expansion of this crop is likely in Europe because of the current shortage of high-protein organic feedstuffs. It has been introduced to other countries partly for the human market because of its reputation as being low in gliadin, the gluten fraction implicated in coeliac disease.

This crop appears to be generally more winter-hardy than soft red winter wheat, but less winter-hardy than hard red winter wheat. The yield is generally lower than that of wheat but equal to wheat when the growing conditions are less than ideal.

Ranhotra *et al.* (1996a, b) reported an average CP value of 166 g/kg in a range of spelt varieties in the USA versus 134 g/kg in a range of wheat varieties. The average lysine content was 29.3 g protein/kg in spelt versus 32.1 g protein/kg in wheat. The nutrient profile of both grains was found to be greatly influenced by cultivar and location. All spelt and wheat samples tested contained gluten. A large study in Poland found that the average CP value was similar at 114.1 g/kg in organic and conventional spelt (Grela and Semeniuk, 2008). The corresponding values for CF were 31.2 and 30.3 g/kg DM.

Triticale (*Triticale hexaploide, tetraploide*)

Triticale is a hybrid of wheat (*Triticum*) and rye (*Secale*) developed with the aim of combining the grain quality, productivity and disease resistance of wheat with the vigour, hardiness and high lysine content of rye. The first crossing of wheat and rye was made in Scotland in 1875 (Wilson, 1876) though the name 'triticale' did not appear in the scientific literature until later. Triticale can be synthesized by crossing rye with either tetraploid (durum) wheat or hexaploid (bread) wheat. It is grown mainly in Poland, China, Russia, Germany, Australia and France, but some is grown in North and South America. It is reported to grow well in regions not suitable for maize or wheat. Globally, triticale is used primarily for livestock feed. Current grain yields in Canada are competitive with the highest-yielding wheat varieties, and may exceed those of barley. Also, the high quality of the protein has been maintained in the newer varieties. Both spring and winter types are now available (including a semi-awnless winter variety) and have provided a new crop option for breaking disease cycles in cereal cropping systems. According to Briggs (2002) the greatest potential for its use as a grain feedstuff is on farms that grow at least part of their own feed grain supply, using lands that are heavily manured from the feed operation. Triticale production in such conditions is generally more productive and sustainable than barley or other cereals for feed-grain. Its greater disease resistance compared with wheat or barley is another advantage. Thus, triticale will be of particular interest to organic poultry producers.

Nutritional features

The use of many varieties and crosses to improve yield and grain quality in triticale, as well as adaptation to local conditions, has resulted in a variation in nutrient composition. The protein content of newer varieties is in the range of 95–132 g/kg, similar to that of wheat (Briggs, 2002; Stacey *et al.*, 2003). A large study in Poland found that the average CP value was 116.9 g/kg in organic triticale compared with 121.2 in conventional triticale (Grela and Semeniuk, 2008). The corresponding values for CF were 26.4 and 24.6 g/kg DM. Typical lysine contents (g/kg) reported by Hede (2001) from work in Mexico and Ecuador are triticale 50.4, barley 29.4, wheat 43.0 and maize 22.7.

ME values have been reported to be generally equal to, or higher than, wheat, Vieira *et al.* (1995) reporting an ME value of 3246 kcal/kg for growing broilers.

According to Beltranena (2016) the newly developed Canadian varieties (× *Triticosecale* Wittmack) possess more of the characteristics of the wheat parent than the rye parent, resulting in an improved energy content and improved palatability and nutritional value. In addition, they are low in ANFs such as ergot which have been found in the older varieties.

Poultry diets

Triticale is used widely in poultry diets in several countries. In a recent study in Poland to compare various grains in the diet of growing broilers, Józefiak *et al.* (2007) found that growth performance was better with diets based on triticale than on wheat or rye. Triticale has been used successfully in diets for growing meat birds at levels up to 400 g/kg (Vieira *et al.*, 1995) and in some trials as the complete grain source (Korver *et al.*, 2004). Final live weight, amount of feed consumed, carcass weight at processing, flock uniformity, percentage of Grade A carcasses and percentage of condemned carcasses were not affected by cereal source (Korver *et al.*, 2004).

Research with turkey males grown to 45 weeks of age showed that replacement of maize in the diet by triticale improved the tenderness of the cooked meat (Savage *et al.*, 1987). Other research with turkeys showed that mature breeding males could be fed diets based on triticale with no effect on semen production (Nilipour *et al.*, 1987).

Some research has shown that feeding low levels of triticale increases microbial diversity in the digestive tract of poultry, discouraging salmonella colonization (Jacob, 2015).

The cultivar 'Bogo', which was developed in Poland, is currently of interest since it has a higher yield than the more traditional varieties of triticale. Trials with this cultivar in the USA (Hermes and Johnson, 2004) found that broiler growth was unaffected by feeding diets containing triticale at levels up to 150 g/kg. In layers, only yolk colour was affected when diets containing up to 300 g/kg triticale were fed, the yolks becoming slightly paler compared with those produced by hens fed no triticale. Other German research (Richter and Lenser, 1993) on layers showed that triticale (cultivar 'Grado') could be included in the diet at levels up to 720 g/kg as a replacement for maize and wheat, with no effect on feed consumption, laying performance, feed efficiency, mortality or weight gain. However, in two of three trials egg weight was reduced with increasing amounts of triticale in the diet, but could be corrected by addition of low levels (10 or 15 g/kg) of sunflower oil to the diet. The effect was attributed to a reduction in linoleic acid content of the diets containing triticale. Therefore, the researchers suggested that layer diets supplemented with oil could contain up to 500 g triticale/kg, but in non-supplemented diets the highest level of triticale should be 200 g/kg. Another finding of this study was that colour intensity of the egg yolks was decreased with increasing amounts of triticale in the diet.

Briggs (2002) reported some negative effects in layer diets when triticale was used at high dietary levels. These findings suggest that triticale should be part of a quality control programme on-farm before high levels are used in poultry diets.

Wheat (*Triticum aestivum*)

Wheat grain consists of the whole seed of the wheat plant. This cereal is widely cultivated in temperate countries and in cooler parts of tropical countries. Several types of wheat are grown in Europe and North America. These include soft white winter, hard red winter, hard red spring and soft red winter wheat, the colour describing the seed colour. Hard wheats contain more CP than soft wheats. The types grown in Europe and Australia include white cultivars. Wheat is commonly used for feeding to poultry when it is surplus to human requirements or is considered not suitable for the human food market. Otherwise it may be too expensive for poultry feeding. Some wheat is grown, however, for feed purposes. By-products of the flour-milling industry are also very desirable ingredients for poultry diets.

During threshing, the husk – unlike that of barley and oats – detaches from the grain, leaving a less fibrous product. As a result, wheat is slightly lower than maize in ME value but contains more CP, lysine and tryptophan. Thus, it can be used as a replacement for maize as a high-energy ingredient and it requires less protein supplementation than maize.

Wheat is very palatable if not ground too finely and can be utilized efficiently by all classes of poultry. Good results are obtained when wheat is coarsely ground (hammer mill screen size of 4.5–6.4 mm) for inclusion in mash diets. Finely ground wheat is not desirable because it is too powdery, causing the birds discomfort in eating unless the diet is pelleted. In addition, finely ground wheat readily absorbs moisture from the air and saliva in the feeder, which can result in feed spoilage and reduced feed intake. Also, feed containing finely ground wheat can bridge and not flow well in feeding equipment. One benefit of wheat as a feed ingredient is that it improves pellet quality due to its gluten content so that the use of a pellet binder may be unnecessary.

Nutritional features

One concern about wheat is that ME and CP contents are more variable than in other cereal grains such as maize, sorghum and barley (Zijlstra et al., 2001). Researchers at the Prairie Swine Centre in Canada analysed a large range of Canadian wheat samples and reported that CP ranged from 122 to 174 g/kg,

NDF from 72 to 91 g/kg, and soluble NSPs from 90 to 115 g/kg. The CF content was low overall and showed little variation. Kernel density was high (77–84 kg/hl) overall. The variation found in composition and nutritive value was related to the different wheat classes and cultivars grown for human consumption and to growing conditions and fertilizer practices. The results indicated a variation in CP content of 50%, though the range in CP in feed wheat is usually about 13–15%. Therefore, periodic testing of batches of wheat for nutrient content is necessary.

A large study in Poland found that the average CP value was 118.7 g/kg in organic wheat compared with 124.2 g/kg in conventional wheat (Grela and Semeniuk, 2008). The corresponding values for CF were 29.9 and 26.7 g/kg DM. Jacob (2007) reported that the CP content of organic wheat was 130.8 g/kg versus 135.0 g/kg in conventional wheat.

Further evidence of the variability in the nutrient content of wheat was obtained in a Danish study by Steenfeldt (2001). The 16 wheat cultivars varied in chemical composition, with protein content ranging from 112 to 127 g/kg DM, starch content from 658 to 722 g/kg DM and NSPs from 98 to 117 g/kg DM. When fed to growing chickens the variability in chemical composition was expressed in differences in growth performance and in intestinal effects, especially when wheat was included at high levels in the diet. Dietary apparent metabolizable energy (AME_n) levels ranged from 12.66 to 14.70 MJ/kg DM with a dietary wheat content of 815 g/kg and from 13.20 to 14.45 g/kg DM with a dietary wheat content of 650 g/kg. Milling-quality wheat resulted in better performance than feed-grade wheat.

Poultry diets

In spite of the recorded variability in nutrient content, there is a large body of evidence that suggests that wheat can be utilized effectively in diets for growing and laying poultry. The efficiency of production can be expected to be lower with complete replacement of maize when the wheat is not supplemented with an appropriate enzyme.

Wheat is the most common cereal used in poultry feed in Australia, Canada and the UK, conventional broiler diets typically containing more than 600 g wheat/kg. Inclusion of xylanase-based enzymes in these diets is now commonplace, to reduce the effects of soluble carbohydrates in wheat on intestinal viscosity and intestinal function. Responses to enzyme supplementation have been shown to depend on age of the bird. More mature birds, because of the enhanced fermentation capacity of the microflora in their intestines, have a greater capacity to deal with soluble carbohydrates in the diet. Replacing maize with wheat reduces the total xanthophyll content of the feed, thus reducing the pigmentation of the broiler skin and egg yolk. Therefore, supplementary sources of xanthophylls may have to be used in broiler and layer diets when wheat is used to replace maize.

Among recent findings Chiang *et al.* (2005) reported that broiler performance in the grower period was best when wheat was substituted for 24% of the maize. During the finisher period and entire experimental period from 0 to 6 weeks, no significant difference was found in growth performance when the replacement was increased to 100% of the maize. However, caecal weight increased linearly with increasing levels of wheat substitution for maize, an effect reduced by supplementation of the diet with xylanase enzyme.

As referred to in the section on rye above, Lázaro *et al.* (2003) fed diets containing wheat or barley at 500 g/kg or rye at 350 g/kg to laying hens and found no significant effects on egg production or feed conversion efficiency in comparison with hens fed a control diet based on maize. However, supplementation of the wheat, barley or rye diets with a fungal β-glucanase–xylanase enzyme complex improved nutrient digestibility, and reduced intestinal viscosity and the incidence of dirty eggs. On the basis of these findings the authors concluded that wheat, barley and rye could replace maize successfully in laying hen diets, and that enzyme supplementation improved digestibility and productive traits. Feeding programmes based on the use of whole, unground wheat have been developed for farm use.

Milling by-products

The main product of wheat milling is flour. Several by-products are available for animal feeding and are used commonly in poultry diets because of their valuable properties as feed ingredients. Their use in this way minimizes the amount of whole grain that has to be used in feed mixtures. These by-products are usually classified according to their CP and CF contents and are traded under a variety of names that are often confusing, such as pollards, offals, shorts, wheatfeed and middlings.

After cleaning (screening), sifting and separating, wheat is passed through corrugated rollers, which crush and shear the kernels, separating the bran and germ from the endosperm. Clean endosperm is then sifted and ground to flour for human consumption. The mill may further separate the remaining product into middlings, bran, germ and mill run. Some bran and germ are used for human consumption as well as animal diets. Mill run includes cleanings (screenings) and all leftover fines and is often used for cattle feeding. In general, the by-products with lower levels of CF are of higher nutritive value for poultry diets. AAFCO (2005) defined the flour and wheat by-products for animal feeding in the USA as follows:

Wheat flour

Wheat flour is defined as consisting principally of wheat flour together with fine particles of wheat bran, wheat germ and the offal from the 'tail of the mill'. This product must be obtained in the usual process of commercial milling and must contain not more than 15 g CF/kg. IFN 4-05-199 Wheat flour less than 15 g fibre/kg.

Wheat bran

Wheat bran is the coarse outer covering of the wheat kernel as separated from cleaned and scoured wheat in the usual process of commercial milling. IFN 4-05-190 Wheat bran. Sometimes screenings are ground and added to the bran. Generally wheat bran has a CP level of 140–170 g/kg; crude fat (oil), 30–45 g/kg and CF 105–120 g/kg. Therefore, while wheat bran may have a CP level as high or higher than the original grain, the higher fibre level results in this product being low in energy. Consequently, wheat bran is low in ME and has limited application in poultry diets other than in adult breeder diets to control excessive consumption of feed.

Wheat germ meal

Wheat germ meal consists chiefly of wheat germ together with some bran and middlings or shorts. It must contain not less than 250 g CP/kg and 70 g CF/kg. IFN 5-05-218 Wheat germ ground.

A wide variety of wheat germ grades is produced, depending on region, type of grain processed and the presence of screenings and other wheat by-products. Generally wheat germ meal has a CP level of 250–300 g/kg, crude fat (oil) 70–120 g/kg and CF 30–60 g/kg. As with other feedstuffs containing a high level of plant oil, a problem that may result on storage is rancidity due to peroxidation of the fat.

A defatted product is also marketed. Defatted wheat germ meal is obtained after the removal of part of the oil or fat from wheat germ meal and must contain not less than 300 g CP/kg. IFN 5-05-217 Wheat germ meal mechanical extracted.

The amount of wheat germ meal available for poultry feeding will usually be very limited, due to availability and cost since there are competing markets for these by-products.

Wheat red dog

Wheat red dog consists of the offal from the 'tail of the mill' together with some fine particles of wheat bran, wheat germ and wheat flour. This product must be obtained in the usual process of commercial milling and must contain not more than 40 g CF/kg. IFN 4-05-203 Wheat flour by-product less than 40 g fibre/kg.

Wheat red dog is a very fine, floury, light-coloured feed ingredient. The colour may range from creamy white to light brown or light red, depending on the type of wheat being milled. Wheat red dog can be used as a pellet binder, as well as a source of protein,

carbohydrate, minerals and vitamins. The average composition (g/kg) is around CP 155–175, crude fat 35–45 and CF 28–40.

Wheat mill run

Wheat mill run consists of coarse wheat bran, fine particles of wheat bran, wheat shorts, wheat germ, wheat flour and the offal from the 'tail of the mill'. This product must be obtained in the usual process of commercial milling and must contain not more than 95 g CF/kg. IFN 4-05-206 Wheat mill run less than 95 g/kg CF.

Wheat mill run usually contains some grain screenings. This by-product may not be available in areas that separate the by-products into bran, middlings and red dog. Wheat mill run generally has a CP level of about 140–170 g/kg, crude fat 30–40 g/kg and CF 85–95 g/kg.

Wheat middlings

Wheat middlings consist of fine particles of wheat bran, wheat shorts, wheat germ, wheat flour and some of the offal from the 'tail of the mill'. This mixture of shorts and germ is the most common by-product of flour mills. It must be obtained in the usual process of commercial milling and must contain not more than 95 g CF/kg. IFN 4-05-205 Wheat flour by-product less than 95 g/kg fibre.

The name 'middlings' derives from the fact that this by-product is somewhere in the middle of flour on one hand and bran on the other. This by-product is known as pollards in Europe and Australia. The composition and quality of middlings vary greatly due to the proportions of fractions included, and also the amount of screenings added and the fineness of grind. A cooperative research study was conducted by members of the US Regional Committee on Swine Nutrition to assess the variability in nutrient composition in 14 samples of wheat middlings from 13 states, mostly in the Midwest (Cromwell et al., 2000). The bulk density of the middlings ranged from 289 to 365 g/l. The middlings averaged (g/kg) 896 DM, 162 CP, 1.2 Ca, 9.7 P and 369 NDF; and 6.6 lysine, 1.9 tryptophan, 5.4 threonine, 2.5 methionine, 3.4 cystine, 5.0 isoleucine, 7.3 valine and 0.53 mg selenium/kg. The variation in nutrient composition was especially high for Ca (0.8–3.0 g/kg) and Se (0.05–1.07 mg/kg). 'Heavy' middlings (high bulk density, ≥ 335 g/l) had a greater proportion of flour attached to the bran, and were lower in CP, lysine, P and NDF than 'light' middlings (≤ 310 g/l).

Steam-pelleting of wheat middlings increases the ME and CP utilization in poultry diets by rupturing the aleurone cells and exposing the content of the cells to attack by digestive enzymes.

In conventional feed manufacturing this by-product is regarded as a combination of bran and shorts and tends to be intermediate between the two in composition. It also may contain some wheat screenings (weed seeds or other foreign matter that is removed prior to milling). It is commonly included in commercial feeds as a source of nutrients and because of its beneficial influence on pellet quality. When middlings (or whole wheat) are included in pelleted feeds, the pellets are more cohesive and there is less breakage and fewer fines.

POULTRY DIETS. Wheat middlings are used commonly in poultry diets in countries such as Australia, as a partial replacement for grain and protein supplement in diets for laying and breeding stock when it can be included at levels up to 450 g/kg (Bai et al., 1992), but lower levels are more usual. Patterson et al. (1988) conducted a series of experiments on wheat middlings as an alternative feedstuff for White Leghorn laying hens. They found that wheat middlings could be included at up to 430 g/kg in the diet without effect on productivity. However, at that level of dietary inclusion feed intake was higher and feed conversion efficiency lower than in hens fed a control diet, unless the diet was supplemented with 50 g fat/kg. A dietary inclusion of 250 g middlings/kg resulted in egg production and feed intake similar to those from hens given a conventional diet.

Although wheat middlings contain a substantial amount of NSP, Im et al. (1999) were able to demonstrate only a slight improvement in the nutritive value of middlings for broilers following xylanase supplementation of the diet. On the other hand, Jaroni et al.

(1999) were able to demonstrate an improvement in egg production in Leghorn layers by supplementation of a middlings-based diet with protease and xylanase enzymes.

Wheat shorts

Wheat shorts consist of fine particles of wheat bran, wheat germ, wheat flour and the offal from the 'tail of the mill'. This product must be obtained in the usual process of commercial milling and must contain not more than 70 g CF/kg. IFN 4-05-201 Wheat flour by-product less than 70 g/kg fibre. (Note: the Canadian feed regulations have the IFNs for middlings and shorts reversed).

As with middlings, the responses to the inclusion of wheat shorts in poultry diets are somewhat variable, related probably to variation in the composition of the product. Shorts can be used in the same way as wheat middlings.

Oilseeds, Their Products and By-products

The major protein sources used in animal production are oilseed meals. Their use in poultry diets was reviewed by Ravindran and Blair (1992). Soybeans, groundnuts, canola and sunflowers are grown primarily for their seeds, which produce oils for human consumption and industrial uses. Cottonseed is a by-product of cotton production, and its oil is widely used for food and other purposes. In the past linseed (flax) was grown to provide fibre for linen cloth production. The invention of the cotton gin made cotton more available for clothing materials and the demand for linen cloth decreased. Production of linseed is now directed mainly to industrial oil production. Thus, soybean is clearly the predominant oilseed produced in the world.

Developments are taking place that allow oilseed crops to be used on-farm for animal feeding, providing high-energy, high-protein feed ingredients. These developments, where warranted technically and economically, are of great relevance to organic animal production.

Moderate heating is generally required to inactivate ANFs present in oilseed meals. Overheating needs to be avoided since it can damage the protein, resulting in a reduction in the amount of digestible or available lysine. Other AA such as arginine, histidine and tryptophan are usually affected to a lesser extent. The potential problems of overheating are usually well recognized by oilseed processors. Only those meals resulting from mechanical extraction of the oil from the seed are acceptable for organic diets.

As a group, the oilseed meals are high in CP content, except safflower meal with hulls. The CP content of conventional oilseed meals is usually standardized before marketing by admixture with hull or other materials. Most oilseed meals are low in lysine, except soybean meal. The extent of dehulling affects the protein and fibre contents, whereas the efficiency of oil extraction influences the oil content and thus the ME content of the meal. Oilseed meals are generally low in Ca and high in P content, though a high proportion of P is present as phytate. The biological availability of minerals in plant sources such as oilseeds is generally low, and this is especially true for P because of the high phytate content (Table 4.3).

Canola (*Brassica* spp., rapeseed)

Canola is a crop belonging to the mustard family, grown for its seed. The leading countries in rapeseed production are China, Canada, India and several countries in the EU. Commercial varieties of canola have been developed from two species, *Brassica napus* (Argentine type) and *Brassica campestris* (Polish type). Rapeseed has been important in Europe for a long time as a source of feed and oil for fuel. Production of this crop became popular in North America during World War II as a source of industrial oil, the crushed seed being used as animal feed. The name 'canola' was registered in 1979 in Canada to describe 'double-low' varieties of rapeseed, i.e. the extracted oil containing less than 20 g erucic acid/kg and the air-dry meal less than 30 µmol of glucosinolates (any mixture of 3-butenyl glucosinolate, 4-pentenyl glucosinolate, 2-hydroxy-3-butenyl glucosinolate or 2-hydroxy-4-pentenyl glucosinolate) per gram of air-dry material. In addition to the above standards for conventional canola,

Table 4.3. Phytate phosphorus contents of grain legumes, oilseed meals and miscellaneous feed ingredients (from Ravindran *et al.*, 1995).

Ingredient	Phytate P, g/100 g	Phytate P, % of total P
Field peas (*Pisum sativum*)	0.19 (0.13–0.24)	46 (36–55)
Cowpeas (*Vigna unguiculata*)	0.26 (0.22–0.28)	79 (72–86)
Green gram (*Vigna radiata*)	0.22 (0.19–0.24)	63 (58–67)
Pigeon peas (*Cajanus cajan*)	0.24 (0.21–0.26)	75 (70–79)
Chickpeas (*Cicer arietinum*)	0.21 (0.19–0.24)	51 (47–58)
Lupins (*Lupinus* spp.)	0.29 (0.29–0.30)	54 (54–55)
Soybean meal	0.39 (0.28–0.44)	60 (46–65)
Cottonseed meal	0.84 (0.75–0.90)	70 (70–83)
Groundnut meal	0.42 (0.30–0.48)	51 (46–80)
Canola meal	0.54 (0.34–0.78)	52 (32–71)
Sunflower meal	0.58 (0.32–0.89)	55 (35–81)
Linseed meal	0.43 (0.39–0.47)	56 (52–58)
Coconut meal	0.27 (0.14–0.33)	46 (30–56)
Sesame meal	1.18 (1.03–1.46)	81 (77–84)
Palm kernel meal	0.39 (0.33–0.41)	66 (60–71)
Grass meal	0.01	2
Lucerne meal	0.02 (0.01–0.03)	12 (5–20)
Ipil-ipil (*Leucaena leucocephala*) leaf meal	0.02	9
Cassava leaf meal	0.04	10
Maize gluten meal	0.41 (0.29–0.63)	59 (46–65)
Maize distillers grain	0.26 (0.17–0.33)	22 (20–43)
Soy protein isolate	0.48	60

the meal is required to have a minimum of 350 g CP/kg and a maximum of 120 g CF/kg. The designation is licensed for use in at least 22 countries. Canola ranks fifth in world production of oilseed crops, after soybeans, sunflowers, groundnuts and cottonseed. It is the primary oilseed crop produced in Canada.

This crop is widely adapted but appears to grow best in temperate climates, being prone to heat stress in very hot weather. As a result, canola is often a good alternative oilseed crop to soybeans in regions not suited for growing soybeans. Some of the canola being grown is from GM-derived seed; therefore, caution must be exercised to ensure the use of non-GM canola for organic poultry production.

Erucic acid is a fatty acid that has toxic properties and has been related to heart disease in humans. Glucosinolates give rise to breakdown products that are toxic to animals. These characteristics make rapeseed products unsuitable as animal feedstuffs, but canola, like soybeans, has both a high oil content and a high protein content and is an excellent feedstuff for poultry.

Canola seed that meets organic standards can be further processed into oil and a high-protein meal, so that the oil and meal are acceptable to the organic industry. In the commercial process in North America, canola seed is purchased by processors on the basis of grading standards set by the Canadian Grain Commission or the National Institute of Oilseed Processors. Several criteria are used to grade canola seed, including the requirement that the seed must meet the canola standard with respect to erucic acid and glucosinolate levels.

Nutritional features

Canola seed contains about 400 g oil/kg, 230 g CP/kg and 70 g CF/kg. A large study in Poland found that the average oil content in organic canola (raps) was 403.2 g/kg compared with 411.5 in conventional canola (Grela and Semeniuk, 2008). The corresponding values for CP and CF were 214.5 and 219.2, and 71.3 and 66.9 g/kg respectively. The oil is high in polyunsaturated fatty acids (oleic, linoleic and linolenic), which makes it valuable for the human food market. It can also be used in animal feed, but the oil is highly unstable due to its content of polyunsaturated fatty acids and, like soybean oil, can result in soft body fat. For organic feed use the extraction has to be done by mechanical methods such as crushing (expeller processing), avoiding the use of solvent used in conventional processing. Two features of expeller processing are important. The amount of residual oil in the meal varies with the efficiency of the crushing process, resulting in a product with a more variable ME content than the commercial, solvent-extracted product. Also, the degree of heating generated by crushing may be insufficient to inactivate myrosinase in the seed. Therefore, more frequent analysis of expeller canola meal for oil and protein contents is recommended and more conservative limits should be placed on the levels of expeller canola meal used in poultry diets.

The extracted meal is a high-quality, high-protein feed ingredient. Canola meal from *B. campestris* contains about 350 g CP/kg, whereas the meal from *B. napus* contains 380–400 g/kg CP. The lysine content of canola meal is lower and the methionine content is higher than in soybean meal. Otherwise it has a comparable AA profile to soybean meal. However, the AA in canola meal are generally 8–10% less available than in soybean meal (Heartland Lysine, 1998); therefore, canola meal must be properly processed to optimize the utilization of the protein.

Because of its high fibre content (> 110 g/kg), canola meal contains about 15–25% less ME than soybean meal. Dehulling can be used to increase the ME content. Compared with soybeans, canola seed is a good source of Ca, selenium and zinc, but it is a poorer source of potassium and copper. Canola meal is generally a better source of many minerals than soybean meal but high phytic acid and fibre contents reduce the availability of many mineral elements. There have been field reports suggesting that the higher sulfur content of canola relative to soybean meal (about three times) may be linked to leg problems in birds. The scientific evidence on this is lacking; however, it is good advice to limit the level of sulfur in poultry diets to less than 5 g/kg.

Canola meal is a good source of choline, niacin and riboflavin, but not folic acid or pantothenic acid. It contains one of the highest levels of biotin found typically in North American feed ingredients. Total biotin in canola meal was found to average 1231 µg/kg with a bioavailability for growing broilers of 0.66 compared with 0.17 for wheat, 0.2 for triticale, 0.21 for barley, 0.39 for sorghum, 0.98 for soybean meal and 1.14 for maize (Blair and Misir, 1989).

The choline level was reported to be approximately three times higher in canola than in soybean meal, but in a less available form (Emmert and Baker, 1997). Although desirable from most respects, the high choline content of canola is a disadvantage with brown-shelled layers based on Rhode Island Red stock. Such stock produces 'fishy eggs' when fed diets containing canola. The explanation appears to be that the high levels of choline and sinapine (see below) compounds in canola are converted to trimethylamine in the gut. This compound has a 'fishy' odour. White-shelled layers have the ability to break down the trimethylamine to the odourless oxide that does not result in the problem, but some strains of brown-shelled layers do not produce enough of the trimethylamine oxidase enzyme to break down the trimethylamine and instead deposit the compound in the egg (Butler et al., 1982). Consequently, canola should not be fed to all strains of brown-shelled laying stock, or only at low levels.

Geese have been shown to appear to digest canola meal more efficiently than other species of poultry (Jamroz et al., 1992) and canola meal and soybean meal have been shown to have similar amino acid digestibility in ducks (Kluth and Rodehutscord, 2006).

Anti-nutritional factors

Glucosinolates represent the major ANF found in canola, occurring mainly in the embryo. This feature limited the use of rapeseed or rapeseed meal in poultry diets in the past. Although glucosinolates themselves are biologically inactive, they can be hydrolysed by myrosinase in the seed to produce goitrogenic compounds that affect thyroid gland function. These cause the thyroid gland to enlarge, resulting in goitre. They can also result in liver damage and can have a negative effect on production and reproduction. Fortunately the modern cultivars of canola contain only about 15% of the glucosinolates found in rapeseed. In addition, heat processing is effective in inactivating myrosinase. Cold-pressing of canola has been tested in Australia (Mullan et al., 2000). This process resulted in a meal with a higher content of oil and glucosinolates than in expeller canola.

Tannins are present in some varieties of canola but only at very low levels (Blair and Reichert, 1984). Canola, rapeseed and soybean hull tannins are not capable of inhibiting α-amylase (Mitaru et al., 1982), in contrast to those in other feedstuffs such as sorghum. Sinapine is the major phenolic constituent of canola and although bitter-tasting (Blair and Reichert, 1984) is not regarded as presenting any practical problems in poultry feeding except possibly for that noted above with brown-shelled layers.

Poultry diets

Canola meal is used in all types of poultry feeds (Table 4.4). However, because of its relatively low energy value for poultry, it is best used in layer and breeder diets rather than in high-energy broiler feeds. Another issue that needs to be addressed is that digestibility of key EAA is lower in canola meal than in soybean meal (Heartland Lysine, 1998). Therefore it is advisable when high inclusion rates are used that the diet should be formulated to digestible AA specifications

Table 4.4. True digestibility coefficients for poultry of some key essential amino acids in canola meal and soybean meal (from Heartland Lysine, 1998).

Amino acid	Digestibility in canola meal (%)	Digestibility in soybean meal (%)
Lysine	79	91
Methionine	90	92
Cystine	73	84
Threonine	78	88
Tryptophan	82	88

rather than to total AA specifications, otherwise flock performance may be affected.

Canola meal has been used effectively in layer diets at levels up to 200 g/kg (Roth-Maier, 1999; Perez-Maldonado and Barram, 2004), though some brown-egg strains lay eggs with a fishy flavour and odour when fed canola-based diets (Butler *et al.*, 1982). Canola and rapeseed are susceptible because they have higher levels of choline and sinapine (precursors of trimethylamine) than other ingredients. In addition, goitrin and tannins are known to inhibit the enzyme. Consequently, in North America, a limit of 30 g canola meal/kg is recommended in diets for brown-shelled layers, unless the brown-egg strain being used does not have the genetic defect that results in the fishy flavour. Higher levels may be used in countries in which high levels of fishmeal have been historically used in feeds, the consumer being accustomed to eggs with a fishy flavour.

Canola meal has also been used successfully in diets for breeding poultry (Kiiskinen, 1989), growing turkeys (Waibel *et al.*, 1992; Zdunczyk *et al.*, 2013) and ducks and geese (Jamroz *et al.*, 1992). The lower ME content of canola meal relative to other protein sources such as soybean meal limits its use in conventional broiler diets, but it can be used in organic broiler diets formulated to a lower ME content than conventional diets.

The glucosinolate levels are continually being improved with the introduction of new cultivars and this has allowed higher levels of canola to be included in poultry diets than in the past (Perez-Maldonado and Barram, 2004; Naseem *et al.*, 2006).

Full-fat canola (canola seed)

A more recent approach with canola is to include the unextracted seed in poultry diets, as a convenient way of providing both supplementary protein and energy. Good results have been achieved with this feedstuff, especially with the lower-glucosinolate cultivars. However, there are two potential problems that need to be addressed,

as with full-fat soybeans. The maximum nutritive value of full-fat rapeseed is only obtained when the product is mechanically disrupted and heat-treated to allow glucosinolate destruction and to expose the oil contained in the cells to the lipolytic enzymes in the gut (Smithard, 1993). Once ground, the oil in full-fat canola becomes highly susceptible to oxidation, resulting in undesirable odours and flavours. The seed contains a high level of α-tocopherol (vitamin E), a natural antioxidant, but additional supplementation with an acceptable antioxidant is needed if the ground product is to be stored. A practical approach to the rancidity problem would be to grind just sufficient canola for immediate use.

Several studies have shown that canola seed could be included in layer, broiler and turkey diets but that at high inclusion levels the performance results were lower than expected unless the diets were steampelleted (Salmon *et al.*, 1988; Nwokolo and Sim, 1989a, b). One interesting finding in the Nwokolo and Sim (1989a) study for organic producers was that there was a linear increase in contents of linoleic acid, linolenic acid and docosahexaenoic acid in the yolk with increasing content of canola seed in the diet. Also, in related work with broilers the same researchers showed that the lipids in skeletal muscle, skin and sub-dermal fat and abdominal fat of birds fed diets containing canola seed had the highest contents of linoleic and linolenic acids (Nwokolo and Sim, 1989b).

Work on canola seed grown in Saudi Arabia showed that a 50–100 g/kg inclusion of whole canola in the diet of Leghorns did not affect hen-day egg production, total egg mass, feed conversion efficiency or egg weight (Huthail and Al-Khateeb, 2004). Talebali and Farzinpour (2005) reported that broilers grew well on diets containing full-fat canola seed at up to 120 g/kg as a replacement for soybean meal.

Work by Meng *et al.* (2006) helps to explain the findings of poorer broiler performance with diets containing high levels of canola seed. Canola seed subjected to hammermilling had a TME_n content of 3642 kcal/kg, which was increased to 4783 kcal/kg following supplementation with enzymes. A similar pattern of increase in digestibility of fat (80.4% versus 63.5%) and NSP (20.4% versus 4.4%) was also observed. Enzyme supplementation of the canola seed diet resulted in an improvement in feed/gain, total tract DM, fat, NSP digestibilities, AME_n content and ileal fat digestibility. The results suggested that poultry are unable to digest canola seed sufficiently to extract all of the lipid from canola seed, and the data indicated the need for enzyme supplementation of poultry diets containing full-fat canola seed in order to maximize the utilization of this ingredient.

On the basis of these and related findings it is suggested that the level of full-fat canola in poultry diets be limited to 50–100 g/kg and that the seed be subjected to some form of heat treatment either before or during feed manufacture. The findings suggest also that supplementation of the dietary mixture by an appropriate enzyme mixture would be beneficial.

Cottonseed meal (*Gossypium* spp.)

Cottonseed is important in world oilseed production, major producing countries being the USA, China, India, Pakistan, Latin America and Europe. It is the second most important protein feedstuff in the USA, produced mainly by solvent extraction. Most of the cottonseed meal is used in ruminant diets, but it can be used in poultry diets when its limitations are taken into account in feed formulation (Ravindran and Blair, 1992).

Nutritional features

The CP content of cottonseed meal may vary from 360 to 410 g/kg, depending on the contents of hulls and residual oil. According to Batal and Dale (2010) mechanically extracted cottonseed meal contains around 410 g CP, 2100 kcal ME, 39 g EE, 1.7 g Ca and 3.2 g available P per kilogram, air-dry basis.

AA content and digestibility of cottonseed meal are lower than in soybean meal. Although fairly high in protein, cottonseed meal is low in lysine and tryptophan. The

fibre content is higher in cottonseed meal than in soybean meal, and its ME value is inversely related to the fibre content. Cottonseed meal is a poorer source of minerals than soybean meal. The content of carotene is low in cottonseed meal, but this meal compares favourably with soybean meal in water-soluble vitamin content, except biotin, pantothenic acid and pyridoxine.

The inclusion of cottonseed meal in poultry diets is further limited by the presence of gossypol found in the pigment glands of the seed. This problem does not occur in glandless cultivars of cottonseed. Gossypol also causes yolk discoloration in laying hens, particularly after storage of the eggs, and the cyclopropenoid fatty acids (CPFAs) present in cottonseed oil result in a pinkish colour in the egg white (Ravindran and Blair, 1992). Gossypol can be classified into bound gossypol, which is non-toxic to non-ruminant species, and free gossypol, which is toxic.

The general signs of gossypol toxicity in broilers are depressed growth, a reduction in plasma iron and haematocrit values, enlarged gallbladder, perivascular lymphoid aggregate formation, biliary hyperplasia and hepatic cholestasis, and increased mortality due to heart failure (Henry *et al.*, 2001). These authors found that toxicity signs became evident when free gossypol level in the diet approached 400 mg/kg.

Anti-nutritional factors

The free gossypol content of cottonseed meal decreases during processing and varies according to the methods used. In new seed, free gossypol accounts for 0.4–1.4% of the weight of the kernel. Screw-pressed materials have 200–500 mg free gossypol/kg. Processing conditions have to be controlled to prevent loss of protein quality owing to binding of gossypol to lysine at high temperatures. Fortunately the shearing effect of the screw press in the expeller process is an efficient gossypol inactivator at temperatures that do not reduce protein quality (Tanksley, 1990).

A general recommendation is that cottonseed meal should replace no more than 50% of the soybean meal or protein supplement in the diet. At this inclusion rate, it is unlikely that the total diet will contain more than the toxic level of free gossypol. Iron salts, such as ferrous sulfate, are effective in blocking the toxic effect of dietary gossypol, possibly by forming a strong complex between iron and gossypol and thus preventing gossypol absorption. It is probable also that gossypol reacts with iron in the liver, and the iron–gossypol complex is then excreted via the bile. A 1:1 weight ratio of iron to free gossypol can be used to inactivate the free gossypol in excess of 100 mg/kg (Tanksley, 1990). However, it is unlikely that this treatment would be acceptable for organic poultry diets. A more acceptable approach would be the use of glandless cotton cultivars devoid of gossypol, if available.

Poultry diets

Limited data are available on the inclusion of expeller cottonseed meal in diets, most of the current data referring to solvent-extracted meal.

Expeller cottonseed cake is used extensively in cotton-producing countries such as Pakistan as a protein supplement in poultry diets. For instance, Amin *et al.* (1986) included expeller or solvent-extracted cottonseed cake and rapeseed cake as the protein source in the diet of starting broilers and found no significant difference among treatments in rate of gain, feed intake, feed conversion efficiency or carcass dressing percentage. In Nigeria, Ojewola *et al.* (2006) conducted a large broiler trial that lasted for 6 weeks and involved cottonseed meal as a replacement for soybean meal at 0, 25%, 50%, 75% and 100%. Results suggested that iron-supplemented cottonseed meal can serve as a substitute for soybean meal in broiler diets. Work on laying hens and growing turkeys in Nigeria indicated that cottonseed meal could replace groundnut meal in layer diets without adverse effects on egg production, feed efficiency, egg weight or egg quality (Nzekwe and Olomu, 1984). The use of cottonseed meal did not result in yolk and albumen discoloration or pink coloration in this study, even after 8 weeks of storage. In turkey diets cottonseed meal replaced

groundnut meal without adverse effect on performance and mortality.

Panigrahi *et al.* (1989) conducted research in the UK on imported screw-press expeller cottonseed meal incorporated into layer diets at levels up to 300 g/kg. Overall performance of hens given a diet with cottonseed meal at 75 g/kg was not significantly different from the control. However, a diet containing 300 g cottonseed meal/kg (providing free gossypol 255 mg/kg and CPFA 87 mg/kg and giving daily intakes per hen of free gossypol 26.2 mg and CPFA 9.0 mg) resulted in reduced feed intake and egg production. A diet containing 150 g cottonseed meal/kg diet (daily intakes of free gossypol 14.6 mg and CPFA 4.8 mg per hen) did not produce adverse effects initially, but egg production was slightly decreased after 10 weeks. Storage of eggs under warm conditions (20°C and 30°C) for up to 1 month did not result in discoloured eggs but did result in yolk mottling, a condition reduced by treatment of the cottonseed meal with iron. Storage of eggs under cold conditions (5°C) for 3 months resulted in brown yolk discoloration and the initial stages of pink albumen discoloration with the diet containing 300 g cottonseed meal/kg diet; the brown yolk discoloration being reduced by treatment of the cottonseed meal with iron salts.

The available findings suggest that, in general, cottonseed meal can be used at low levels in organic diets for growing poultry and that it is utilized most effectively when its reduced digestibility and gossypol content are taken into account in formulation. Where iron supplementation is acceptable under the prevailing regulations, higher levels may be used and this ingredient may be suitable for use in layer diets. In developing countries in which cottonseed meal is the only available source of supplementary protein, higher levels may be used together with iron supplementation of the diet.

Linseed (flax) (*Linum usitatissimum*)

Flax is grown mainly to produce linseed oil for industrial applications, western Canada,

China and India being leading producers. Other important areas of production are the Northern Plains of the USA (Maddock et al., 2005), Argentina, the former USSR and Uruguay. Flax is grown typically under dryland conditions. In Canada, flax is produced only as an industrial oilseed crop and not for textile use as in some countries.

The oil content of flaxseed ranges from 400 to 450 g/kg and the by-product of mechanical oil extraction – flaxseed (or linseed) meal – can be used in organic poultry feeding. There is also interest in feeding the ground whole oil-containing seed to poultry for two main reasons: to produce meat with a fatty acid profile in the fat that confers health benefits to the consumer (Conners, 2000) and to impart an enhanced flavour to the meat.

Nutritional features

As with most grains and oilseeds, the composition of flax varies, depending on cultivar and environmental factors. Typical values are 410 g oil/kg and 200 g CP/kg (DM basis; DeClercq, 2006). Reported CP values range from 188 to 244 g/kg (Daun and Pryzbylski, 2000; Singh et al., 2011; Jacob, 2013). A large study in Poland found that the average EE content was 321.4 g/kg in organic flax compared with 324.2 g/kg in conventional flax (Grela and Semeniuk, 2008). The corresponding values for CP and CF were 221.2 and 229.1 g/kg, and 106.5 and 105.8 g/kg DM, respectively. Jacob (2007) reported similar values for CP in organic and conventional flaxseed, i.e. 218.1 and 220 g/kg, respectively.

As with other oilseeds, mechanical extraction results in a higher residual oil content in the meal than in the solvent-extracted product. As reviewed by Maddock et al. (2005) several reports have indicated possible human health benefits associated with consumption of flaxseed. The oil contains α-linolenic acid (ALA), an essential omega-3 fatty acid that is a precursor for eicosapentaenoic acid (EPA), at about 230 g/kg, and linoleic acid at about 65 g/kg. EPA is a precursor for the formation of eicosanoids, which are hormone-like compounds that play an essential role in the immune response. Additionally, some evidence suggests EPA can be converted to docosahexaenoic acid (DHA), an omega-3 fatty acid that is essential for cell membrane integrity, as well as brain and eye health. Flax is the richest plant source of the lignan precursor secoisolariciresinol diglycoside (SDG), which is converted by microorganisms in the hind-gut of rats and other mammals to mammalian phyto-oestrogens (Begum et al., 2004). It does not appear to have been established that this process occurs in avian species. Phyto-oestrogens are believed to have potential uses in hormone replacement therapy and cancer prevention in humans (Harris and Haggerty, 1993).

Recent research indicates that products from animals fed flax have increased levels of omega-3 fatty acids (Scheideler et al., 1994; Maddock et al., 2005), considered beneficial in the human diet. Eggs are a convenient way of providing omega-3 fatty acids in the human diet. For instance, Farrell (1995) conducted research with human volunteers who consumed ordinary eggs or omega-3 enriched eggs at a rate of seven eggs per week. After 20 weeks, the plasma levels of omega-3 fatty acids in volunteers consuming the enriched eggs were significantly higher than in those consuming the ordinary eggs and the ratio of omega-6 to omega-3 fatty acids was reduced. There were only small differences in the plasma cholesterol. Farrell concluded that an enriched egg could supply approximately 40–50% of the daily human requirement for omega-3 polyunsaturated fatty acids. In another study Marshall et al. (1994) found that 65% of consumers surveyed reported a willingness to purchase omega-3 enriched eggs.

The nutrient content of linseed meal has been reviewed by Chiba (2001) and Maddock et al. (2005). The CP content averages 350–360 g/kg, but may vary from 340 to 420 g/kg. Linseed meal is deficient in lysine and contains less methionine than other oilseed meals (Ravindran and Blair, 1992). Because of the hulls, which are coated with high quantities of mucilage, the CF content of linseed meal is relatively high. The mucilage contains a water-dispersible carbohydrate, which has low digestibility for non-ruminant species (Batterham et al., 1991). The major macrominerals in linseed meal are comparable with those in other oilseed meals, although the levels of Ca, P and magnesium are higher than the levels found in soybean meal. Although microminerals in linseed meal vary widely, it is a good source of Se, possibly because it has been grown in Se-adequate areas. The water-soluble vitamin content of linseed meal is similar to that of soybean meal and most other oilseed meals.

Anti-nutritional factors

Flaxseed or flaxseed meal contains a number of ANFs for livestock, the main ones being linamarin and linatine. Linamarin is a cyanoglycoside, which has the potential to cause cyanide poisoning by the action of the enzyme linamarase (linase). The mature seed contains little or no linamarin, and linamarase is normally destroyed by heat during oil extraction. Linatine is a dipeptide that can act as an antagonist for pyridoxine.

Because it is deficient in lysine, linseed meal should be used in combination with a complementary protein source(s). As with other oilseeds containing oil that is subject to rancidity, the ground seed should be mixed into diets and used quickly after processing.

Poultry diets

Early work showed that flaxseed meal could be used satisfactorily as a protein supplement for poultry, following soaking in water and drying or supplementation with pyridoxine to counteract the pyridoxine antagonist linatine (Kratzer and Vohra, 1996) and supplementation with lysine. The mucilage in the meal caused sticky droppings, but did not affect the performance of the birds. This knowledge allowed water-treated flaxseed meal to supply 50–75% of the protein in broiler diets, with successful results (Madhusudhan et al., 1986). Expeller flaxseed meal was found to be a suitable replacement for half of the protein supplied by soybean meal in diets for growing pullets, provided that the diet was supplemented with methionine and pyridoxine (Wylie et al., 1972).

However, flaxseed meal is unlikely to be widely available to organic poultry producers, most of the meal for animal feeding being used with ruminants. The use of flaxseed in poultry feeding that is of most interest to organic producers is the inclusion of home-grown whole seed in poultry diets to provide protein and energy and to confer health benefits in poultry meat and eggs to the human consumer. Thacker et al. (2005) fed broilers on diets containing 125 g flaxseed/kg in combination with peas or canola as a partial replacement for soybean meal and found no differences in growth rate or feed conversion efficiency when these protein supplements were extruded. Ajuyah et al. (1993) showed that breast meat from broilers fed diets containing full-fat flaxseed at 150 g/kg and antioxidants had higher levels of polyunsaturated fatty acids (C18:3n3, C20:5n3, C22:5n3 and C22:6n3) and lower levels of n-6:n-3 and total saturated fatty acids, than breast meat from birds fed diets containing full-fat flaxseed and no antioxidant. However, the beneficial effect of feeding flaxseed may be at the expense of productivity; for instance, Najib and Al-Yousef (2011) found that while feeding a diet containing 150 g flaxseed/kg increased the omega-3 fatty acid level of dark chicken meat, inclusion levels as low as 50 g/kg resulted in reduced body weight gain and feed efficiency.

Caston and Leeson (1990) reported a large increase in the omega-3 fatty acid content of the eggs of hens fed diets containing 100, 200 or 300 g flaxseed/kg. Cherian and Sim (1991) fed diets containing ground flaxseed at 80 and 160 g/kg and supplemented with pyridoxine and found an increase in the omega-3 fatty acid content of the eggs, also in the brain tissue of embryos and chicks from the hens fed the flaxseed. The increase in linolenic acid in eggs from hens fed flaxseed was mainly in the triglycerides. The longer-chain omega-3 fatty acids were deposited exclusively in the phospholipids. Aymond and Van Elswyk (1995) reported that diets containing 50 and 150 g flaxseed/kg resulted in an increase in total omega-3 fatty acids in the eggs and that the ground seeds caused a greater level of these fatty acids at the 150 g/kg level than the whole seed. The content of yolk thiobarbituric acid reactive substances (a measure of rancidity) was not influenced by inclusion of flaxseed at either level.

Basmacioğlu et al. (2003) reported a significant decrease in yolk cholesterol content in eggs from hens fed diets containing 15 g fish oil/kg or 86.4 g flaxseed/kg and a significant increase in total omega-3 fatty acid content of eggs at 28 (phase 1) and 56 (phase 2) days of the trial. Linolenic acid content of eggs was highest in eggs from hens fed the diet containing flaxseed. Serum cholesterol content was lower in hens fed diets containing 15 g fish oil/kg and 43.2 g flaxseed/kg or 86.4 g flaxseed/kg than in hens fed the control diet. Dietary inclusion of flaxseed did not result in any negative effects on egg weight, yolk weight, yolk ratio, albumen weight, albumen ratio, shell weight, shell ratio, shell strength or shell thickness. Egg production of hens fed the diet containing 43.2 g flaxseed/kg was significantly higher than with the control diet. Feed intake and feed conversion were not affected by diet.

One possible problem that may occur in feeding flaxseed diets to layers is a fishy flavour in the eggs. Jiang et al. (1992) reported a 36% incidence of fishy or fish-related flavour in eggs from hens fed flaxseed. This was not found in eggs from hens fed a diet containing no flaxseed or diets containing high-oleic acid or high-linoleic acid sunflower seeds. Another possible problem is in relation to coccidiosis. Allen et al. (1997) found that the feeding of flaxseed meal was beneficial in reducing the effects of *Eimeria tenella* infections but it was not beneficial in reducing effects of *Eimeria maxima* infections and might exacerbate lesions at high infection levels.

Mustard

Mustard is both a condiment and an oilseed crop. It grows well in temperate and in high-altitude, subtropical areas and is moderately drought-resistant. Two species are grown: *Brassica juncea* (brown and oriental mustard) and *Sinapis alba* (yellow or white mustard). *B. juncea* is grown as an edible oil crop in China,

India, Russia and Eastern Europe. Canada is the world's largest supplier of mustard seed, exporting the seed to Japan, the USA, Europe and Asia. Newer mustard cultivars appear to have several advantages over some of the canola cultivars grown, being higher yielding, early maturing, more resistant to late spring frosts, more heat- and drought-tolerant, more resistant to seed shattering and more resistant to disease.

Nutritional features

Mustard meal contains more glucosinolates than rapeseed, though of different kinds (Ravindran and Blair, 1992), the high content of glucosinolates determining the value of mustard as a condiment in the human diet. Several methods of detoxification have been developed, including ammoniation and sodium carbonate treatment, which would render the meal unacceptable for use in organic diets.

Some of the new low-glucosinolate cultivars can be considered as an alternative to canola as an oilseed crop, since the meal appears to be similar to canola meal in nutrient content. The new cultivars of mustard are being grown in several countries, including Australia. These cultivars are of interest to organic producers since they have the potential to be grown and used on-farm as ingredients in organic diets.

Information on the nutritional value of unextracted, low-glucosinolate mustard seed for poultry is very limited. Therefore, the available data on extracted mustard meal have to be used as a guide. Blair (1984) found the AME value of ammoniated, solvent-extracted mustard meal for starting broilers to be 2648 kcal/kg DM basis (2383 kcal/kg, air-dry basis). Bell et al. (1998) reported that the CP contents of canola and mustard meal were as follows: 'Excel' canola 418 g/kg, 'Parkland' canola 401 g/kg and *B. juncea* 439 g/kg. Cheva-Isarakul et al. (2003) studied the composition of mustard meal grown in Thailand and reported that it contained 300–320 g CP/kg, 190–220 g lipid/kg, 120–130 g CF/kg, 50–60 g ash/kg and 280–310 nitrogen-free extract (NFE) g/kg (DM basis). AME and true ME of sun-dried mustard meal were 2888 and 3348 kcal/kg DM (2724 and 3161 kcal/kg air-dry basis), respectively, while those of gas-dried meal were lower (2435 and 2892 kcal/kg DM, 2328 and 2765 kcal/kg air-dry basis, respectively).

Newkirk *et al.* (1997) conducted a nutritional evaluation of low-glucosinolate mustard meals. Samples of brassica seed (four *B. juncea*, one *B. napus* and one *B. rapa*) were processed to produce oil-extracted meals, which were then fed to broiler chickens. Meals derived from *B. juncea* contained more CP and less TDF on a dry basis than *B. napus* or *B. campestris*: 459 versus 446 and 431 g CP/kg and 272.2 versus 294.7 and 296.7 g TDF/kg, respectively. ADF and NDF levels for *B. juncea* and *B. campestris* meals were similar to each other, but lower than those of *B. napus*: 127.9 and 132.0 versus 206 g ADF/kg, and 211.5 and 195.8 versus 294.7 g NDF/kg, respectively. *B. juncea* meals contained more glucosinolates than *B. napus* and *B. campestris*: 34.3 versus 21.8 and 25.5 µmol total glucosinolates/g, respectively. *B. juncea* meals were equal or superior to *B. napus* and *B. campestris* meals for AME_n and apparent ileal protein digestibility.

Poultry diets

Blair (1984) studied the value of ammoniated, solvent-extracted mustard meal for broiler chicks aged 0–4 weeks and found the AME value to be 2648 kcal/kg DM basis (2383 kcal/kg, air-dry basis). Results showed that ammoniated mustard meal could be included in the diet successfully at up to 100 g/kg, though thyroid size was increased. A 200 g/kg inclusion rate resulted in poorer growth. Up to 200 g/kg could successfully be included in the diet provided that the diet was supplemented with L-lysine.

Rao *et al.* (2005) compared low-glucosinolate and conventional mustard oilseed cakes in commercial broiler chicken diets and concluded that soybean meal could be replaced completely with the low-glucosinolate product (535.0 and 466.5 g/kg in starter and finisher diets, respectively) or up to 50% of the conventional product (215.0 and 186.7 g/kg in starter and finisher diets, respectively). Choline supplementation at the 1 g/kg level in broiler diets containing the conventional mustard product was found to be beneficial during the starter phase.

Newkirk *et al.* (1997) found that broilers fed on *B. juncea* meals grew as quickly and converted feed to gain as efficiently to 21 days of age as those fed on *B. napus* and *B. campestris* meals. Feeding meal from *B. campestris* reduced growth rate and gain/feed ratio. They concluded that the nutritional value of meals from low-glucosinolate mustard is equal or superior to that of rapeseed meal samples derived from *B. napus* and *B. campestris* cultivars.

Cheva-Isarakul *et al.* (2001) studied the effects of including mustard meal from a mustard-processing plant in layer diets. The meal was sun-dried or was dehydrated in a gas-heated pan and incorporated into diets at 0, 100, 200 and 300 g/kg as a replacement for soybean meal. Egg production, feed intake, body weight gain and egg weight decreased significantly with increased level of mustard meal. Fat deposition of the birds fed diets containing mustard meal decreased significantly, while kidney weight increased in comparison with the control group. Weight of thyroid and spleen tended to be heavier in the mustard-fed groups but the effect was not statistically significant. It was concluded that mustard meal from either drying methods could be incorporated in laying hen diets at a concentration of 100 g/kg without any adverse effects.

Groundnuts (peanuts) (*Arachis hypogaea* L.)

Groundnuts (also known as peanuts) are not included as an approved feedstuff in either the EU or New Zealand lists but should be acceptable for organic poultry diets if grown organically. The reason for omission may be that groundnuts are grown mainly for the human market. This crop is grown extensively in tropical and subtropical regions and is too important to be rejected for use in organic poultry diets. However, this issue should be clarified with the local certifying agency. India and China are the major producing countries. Groundnut occupies an important position as an oilseed in many developing countries. Groundnuts not suitable for human consumption are used in the production of groundnut oil. The by-product of oil extraction, groundnut meal, is widely used as a protein supplement in livestock diets.

Nutritional features

Raw groundnuts contain 400–550 g oil/kg. Groundnut meal is the ground product of shelled groundnuts, composed principally of the kernels, with such portion of the hull, or fibre and oil remaining after oil extraction by a mechanical extraction process. Mechanically extracted meal may contain 50–70 g oil/kg and thus tends to become rancid during storage, especially during summer. The CP content of extracted meal ranges from 410 to 500 g/kg. In the USA, the meal is usually adjusted to a standard protein level with ground groundnut hulls. The traded product in the USA must contain not more than 70 g CF/kg and only such amount of hulls as is unavoidable in good manufacturing practice. Groundnut protein is deficient in methionine, lysine, threonine and tryptophan, with lysine being the most limiting AA, followed by methionine and then threonine. Thus, groundnut is not suitable as the sole supplemental protein for poultry diets and needs to be mixed with other protein sources. Groundnut meal is low in Ca, sodium and chloride and much of the P occurs as phytate. It is a good source of niacin, thiamin, riboflavin, pyridoxine, pantothenic acid and choline.

Batal *et al.* (2005) surveyed the nutrient composition of solvent-extracted groundnut meal available in the USA and found that N-corrected ME ranged from 2273 to 3009 kcal/kg with a mean of 2664 kcal/kg, whereas CP ranged from 401 to 509 g/kg with a mean of 456 g/kg. Mean values for fat, fibre and ash were 25, 83 and 50 g/kg, respectively. Total concentration (g/kg) and percentage availability of several critical AA were lysine 15.4 (85), methionine 5.2 (87), cystine 6.4 (78), threonine 11.7 (81) and arginine 50.4 (90). Average levels of Ca, P, sodium and potassium were 0.8, 5.7, 0.1 and 12.2 g/kg respectively. The variation observed among samples strongly indicates that confirmatory analyses should be conducted prior to use of samples from a new supplier.

Anti-nutritional factors

Groundnuts are subject to contamination with moulds. *Aspergillus flavus*, which produces aflatoxin, can grow in groundnuts and occur in groundnut meal. Aflatoxin is carcinogenic and acutely toxic to animals and humans, depending on the level of contamination (see section on 'Mycotoxins' in Chapter 7). Their presence in the diet has implications for livestock feeding (Ravindran and Blair, 1992). The primary factors leading to high contamination levels are shell damage and kernel splitting, which are usually induced by insects, poor harvesting and poor drying conditions. Contamination can be minimized by prompt harvesting and drying and proper storage.

Groundnuts also contain protease inhibitors and tannins, but generally not at levels high enough to cause concern. This trypsin inhibitor can be destroyed by mild heating.

Poultry diets

Both groundnuts and groundnut meal have been used successfully in poultry diets. However, productivity of growing and laying birds has generally been reduced when more than 50% of the protein from soybean meal has been replaced by groundnut meal, unless supplemented with protein sources rich in lysine and methionine (El-Boushy and Raterink, 1989; Ravindran and Blair, 1992; Amaefule and Osuagwu, 2005; Lu *et al.*, 2013).

Offiong *et al.* (1974) reported that broiler chicks fed diets containing raw groundnuts and fishmeal at 50, 100 or 125 g/kg were significantly heavier than those given diets based on soybean meal. Efficiency of feed conversion was also higher with the diets containing 100 or 250 g groundnuts/kg.

Pesti *et al.* (2003) compared groundnut meal and soybean meal as protein supplements in maize-based diets for laying hens and examined the effects on egg quality. The diets contained three CP levels: 160, 185 and 210 g/kg. Increasing dietary protein level had no consistent effect on productivity but significantly improved body weight gains and egg weights (1.2–2.5 g per egg). Hens fed the diets containing groundnut meal laid slightly smaller eggs initially. Their eggs were found to have better interior quality after 2 weeks of storage, Haugh units remaining better for eggs from hens fed groundnut meal than soybean meal when kept refrigerated (4°C) or at room temperature (20°C). Egg specific gravity was slightly lower for hens fed groundnut meal.

Safflower meal (*Carthamus tictorius*)

Safflower is an oilseed crop cultivated mainly in tropical regions. The oil is high in poly-unsaturated fatty acids, particularly linoleic acid, making it an important oil for human use, like canola and olive oils. India, the USA and Mexico are major producers of safflower, but it can be grown in cooler areas. Because it is a long-season crop, safflower extracts water from the soil for a longer period than cereal crops and the long taproot can draw moisture from deep in the subsoil. These properties can help to prevent the spread of dryland salinity in areas such as the Canadian prairies, using up surplus water from areas that otherwise would contribute to the development or expansion of salinity. Safflower meal is not included in the lists of approved feedstuffs but should be acceptable if produced organically.

Nutritional features

Safflower seeds resemble those of sunflower and have an average composition of 400 g hull/kg, about 170 g CP/kg and 350 g crude fat/kg (Seerley, 1991). The seed is composed of a kernel surrounded by a thick fibrous hull that is difficult to remove. As a result, much safflower meal is made from unhulled seed, suitable only for feeding to ruminants. Australian researchers (Ashes and Peck, 1978) described a simple mill and screening device for dehulling safflower seed and other seed and grains. The device operates by bouncing the seed between a 'squirrel cage'-type rotor and a ripple plate, thus forcing the hull from the kernel in contrast to conventional milling or rolling. Thirteen seed and grain types were dehulled during a single pass through the mill and screening device. Safflower seed was effectively dehulled by the device but

required two passes through the mill. The efficiency of dehulling with other seed and grains after one pass varied, but was 90% with sunflower seed and 95% with cottonseed. The extent of the dehulling was proportional to the velocity of the rotor tips and could be varied readily. Results showed that the mill was able to process a wide variety of seed and also other ingredients such as dry lucerne hay. The Australian device may be of interest to organic poultry producers who wish to grow their own protein feedstuffs such as safflower.

Safflower meal may contain CP levels ranging from 200 to 600 g/kg, with the composition being dependent on the proportion of hulls in the meal and the processing method. Although complete dehulling could result in a meal with about 600 g CP/kg, in practice it is very difficult to effect a complete separation because of the hardness of the seedcoat and the extreme softness of the kernel (Ravindran and Blair, 1992). Meals produced from partially dehulled seeds typically contain over 400 g CP/kg and about 150 g CF/kg. The quality of safflower protein is low due to the low content and poor availability of lysine in the meal. In an assessment of the nutritive value of oilseed proteins, Evans and Bandemer (1967) ranked safflower protein at 50 and soybean protein at 96 relative to casein.

Oil extraction produces an undecorticated (hulled) safflower meal with approximately 200–220 g CP/kg and 400 g CF/kg. The undecorticated meal is also called whole pressed seed meal, whereas the decorticated meal is referred to as safflower meal. Decortication of the hulled meal yields a high protein (420–450 g CP/kg), less fibrous (150–160 g CF/kg) meal, which is more suitable for inclusion in poultry or pig diets (Darroch, 1990).

The mineral content of safflower meal is generally less than that of soybean meal, but safflower meal is a comparable source of Ca and P. Safflower meal is a rich plant source of iron (Darroch, 1990). Compared with other oilseed meals, safflower meal has a relatively poor vitamin profile, but is a good source of biotin, riboflavin and niacin compared with soybean meal (Darroch, 1990).

Anti-nutritional factors

The palatability of safflower meal to poultry tends to be low (Ravindran and Blair, 1992) due to the presence of two phenolic glucosides: matairesinol-β-glucoside, which imparts a bitter flavour, and 2-hydroxyarctiin-β-glucoside, which also has a bitter flavour and has cathartic properties (Darroch, 1990). Both glucosides are associated with the protein fraction of the meal, and they can be removed by extraction with water or methanol, or by the addition of β-glucosidase.

Poultry diets

A review of early studies showed that poultry performed poorly when fed diets containing unhulled safflower meal (Ravindran and Blair, 1992). Partially dehulled safflower meal, however, can be utilized in balanced diets for broilers and layers at levels limited by its fibre content. Satisfactory performance is achieved only with supplementation with other proteins to improve the protein quality of the diet.

Data on the use of expeller safflower meal in poultry diets are limited. Thomas et al. (1983) found the TME value of this product for adult roosters to be 2402 kcal/kg. Chickens fed diets based on safflower meal replacing 25% or 50% of the soybean protein ate significantly more and gained significantly more weight than those fed diets based on rapeseed meal replacing 25% or 50% of soybean protein. Safflower oil is used to impart desirable nutritional qualities in eggs, because of its high content of polyunsaturated fatty acids (Kim et al., 1997).

Sesame meal (*Sesamum indicum*)

Sesame is grown mainly in China, India, Africa, South-east Asia and Mexico as an oil crop. It is known as the 'queen of the oilseed crops' because of the excellent culinary properties of the oil (Ravindran and Blair, 1992). Due to an increasing demand for sesame oil, world production of sesame has increased. Sesame is a crop of the tropics, but its extension into the temperate zone has been made possible through breeding of suitable varieties. After oil extraction the meal can be used for animal feeding. Sesame meal is not of significant importance for poultry feeding, however.

Nutritional features

The nutrient composition of sesame meal compares closely with that of soybean meal (Ravindran and Blair, 1992). An average CP content of 400 g/kg and CF content of 65 g/kg are typical for expeller-extracted sesame meal, but these values may vary widely depending on the variety used, degree of decortication and method of processing. The ME content of sesame meal is lower than that of soybean meal and this appears to be related largely to its high ash content. Sesame meal is an excellent source of methionine, cystine and tryptophan but is deficient in lysine, indicating that sesame meal cannot be used effectively as the sole protein supplement in poultry diets. As suggested by its high ash content, sesame meal is a rich source of minerals. The Ca content, in particular, is ten times higher than that of soybean meal. However, mineral availability from sesame meal may be lower due to the presence of high levels of oxalates (35 mg/100 g) and phytate (5 g/100 g) in the hull fraction of the seed. Removal of hull not only improves mineral availability, but also reduces the fibre content and increases the protein level and palatability. Complete dehulling, however, is not always possible due to the small size of the seeds. Vitamin levels in sesame meal are comparable with soybean meal and most other oilseed meals (Ravindran and Blair, 1992).

Anti-Nutritional Factors

Although sesame seed is not known to contain any protease inhibitors or other ANFs, high levels of oxalic and phytic acids may have adverse effects on palatability (Ravindran and Blair, 1992) and on availability of minerals and protein (Aherne and Kennelly, 1985). Decortication of seeds almost completely removes oxalates, but it has little effect on phytate (Ravindran and Blair, 1992).

Poultry Diets

Use of sesame meal in starter diets should be limited because of its high fibre content and possible palatability problems associated with phytates and oxalates (Ravindran and Blair, 1992).

Good-quality sesame meal can be included up to 150 g/kg in poultry diets, but for optimal growth and feed conversion efficiency it should be supplemented with high-lysine ingredients such as soybean meal or fishmeal (Mohan *et al.*, 1984; Ravindran and Blair, 1992; Jacob *et al.*, 1996).

Sesame meal may have a role as a natural antioxidant in feeds for small-scale farmers using diets based on rice bran, which is very unstable and can become rancid on storage (Yamasaki *et al.*, 2003). In areas such as the Mekong Delta area of Vietnam, rice bran, broken rice, protein concentrate, vegetables, etc. are used by small-scale farmers for livestock feeding. In this environment rice bran, which is the major regional feed resource, must be used within a few weeks of production because it is generally not defatted and the oil is prone to peroxidation and loss of palatability. Yamasaki *et al.* (2003) tested the inclusion of ground white sesame at 10–35 g/kg into the diet of growing pigs and reported an improvement in feed intake and feed conversion efficiency. These researchers recommended the use of small amounts of sesame meal as a natural antioxidant for use with rice bran diets, but only when the sesame meal was fresh. Presumably the sesame meal acted in this way due to its content of vitamin E. These findings provide a guide to the use of sesame meal in poultry diets.

Soybeans (*Glycine max*) and soybean products

Soybeans and soybean meal are now used widely in animal feeding. The crop is grown as a source of protein and oil for the human market and for the animal feed market, the USA, Brazil, Argentina and China being the main producers of soybeans. Soybean meal is generally regarded as the best of plant protein source in terms of its nutritional value.

Also, it has a complementary relationship with cereal grains in meeting the AA requirements of farm animals. Consequently, it is the standard with which other plant protein sources are compared.

Several genetically modified strains of soybeans are now grown; therefore, organic producers have to be careful to select non-GM products. The major GM crops grown in North America are soybeans, maize, canola and cotton.

Whole soybeans contain 150–210 g oil/kg, which is removed in the oil-extraction process. When the North American industry started in the 1930s, soybeans were processed mechanically using hydraulic or screw presses (expellers), to remove much of the oil. Later most of the industry converted to the solvent-extraction process. Disadvantages of the mechanical process are that it is less efficient than the solvent process in extracting the oil and that the heat generated by friction of the screw presses, while inactivating ANFs present in raw soybeans, subjects the product to a higher processing temperature than in the solvent-extraction process, which makes the protein more difficult to digest. Expeller soybean meal is thus favoured for dairy cow feeding since the higher content of rumen bypass protein results in improved milk production. Consequently, most of the expeller soybean meal available in North America is used in the dairy feed industry and may be difficult to obtain by the organic poultry producer.

Fig. 4.2. Faba beans: an example of protein crop that can be grown for self-sufficiency on organic poultry farms.

More recently a new process known as extruding-expelling has been developed. Extruders are machines in which soybeans or other oilseeds are forced through a tapered die. The frictional pressure causes heating. In the extruding-expelling process a dry extruder in front of the screw presses eliminates the need for steam. These plants are relatively small, typically processing 5–25 t of soybeans per day in the USA. The dry extrusion-expelling procedure results in a meal with greater oil content than in conventional solvent-extracted meal, but with similar low trypsin-inhibitor values. The nutritional characteristics of extruded-expelled meal have been shown to be similar to those of screw-pressed meal, with Woodworth *et al.* (2001) showing that extruded-expelled soybean meal was more digestible than conventional soybean meal. These researchers also showed that extruded-expelled soybean meal could be used successfully in pig diets. This process should be of interest, therefore, to organic poultry producers since the soy product qualifies for acceptance in organic diets.

Yet another process used in small plants is extrusion, but without removal of the oil, the product being a full-fat meal. Often these plants are operated by cooperatives and should be of interest to organic poultry producers, since the product also qualifies for acceptance in organic diets. Another interesting development with soybeans is the introduction of strains suitable for cultivation in cooler climates, for instance the Maritime region of Eastern Canada. This development together with the installation of extruder plants allows the crop to be grown and utilized locally, holding the promise for regions that are deficient in protein feedstuffs to become self-sufficient in feed needs. Developments such as this may help to solve the ongoing problem of an inadequate supply of organic protein feedstuffs in Europe.

Nutritional features

Whole soybeans contain 360–370 g CP/kg, whereas soybean meal contains 410–500 g CP/kg depending on efficiency of the oil-extraction process and the amount of residual hulls present. Jacob (2007) reported a slightly higher content of CP in organic soybeans and in expeller soybean meal than in conventional soybeans and expeller meal: 398.5 versus 380 g/kg and 431.3 versus 420 g/kg, respectively. The oil has a high content of the polyunsaturated fatty acids, linoleic (C18:2) and linolenic (C18:3) acids. It also contains high amounts of another unsaturated fatty acid, oleic (C18:1) and moderate amounts of the saturated fatty acids, palmitic (C16:0) and stearic (C18:0).

Conventional soybean meal is generally available in two forms: 440 g CP/kg meal; and dehulled meal, which contains 480–500 g CP/kg. Because of its low fibre content, the ME content of soybean meal is higher than in most other oilseed meals. Soybean meal has an excellent AA profile. The content of lysine in soybean protein is exceeded only in pea, fish and milk proteins. Soybean meal is an excellent source of tryptophan, threonine and isoleucine, complementing the limiting AA in cereal grains. In addition, the AA in soybean meal are highly digestible in relation to other protein sources of plant origin. Apparent digestibility of N and most AA has been shown to be similar for both types of soybean meal. These features allow the formulation of diets that contain lower total protein than with other oilseed meals, thereby reducing the amount of N to be excreted by the animal and reducing the N load on the environment.

Soybean meal is generally low in minerals and vitamins (except choline and folic acid). About two-thirds of the P in soybeans is bound as phytate and is mostly unavailable to animals. This compound also chelates mineral elements, including Ca, magnesium, potassium, iron and zinc, rendering them unavailable to poultry. Therefore, it is important that diets based on soybean meal contain adequate amounts of these trace minerals. Another approach to the phytate problem is to add phytase, a phytic acid degrading enzyme, to the feed to release phytin-bound P. A benefit of this approach is that less P needs to be added to the diet, reducing excess P loading into the environment.

Anti-nutritional factors

ANFs are found in all oilseed proteins. Among these in raw soybeans are protease inhibitors. These are known as the Kunitz inhibitor and the Bowman-Birk inhibitor, which are active against trypsin, while the latter is also active against chymotrypsin (Liener, 1994). These protease inhibitors interfere with the digestion of proteins, resulting in decreased animal growth rate. They are inactivated when the beans are toasted or heated during processing. However, care has to be taken that the beans are not overheated. When proteins are heat-treated at too high a temperature the bioavailability of protein and AA may be reduced because of a substantial loss of available lysine as a result of the browning or Maillard reaction. The browning reaction involves a reaction between the free amino group on the side chain of lysine with a reducing sugar, to form brown, indigestible polymers.

Lectins (haemagglutinins) in raw soybeans can inhibit growth and cause death in animals. They are proteins that bind to carbohydrate-containing molecules and cause blood clotting. Fortunately lectins are degraded rapidly by heating. Soybeans also contain growth inhibitors that are not easily deactivated by heat treatment.

Conventional soybean meal is one of the most consistent feed ingredients available to the feed manufacturer, with the nutrient composition and physical characteristics varying very little between sources. Suppliers of organic soybean meal need to adopt similar quality control measures to ensure similar consistency in composition.

Proper processing of soybeans requires precise control of moisture content, temperature and processing time. Adequate moisture during processing ensures destruction of the ANFs. Both over- and under-toasting of soybean meal can result in a meal of lower nutritional quality. Under-heating produces incomplete inactivation of the ANFs and over-toasting can reduce AA availability.

The industry monitors soybean meal quality by using urease activity to detect under-heating and potassium hydroxide (KOH) solubility to detect overheating. The urease assay measures urease activity based on the pH increase caused by ammonia release from the action of the urease enzyme. Destruction of the urease activity is correlated with destruction of trypsin inhibitors and other ANFs. To measure KOH solubility, soy products are mixed with 0.2% KOH, and the amount of N solubilized is measured. The amount of N solubilized decreases as heating time increases, indicating decreased AA availability.

Poultry diets

An appropriate combination of soybean meal and maize (or most other cereal grains) provides an excellent dietary AA pattern for all classes of poultry.

Full-fat soybeans

Several reports show that properly processed whole soybeans can be used effectively in broiler (Pârvu et al., 2001) and layer (Sakomura et al., 1998) diets as a partial or complete replacement for soybean or other protein meals, though the digestibility of this ingredient may be more variable than with conventional soybean meal (Opapeju et al., 2006).

Extruded soybeans are whole soybeans that have been exposed to a dry or wet (steam) friction heat treatment without removing any of the component parts. Roasted soybeans are whole soybeans that have been exposed to heat or micronized treatment without removing any of the component parts. Since soybeans contain a high-quality protein and are a rich source of oil they have the ability to provide major amounts of protein and energy in the diet. Use of full-fat beans is a good way of increasing the energy level of the diet, particularly when they are combined with low-energy ingredients. Also, this is an easier way to blend fat into a diet than by the addition of liquid fat.

Sakomura et al. (1998) reported that the processing method led to differences in the performance of laying hens, extruded soybeans being superior to toasted soybeans. In addition, these researchers showed higher utilization of oil when included in the diet as extruded soybeans or soybean meal plus added soybean oil than as toasted soybeans. The AME_n value for extruded soybeans was about 12% higher than for soybean plus oil or toasted soybeans.

However, whole soybeans may have adverse effects on the carcass fat of broilers. A standard recommendation is that the soybean product should be limited to supplying a dietary addition of 20 g soybean oil/kg in order to ensure an acceptable carcass fat quality and good pellet quality. This may require that the incorporation rate of soybeans in the feed of finishing broilers should not exceed 100 g/kg.

Including full-fat soybeans in poultry diets reduces aerial dust levels, and is likely to benefit the health of animals and workers in buildings. Because of possible rancidity problems, diets based on full-fat soybean should be used immediately and not stored. Otherwise, an approved antioxidant should be added to the dietary mixture.

Soy protein isolates

Soybean protein concentrate (or soy protein concentrate) (IFN 5-08-038) is the product obtained by removing most of the oil and water-soluble non-protein constituents from selected, sound, cleaned, dehulled soybeans. The traded product in North America contains no less than 650 g CP/kg on a moisture-free basis. Soybean protein isolate (or soy protein isolate) (IFN 5-24-811) is the dried product obtained by removing most of the non-protein constituents from selected, sound, cleaned, dehulled soybeans. The traded product contains no less than 900 g CP/kg on a moisture-free basis. Both soy protein concentrate and isolate have the potential to be used in poultry diets as a source of protein and AA, though not listed specifically as approved organic feedstuffs. Several agencies regard these soy products as 'synthetic' or possibly derived from GM soybeans. The matter should be checked with the local certifying agency.

Sunflower seeds and meal (*Helianthus annus*)

Sunflower is an oilseed crop of considerable potential for organic poultry production since it grows in many parts of the world (Fig. 4.3). The main producers are Europe (France, Russia, Romania and Ukraine), South America, China and India. Sunflower is grown for oil production, leaving the extracted meal available for animal feeding.

Fig. 4.3. Sunflowers: another example of protein crop that can be grown for self-sufficiency on organic poultry farms.

Sunflower oil is highly valued for its high content of polyunsaturated fatty acids and stability at high temperatures. Sunflower seed surplus to processing needs, and seed unsuitable for oil production, may also be available for feed use. On-farm processing of sunflower seed is being done in countries such as Austria (Zettl et al., 1993).

Nutritional features

Senkoylu and Dale (1999) reviewed the nutritional value of sunflower seed and meal for poultry. The seeds contain approximately 380 g oil/kg, 170 g CP/kg and 159 g CF/kg and are a good source of dietary lipid. Sunflower meal is produced by extraction of the oil from sunflower seeds. The nutrient composition of the meal varies considerably, depending on the quality of the seed, method of extraction and content of hulls. The CF content of whole (hulled) sunflower meal is around 300 g/kg and with a complete decortication (hull removal) the fibre content is around 120 g/kg. Sunflower meal is lower in lysine and higher in sulfur-containing AA than soybean meal. However, its ME value is considerably lower than in soybean meal. Ca and P levels compare favourably with those of other plant protein sources. Sunflower meal tends to be lower in trace elements, compared with soybean meal. In general, sunflower meal is high in B vitamins and β-carotene. Because of its high fibre content it should be used in limited quantities in poultry diets.

Anti-nutritional factors

In contrast to other major oilseeds and oilseed meals, sunflower seeds and meals appear to be relatively free of ANFs. Chlorogenic acid, a polyphenolic compound occurring in sunflower meal (12 g/kg), has been reported to inhibit the activity of hydrolytic enzyme, with possible adverse effects on poultry production (Ravindran and Blair, 1992). These effects may be overcome by supplementing the diet with a methyl donor such as choline.

Poultry diets

Senkoylu and Dale (1999) reviewed findings on the use of sunflower meal in poultry diets and concluded that it could be used successfully in layer, broiler and waterfowl diets to replace 50–100% of soybean meal, depending on the type of diet and the nature of the other ingredients. However, results were inconsistent, due to differences in cultivar types, hull content of the meal, processing method and composition of the basal diet used in the various studies. Ravindran and Blair (1992) concluded that dehulled sunflower meal could replace up to 50% of the soybean in broiler and layer diets, provided that the dietary lysine content was adequate.

High dietary levels of sunflower meal require adequate supplementation with lysine, methionine and energy sources, and pelleting of the diet has been found to be beneficial in helping to overcome the bulkiness of the diet due to the high content of fibre in this ingredient. Another approach to addressing the low energy in sunflower meal is to utilize high-oil cultivars (Senkoylu and Dale, 2006), which may require dietary supplementation with vitamin E to prevent the susceptibility of the lipids in broiler meat to oxidation during refrigerated storage (Rebolé et al., 2006).

However, it is unlikely that maximal levels of sunflower meal can be achieved in organic diets since supplements of pure lysine and methionine are not acceptable in the organic regulations in most jurisdictions.

Jacob et al. (1996) obtained excellent broiler growth and feed conversion when Kenyan sunflower meal replaced about one-third of the soybean meal in mash-type diets. In another study Rao et al. (2006) conducted research on sunflower seed meal as a substitute for soybean meal in commercial broiler chicken diets. Soybean meal in the starter (318 g/kg) and finisher (275 g/kg) diets was replaced with sunflower meal at 33%, 67% and 100% on an isoenergetic and isonitrogenous basis, using supplementary sunflower oil to balance the energy content of the diets. However, even with oil supplementation, the dietary ME levels used were considerably lower than recommended for broilers. Digestibility of DM decreased with increasing level of dietary sunflower seed meal. Body weight gain to 42 days of age was not affected by total replacement of soybean meal with sunflower seed meal. Based on feed conversion efficiency, carcass yield, serum lipid profile and level of supplemental fat required to provide a satisfactory level of energy in the diet, it was concluded that sunflower seed meal could replace up to 67% of soybean meal, corresponding to inclusion rates of 345 g and 296 g sunflower seed meal/kg in starter and finisher diets, respectively.

Jacob *et al.* (1996) reported that egg production, feed intake, feed conversion efficiency and egg weight were lower in hens receiving a diet containing Kenyan sunflower seed cake replacing 50% of the soybean meal (about 135 g/kg diet). There was no effect of protein source on the incidence or severity of yolk mottling. On the other hand, the results of Šerman *et al.* (1997) suggested that sunflower meal could be used as the protein source in layer diets provided that the lysine and energy values were maintained. Karunajeewa *et al.* (1989) found that a diet containing undecorticated, oil-extracted sunflower seed meal up to 190 g/kg with or without added sunflower oil, or full-fat sunflower seeds at 23.5 g/kg with or without sunflower oil had no effect on rate of lay, total egg mass, feed conversion efficiency, body weight gain, mortality, egg specific gravity or faecal moisture content. Egg weight and feed intake increased and Haugh unit score decreased with diets containing sunflower seed meal at 122 and 190 g/kg. Casartelli *et al.* (2006) found that sunflower seed meal could be used at up to 120 g/kg (the highest level tested) in layer diets without affecting performance or egg quality parameters. Diets formulated on the basis of digestible AA gave improved eggshell percentage and egg specific gravity without decreasing egg weight.

Because of its high oil content, unextracted sunflower seed provides a convenient method of providing additional energy to poultry diets in the form of sunflower oil. Elzubeir and Ibrahim (1991) fed starting broiler chickens on diets containing unprocessed ground, hulled sunflower seed at 0, 75, 150 and 225 g/kg as a partial replacement for groundnut meal and sesame meal. No significant differences in weight gain, feed intake, feed efficiency, mortality, skin colour, or liver and pancreas weight were observed. Rodríguez *et al.* (1997) evaluated the nutritive value of hulled full-fat sunflower seed for broiler chicks. The crude fat digestibility coefficient and AME value were 0.84 and 18.71 MJ/kg DM, respectively, and mean AA digestibility values were similar for a maize–soybean control diet and for diets containing sunflower seed meal. No significant differences in feed intake, weight gain and feed efficiency were found with graded levels of sunflower seed (0, 50, 150 and 250 g/kg diet). Tsuzuki *et al.* (2003) studied the effects of sunflower seed inclusion (0, 14, 28, 42 and 56 g/kg) in layer diets containing maize, soybean meal, wheat meal and soybean oil as the main ingredients. Dietary inclusion of sunflower seed had no effect on daily feed intake, average egg weight, feed conversion efficiency, eggshell percentage, yolk colour or Haugh unit score.

Fat sources

Oils are often used as dietary supplements to provide additional energy. They have other benefits such as increasing palatability of the diet and reducing dustiness of the feed mixture. Animal fats are not permitted in organic diets but plant oils should be acceptable. These oils are concentrated sources of energy, with an average ME value of 7800–8000 kcal/kg. Thus, they provide about 2.25 times as much energy as starch on an equivalent weight basis. This makes them very useful ingredients when the energy level of the diet needs to be increased. The energy values are generally higher for birds over 3 weeks than for younger birds, due to a higher digestibility.

As outlined in Chapter 3, adding high levels of fat to diets often leads to a higher dietary ME than can be accounted for from the summation of ingredients (NRC, 1994). High-level fat feeding increases the intestinal retention time of feed and so allows for more complete digestion and absorption of the non-lipid constituents.

Normally, low levels of oil can be added to diets, otherwise the diets become soft, difficult to pellet and subject to rancidity. All of the oils are unstable and subject to rancidity; therefore, they should be used rapidly after delivery and not stored for extended periods. Purchased oils may be stabilized by the addition of an antioxidant which should meet organic standards.

Legume Seeds, Their Product and By-products

As reviewed by Gatel (1993), the AA profile of legume seeds for non-ruminant animals is characterized by a high lysine content (7.3% of CP in peas) and a relative deficiency in sulfur AA (methionine, cystine) and tryptophan. Protein digestibility is slightly lower than in soybean meal, especially for pigs (0.74 in peas versus 0.80 in soybean meal) and for young animals and appears variable between and within species. This lower digestibility can partly be explained by the presence in some plant species or cultivars of ANFs (e.g. protease inhibitors, lectins, tannins) and/or fibrous material leading to low accessibility of legume seed protein to digestive enzymes. This researcher recommended that more work was needed to separate the effects of these factors and their practical importance. Because of the high variability of ANF activity, it was also necessary to develop reliable and quick assays that allowed a check and selection of batches by plant breeders or feed manufacturers. In the short term, it was concluded that technological treatments could lead to improved utilization, especially in poultry, but attention had to be paid to the cost of treatment.

Faba beans (*Vicia faba* L.)

The faba bean, also known as field bean, horse bean and broad bean, is an annual legume that grows well in cool climates (Fig. 4.2). It is well established as a feedstuff for horses and ruminants and is now receiving more attention as a feedstuff for poultry, particularly

in Europe, because of the deficit in protein production. At the current time, the EU uses over 20 million tonnes of protein feeds annually, but produces only about 6 million tonnes. The most suitable expansion in locally produced protein feedstuffs may be from crops of the legume family (beans, peas, lupins and soybeans). Field beans grow well in regions with mild winters and adequate summer rainfall and the beans store well for use on-farm.

Nutritional features

Field beans (faba beans) are often regarded nutritionally as high-protein cereal grains. They contain about 240–300 g CP/kg, the protein being high in lysine and (like most legume seeds) low in sulfur AA. The ME content is intermediate between that of barley and wheat. The CF content is around 80 g/kg, air-dry basis. A large study in Poland found that the average CP value was 235.2 g/kg in organic faba beans compared with 247.5 in conventional field beans (Grela and Semeniuk, 2008). The corresponding values for CF were 79.1 and 72.6 g/kg DM. The oil content of the bean is relatively low (10 g/kg DM), with a high proportion of linoleic and linolenic acids. This makes the beans very susceptible to rancidity if stored for more than about 1 week after grinding. When fresh they are very palatable.

As with the main cereals, faba beans are a relatively poor source of Ca and are low in iron and manganese (Mn). The P content is higher than in canola. Faba beans contain lower levels of biotin, choline, niacin, pantothenic acid and riboflavin, but a higher level of thiamin, than soybean meal or canola meal.

Anti-nutritional factors

Faba beans contain several ANFs such as tannins, protease inhibitors (vicin/convicin) and lectins. Use of low-vicin/convicin cultivars may allow substantial levels of faba beans to be included in poultry diets (Dänner, 2003). The levels of trypsin inhibitor and lectin activities are low compared with other legume seeds and do not pose problems in poultry diets when faba beans are incorporated into diets at the levels shown below. Of most concern for poultry is the tannin fraction, which has been shown to depress digestibility of the protein and AA (Ortiz *et al.*, 1993). Tannins in whole faba beans are associated with the seedcoat (testa), and the tannin content is related to the colour of the seedcoat (and flowers). Tannins are lower in white-than in the colour-seeded varieties.

As a result of the possible presence of ANFs in the diet the small-seed varieties with low tannin, low vicine/convicine and low trypsin inhibitor contents are preferred for poultry feeding.

Various processing methods have been investigated in an attempt to deal with ANFs in faba beans. For instance, Castanon and Marquardt (1989) tested the effects of adding different enzyme preparations to diets containing raw, autoclaved or fermented (water-soaked) faba beans on weight gain, feed/gain ratio and DM retention in young Leghorn chickens. They found that addition of enzymes to raw beans did not affect growth during week 1, but improved weight gain during week 2 by more than 10% with protease, 6–7% with cellulase and 8% with cellulase plus protease, and improved feed/gain ratio by more than 5%. Autoclaving of beans improved performance by more than 13%. Fermentation of beans gave a decrease in feed intake.

Poultry diets

Faba beans can be included successfully at moderate levels in poultry diets. For instance, Blair *et al.* (1970) fed broilers on mash diets containing a mixture of equal parts of spring and winter beans at 0, 150, 300 or 450 g/kg and reported body weights at 4 weeks of age of 557, 535, 503 and 518 g and intakes per kilogram gained of 1.94, 1.98, 2.15 and 2.06 kg. Castro *et al.* (1992) fed broilers a diet based on maize and soybean meal without or with 100, 200 or 300 g/kg faba beans from 4 to 46 days old and found no significant differences in mean daily gain, feed intake or feed conversion efficiency. Farrell *et al.* (1999) conducted a study in Australia to determine the optimum inclusion rates of various

protein supplements in broiler starter and finisher diets at levels up to 360 g/kg. Birds grown to 21 days on diets containing faba beans, field peas or sweet lupin seed gave better growth rates and feed conversion efficiency than those given chickpeas. In another experiment all protein supplements resulted in similar growth performance, at all levels of inclusion. Steam-pelleting improved growth rates and feed conversion efficiency. Based on the results these authors recommended an optimal inclusion rate of 200 g faba beans/kg diet. Moschini *et al.* (2005) conducted a similar study in Italy. Raw faba beans were included in broiler diets at 480 g/kg (1–10 days old) and 500 g/kg (11–42 days). Over the whole growth period the broilers fed diets containing faba beans had similar growth rates to a control group fed a soybean meal-based diet at both levels of faba bean inclusion. No effects were observed on dressing percentage or breast and leg quarter cuts.

Castro *et al.* (1992) fed laying hens on diets based on maize and soybean without or with faba beans at 100, 150 or 200 g/kg. Total weight of eggs laid was similar. Feed intake did not differ among the groups and feed conversion efficiency was improved by dietary inclusion of faba beans. A more recent study with layers was conducted by Fru-Nji *et al.* (2007), who investigated the effect of graded replacement of soybean meal and wheat by faba beans. Increasing the level of dietary field beans to 400 g/kg as a complete replacement for soybean meal decreased hen-day egg production from 85% to 75%, average egg mass from 50.8 to 43.5 g per hen per day and increased feed consumption from 110 to 113 g per hen per day. Most traits studied were unaffected with up to 160 g field beans/kg feed. For use in poultry diets, the beans are usually ground to pass through a 3 mm screen.

Field peas (*Pisum sativum* L)

Field peas are grown primarily for human consumption, but they are now used widely in poultry feeding also, especially in Canada, the northern USA and Australia.

Some producers grow peas in conjunction with barley, as these two ingredients can be successfully incorporated into a poultry feeding programme. Peas are a good cool-season alternative crop for regions not suited to growing soybeans. They may be particularly well suited for early planting on soils that lack water-holding capacity and they mature early. There are green and yellow varieties, which are similar in nutrient content. Those grown in North America and Europe, both green and yellow, are derived from white-flowered varieties. Brown peas are derived from coloured-flower varieties: they have higher tannin levels, lower starch, higher protein and higher fibre contents than green and yellow peas. These varietal differences account for much of the reported variation in nutrient content. White-flowered cultivars are preferred for poultry feeding. Pea protein concentrate from starch production may also be available as a feed ingredient.

Nutritional features

Peas are slightly lower in energy content than high-energy grains such as maize and wheat, with about 2600 kcal ME/kg, but have a higher CP content (about 230 g/kg) than grains. Thus, they are regarded primarily as a protein source. Jacob (2007) reported a lower CP value in organic peas (216.1 g/kg) than in peas grown conventionally (238 g/kg). A similar finding was reported in a large study in Poland, the average CP value being 215.7 g/kg in organic peas compared with 221.3 g/kg in conventional peas (Grela and Semeniuk, 2008). The corresponding values for CF were 68.1 and 64.8 g/kg DM. Pea protein is particularly rich in lysine, but relatively deficient in tryptophan and sulfur AA. The ME content of peas is higher than that of soybean meal and is due to the high starch content, which is highly digestible. The starch type is similar to that in cereal grains. As with most crops, environmental conditions can affect the protein content. Hot, dry growing conditions tend to increase CP content. Starch content has been found to be correlated inversely with CP content. At 230 g CP/kg the starch content is approximately 460 g/kg; therefore, a correction for starch and energy content should be made if the peas differ significantly from 230 g CP/kg. AA content is correlated with CP content and prediction equations have been produced.

Ether extract (oil content) of peas is around 14 g/kg. The fatty acid profile of the oil in peas is similar to that of cereal grains, being primarily polyunsaturated. The proportions of major unsaturated fatty acids are linoleic (50%), oleic (20%) and linolenic (12%) (Carrouée and Gatel, 1995).

Field peas, like cereal grains, are low in Ca but contain a slightly higher level of P (about 4 g/kg). They contain about 12 g phytate/kg, similar to that in soybeans (Reddy et al., 1982). The levels of trace minerals and vitamins in peas are similar to those found in cereal grains.

The CF content of peas is around 60 g/kg, the fibre being found mainly in the cell walls. Cellulose and lignin levels are relatively low. Appreciable levels of galactans are found in peas. Peas contain approximately 50 g oligosaccharides/kg, made up mainly of sucrose, stachyose, verbascose and raffinose. Compared with some other pulses such as lupins and beans, the levels of gas-producing oligosaccharides are fairly low and not enough to create sufficient gas production in the hindgut to result in flatulence.

Anti-nutritional factors

The factors that may be present in peas include amylase, trypsin and chymotrypsin inhibitors; tannins (proanthocyanidins); phytate; saponins; haemagglutinins (lectins); and oligosaccharides. However, they generally do not present any major problems in feeding peas to poultry, except that it is advisable to restrict the level of peas in diets for young poultry. Some research has shown that smooth peas have higher trypsin inhibitor activity (TIA) than wrinkled varieties (Valdebouze et al., 1980), but that TIA was significantly affected also by cultivar and environmental conditions (Bacon et al., 1991). Tannins are polyphenolic compounds that may inhibit the activity of several digestive enzymes. They are found in the hull, particularly in brown-seeded cultivars, but are unlikely to pose a problem if brown-seeded cultivars are avoided. Most of the field peas grown in Europe and Canada have a zero tannin content. Saponins and other ANFs that occur in peas appear to have insignificant effects in poultry.

Poultry diets

Field peas can be used successfully in poultry diets when included at levels that allow the dietary ME and AA levels to be maintained (Harrold, 2002; Nalle *et al.*, 2011). High inclusion levels may require supplementation of the diet with a high-energy source such as oil.

Castell *et al.* (1996) suggested that the maximal level in broiler diets be 200 g/kg, on the basis of published findings. Subsequent research showed that inclusion of dehulled peas at 700 g/kg diet resulted in acceptable performance of broilers (Daveby *et al.*, 1998) and a response to alpha-galactosidase was observed with a small particle size produced by grinding. No response to the enzyme addition was obtained with the larger particle size of crushed peas. Farrell *et al.* (1999) reported that steam-pelleting of diets containing field peas improved growth rate and feed conversion efficiency of broilers and recommended an upper inclusion rate of 300 g/kg diet for broilers. Research by Richter *et al.* (1999) showed that supplementation with enzyme mixtures improved weight gain of broilers by 2.5% when starter and finisher diets contained 290 g peas/kg. In contrast, Igbasan *et al.* (1997) reported that pectinase or a combination of pectinase and alpha-galactosidase enzymes did not significantly improve growth rate, feed intake or feed conversion. Nimruzi (1998) reported that fig powder acted as a source of enzymes that reduced intestinal viscosity and improved digestibility when diets containing peas were fed to broilers – an approach that may be suitable for organic production. Moschini *et al.* (2005) conducted a study in which raw field peas were included in broiler diets at 350 g/kg and reported a lower growth rate than with a control diet.

Diets for layers frequently contain up to 100 g field peas/kg since a slight decrease in the rate of lay (typically 2.5–3%) is observed when higher levels of peas are fed (Harrold, 2002). However, good results have been obtained with layer diets containing up to 300 g peas/kg (Castell *et al.*, 1996). In another study yellow, green, and brown-seeded peas were fed at levels of 0, 200, 400 and 600 g/kg diet as a replacement for wheat and soybean meal (Igbassen and Guenter, 1997a). Egg yolk colour was improved as the level of peas was increased. Shell quality was reduced by increasing levels of yellow- and brown-seeded peas but not by green-seeded peas. These authors concluded that yellow-, green- or brown-seeded peas could not be used to replace all of the soybean meal but could be included at levels up to 400 g/kg diet without adverse influence on egg production or efficiency of feed conversion. In a related study Igbassen and Guenter (1997b) reported that micronization, an infrared heat treatment, improved the feeding value of peas for laying hens but that dehulling or supplementation with a pectinase enzyme was not effective in improving hen performance. Perez-Maldonado *et al.* (1999) reported that the viscosity of intestinal contents from hens fed diets containing 250 g field peas/kg was higher than with a similar content of faba beans.

Field peas have been used in turkey diets. Czech researchers (Mikulski *et al.*, 1997) reported that 200–240 g peas/kg could be used to replace part of the soybean meal in diets for growing turkeys without adversely affecting performance, dressing percentage or meat quality. These values are similar to the recommendations of Castell *et al.* (1996) on the use of peas as 250 g/kg diet for turkeys. Diets for turkeys may contain higher levels of field peas as the birds mature and approach market weights (Harrold, 2002).

Harrold (2002) recommended that attention should be paid to particle size when peas are being processed for inclusion in poultry diets since feed intake in poultry is influenced greatly by particle size. Consequently, the production of 'fines' is to be avoided during grinding unless the diets are to be pelleted. Grinding to an extremely small particle size is not economical and may lead to interference with feed intake due to build-up of material in the beak. As pointed out by Harrold (2002), beak necrosis has been suggested to be one result of impaction of small particles in the beak. A related issue is that diets containing field peas at more than 200 g/kg may be difficult to pellet, although other reports do not confirm a problem in pelleting diets containing field peas.

The use of high levels of field peas (200–300 g/kg) may result in a slight increase in viscosity of the digesta, but this increase is considerably lower than that associated with feeding several other feedstuffs (Harrold, 2002). Some commercial enzyme supplements containing xylanases and β-glucanases have been reported to reduce the viscosity of intestinal contents and increase protein digestibility when diets containing high levels of peas have been fed. At the highest levels of recommended use the droppings may be slightly wetter than usual, which should not present a problem with poultry given access to outdoors.

Lentils (*Lens culinaris*)

Lentils are a legume seed, grown primarily as a human food crop. They do not appear on any approved list of organic feedstuffs but should be acceptable if grown organically. Surplus and cull lentils may be available economically for inclusion in poultry diets. The main producing countries are India, Turkey and Canada. Canadian cultivars are mostly green with yellow cotyledons, unlike the red-cotyledon types grown in other countries.

Nutritional features

Castell (1990) reviewed findings on the nutrient content of lentils and concluded that they are similar to peas in nutrient content. However, the CP content may be slightly higher than in field peas. As with other legume seeds, lentils have a low sulfur AA content. As a result, lentils need to be combined with other protein sources in the diet to provide a satisfactory AA profile. The oil content of lentils is low at around 20 g/kg, with linoleic (C18:2) and linolenic (C18:3) accounting for 44% and 12%, respectively, of the total fatty acids (Castell, 1990). The level of lipoxygenase present in the seed suggests the potential for rapid development of rancidity after grinding of lentil seeds (Castell, 1990).

Anti-nutritional factors

In common with many other legume seeds, raw lentils contain some undesirable constituents, though the levels of these are not likely to be of concern in poultry feeding. Weder (1981) reported the presence of several protease inhibitors in lentils. Marquardt and Bell (1988) also identified lectins (haemagglutinins), phytic acid, saponins and tannins as potential problems but could find no evidence that these had adversely affected performance of pigs fed lentils. It is known that cooking improves the nutritive value of lentils for humans but the effects of consumption of raw lentils by non-ruminants have not been well documented (Castell, 1990).

Poultry diets

There is a lack of published data on the use of lentils in poultry diets, though they are being used in pig diets in some countries. In a study in Iran, Parviz (2006) included raw or processed (heated followed by boiling) lentils in broiler diets at concentrations of 100, 200 and 300 g/kg. Growth was best with the 100 g/kg level and lowest with the 300 g/kg level. Processing did not confer any advantages. The author suggested that lentil seeds could be used in broiler diets up to 200 g/kg but not as the sole source of protein. The recommendations applying to the inclusion of peas in poultry diets can probably be adopted for lentils.

Lupins

Lupins are becoming increasingly important as an alternative protein feed ingredient to soybean meal, especially in Australia, which is also an exporter of this product. Benefits of this crop for the organic producer are that the plant is an N-fixing legume and, like peas, can be grown and utilized on-farm, with minimal processing. Another advantage of lupins is that the seed stores well. The shortage of organic protein feedstuffs in Europe has stimulated interest in lupins as an alternative protein source.

The development of low-alkaloid (sweet) cultivars in Germany in the 1920s allowed the seed to be used as animal feed. Prior to that the crop was unsuitable for animal feeding because of a high content of toxic alkaloids in the seed. In Australia, where much of the research on lupins as a feedstuff has been done, the main species of lupins used in poultry diets are *Lupinus angustifolius* and *Lupinus luteus* (yellow lupin). *L. luteus* is regarded as having significant potential as a feedstuff in Australia. This lupin is native to Portugal, western Spain and the wetter parts of Morocco and Algeria.

Nutritional features

Lupin seed is lower in protein and energy contents than soybean meal. A factor related to the relatively low energy level is the high content of fibre, 130–150 g/kg in *L. luteus* and *L. angustifolius* and slightly lower in *L. albus* (Gdala *et al.*, 1996). Another factor is the type of carbohydrate in lupins, which is different from that in most legumes with negligible levels of starch and high levels of soluble and insoluble NSPs and oligosaccharides (up to 500 g/kg seed; van Barnveld, 1999). These compounds influence the transit of feed through the gut, the utilization of nutrients and also the microflora and the morphology of the gastrointestinal tract. They also influence water intake and reduce the quality of the litter. One of the main ways of addressing the adverse effects of the NSPs and improving the utilization of lupin seed is to include a supplement of exogenous enzymes in the diets. Dehulling the seed can produce a feedstuff more comparable to soybean meal.

The CP content of *L. angustifolius* seed has been reported as ranging from 272 to 372 g/kg and of *L. albus* from 291 to 403 g/kg (air-dry basis) (van Barnveld, 1999). Recent selections of *L. luteus* have been found to have a higher CP content (380 g/kg, air-dry basis) than either *L. angustifolius* (320 g/kg, air-DM basis) or *L. albus* (360 g/kg, air-dry basis; Petterson *et al.*, 1997) and to yield better than *L. angustifolius* on acid soils of low fertility (Mullan *et al.*, 1997). Roth-Maier (1999) reported on the nutrient composition of white and yellow lupin cultivars originating from different regions of Germany. The composition of white lupin seed was 340–380 g CP/kg and 7.7 MJ ME_n/kg (DM basis) for poultry.

Zduńczyk *et al.* (1996) studied the nutritive value of low-alkaloid varieties of white lupin and found that the lysine contents were relatively low (4.70–5.25 g/16 g N), with methionine as the limiting AA. Roth-Maier and Paulicks (2004) studied the digestibility and energy contents of the seeds of sweet blue lupins (*L. angustifolius*) and found digestibility coefficients of 0.43–0.5 for organic matter, 0.36–0.43 for protein, 0.69–0.83 for fat, 0.46–0.58 for NFE and an ME concentration of 7.54–8.22 MJ/kg.

The crude fat content of lupins appears to vary within and between species. The range of values for common species grown in Australia has been reported as 49.4–130.0 g/kg (van Barnveld, 1999), the main fatty acids being linoleic 483, oleic 312, palmitic 76 and linolenic 54 g/kg. Petterson (1998) reported that extracts of *L. angustifolius* oil were stable for 3 months at 51°C, indicating a high level of antioxidant activity in this material, helping to explain the good storage characteristics of lupins. Gdala *et al.* (1996) reported that the oil content in *L. albus* was almost twice as high (104 g/kg) as in *L. angustifolius* and more than twice as high as in *L. luteus*.

Lupins are low in most minerals, with the exception of manganese (Mn). *Lupinus albus* is known to be an accumulator of Mn and it has been suggested that high Mn might be the explanation for a reduced voluntary feed intake in some farm animals fed diets containing this species of lupin. However, excessive Mn levels in lupins do not appear to be the cause of reduced feed intake.

Anti-nutritional factors

Non-ruminant animals are known to be sensitive to alkaloids in lupins, but the average alkaloid content of cultivars of *L. angustifolius* and *L. albus* now used is generally low (< 0.04 g/kg; van Barnveld, 1999). Cuadrado *et al.* (1995) reported levels of saponins in *L. albus* of less than 12 mg/kg and 55 mg total saponins/kg in *L. luteus*, leading these authors to conclude that it was unlikely that saponins were responsible for reduced feed intakes of diets containing *L. albus*. The levels of tannins in *L. angustifolius* are considered to be low enough for this not to be a problem in poultry diets. Data on the tannin levels in *L. albus* and *L. luteus* appear to be more limited. Low-alkaloid lupins appear to contain zero or very low levels of trypsin inhibitors and haemagglutinins (van Barnveld, 1999).

Poultry diets

Farrell *et al.* (1999) conducted a study in Australia to determine the optimum inclusion rates of various protein supplements in broiler starter and finisher diets at levels up to 360 g/kg. Birds grown to 21 days on diets containing faba beans, field peas or sweet lupin seed gave better growth rate and feed efficiency than those given chickpeas. In another experiment all protein supplements resulted in similar growth performance, at all levels of inclusion. Steam-pelleting improved growth rate and feed conversion efficiency. Digesta viscosity and excreta stickiness scores were much higher on diets with sweet lupin seed. Based on the results the authors recommended an optimal inclusion rate of less than 100 g sweet lupin seed/kg diet.

Rubio *et al.* (2003) investigated the effects of whole or dehulled lupin (*L. angustifolius*) seed meal in broiler diets. Raw whole (not heat-treated) or dehulled sweet (low in alkaloids) lupin seed meal (400 and 320 g/kg, respectively) was used. Final body weight and feed intake of chickens fed diets containing whole lupin seed meal (400 g/kg) were lower than controls, but gain/feed ratios were not different. Birds fed dehulled (320 g/kg) instead of whole lupin seed meal

had similar body weight, feed intake and gain/feed values to controls. The addition of a commercial protease at 1 g/kg tended to increase feed intake and final weight of the birds. Apparent ileal digestibility of AA was not different from controls for the different lupin diets. The relative weight of the liver was higher in lupin-fed birds than in controls, but not in those given enzyme-supplemented lupin-based diets.

Roth-Maier and Paulicks (2004) studied the growth of broilers fed diets containing seed from sweet yellow lupin and sweet blue lupin in isoenergetic (12.5 MJ AME/kg) and isonitrogenous (220 CP, 12 lysine and 9 methionine/kg) diets at concentrations of 200 and 300 g/kg, respectively. It was found that feed intake (64.5 g/day without lupin seed) was increased with lupin seed to 70 g/day, but growth performance was similar for all groups. Based on the findings, these researchers suggested that yellow lupin seed at up to 300 g/kg can be included in broiler diets in place of soybean meal without impairing growth performance, provided that AA supplementation was used. With blue lupin seed at 300 g/kg, however, an impairment of growth performance of up to 4% might occur.

Moschini *et al.* (2005) conducted a study in which sweet white lupin seed was included in broiler diets at 50 and 100 g/kg as a replacement for soybean meal. Growth rate was improved with the lower level of lupin seed but was reduced with the higher level.

Perez-Maldonado *et al.* (1999) studied the inclusion of sweet lupin seed on the productivity of laying hens. An inclusion rate of 250 g/kg was used, the lupin seed being either raw or steam-pelleted, then ground. Their results showed comparable egg production with chickpeas, field peas or lupin seed, though feed intake was higher with lupin seed. Steam-pelleting of the lupin seed resulted in a decrease in feed intake and improvement in egg production compared with raw lupin seed. These researchers concluded that sweet lupin seed could be included successfully in layer diets at a level up to 250 g/kg, but the resultant increase in gut viscosity was of some concern. Based on research conducted in Germany, Roth-Maier (1999) recommended that a dietary concentration of stored lupin seed at up to 300 g/kg yielded favourable results with layers but that freshly harvested seed should be used only up to 200 g/kg in the diet. Cholesterol, triglyceride and phospholipid concentrations were determined in serum, eggs and meat. A high dietary inclusion rate of white lupin seed at 300 g/kg resulted in only a slight influence on these parameters, not requiring a restriction on the inclusion of lupin seed in poultry.

Lee *et al.* (2016) reported that inclusion of blue lupins in the diet of laying hens at a rate of 150 g/kg DM resulted in no adverse effects in production or hen health and could be used as part of a balanced diet with inclusion of NSP-degrading enzymes as a replacement for soybean meal.

The above findings suggest that lupin seed can be a useful alternative to soybeans in poultry diets. However, their inclusion, like that of several other alternative protein sources, will be constrained by the availability of a supplementary source of AA. Also, supplementation with NSP-degrading enzymes can be beneficial.

Tuber Roots, Their Products and By-products

Examples are sugarbeet pulp, dried beet, potato, sweet potato as tuber, manioc as roots, potato pulp (by-product of the extraction of potato starch), potato starch, potato protein and tapioca.

Cassava

Cassava (tapioca, manioc; *Manihot esculentis crantz*) is a perennial woody shrub that is grown almost entirely in the tropics. It is one of the world's most productive crops, with possible yields of 20–30 t/ha of starchy tubers (Oke, 1990). Cassava is an approved ingredient in organic poultry diets, though in many countries it will represent an imported product not produced regionally.

Nutritional features

Oke (1990) reviewed the nutritional features of cassava. Fresh cassava contains about 65% moisture. The DM portion is high in starch and low in protein (20–30 g/kg of which only about 50% is in the form of true protein). Cassava can be fed fresh, cooked, ensiled or as dried chips or (usually) dried meal. The meal is quite powdery and tends to produce a powdery, dusty diet when included at high levels. Other features of cassava are high levels of potassium and usually of sand. Cassava meal is an excellent energy source because of its highly digestible carbohydrates (700–800 g/kg), mainly in the form of starch. However, its main drawback is the negligible content of protein and micronutrients, including carotenoids.

Anti-nutritional factors

Fresh cassava contains cyanogenic glucosides (mainly linamarin), which on ingestion are hydrolysed to hydrocyanic acid, a poisonous compound. Boiling, roasting, soaking, ensiling or sun-drying can be used to reduce the levels of these compounds (Oke, 1990). Sulfur is required by the body to detoxify cyanide; therefore, the diet needs to be adequate in sulfur-containing compounds such as methionine and cystine. The normal range of cyanide in fresh cassava is about 15–500 mg/kg fresh weight. A dietary cyanide content in excess of 100 mg/kg diet appears to adversely affect broiler performance but laying hens may be affected by levels as low as 25 mg total cyanide/kg diet (Panigrahi, 1996).

Poultry diets

According to several reviews, low-cyanide cassava root meal can be included in balanced poultry diets at levels up to 250 g/kg (Garcia and Dale, 1999) or 600 g/kg (Hamid and Jalaludin, 1972) without any reduction in weight gain or egg production. Morgan and Choct (2016) conducted an extensive review on the utilization of cassava meal in poultry diets and concluded that it could potentially be substituted quantitatively for up to 50% of the maize in poultry diets without adverse effects on bird performance. However, it had to be processed correctly, by methods such as drying, boiling and fermentation, to reduce the HCN content to non-toxic levels. Also, diets containing cassava had to be formulated with care, particularly with respect to the balance of limiting amino acids, vitamins and minerals and essential fatty acids. It is unlikely that organic diets can be formulated to contain such high levels, due to difficulties in maintaining the correct concentration of supplementary protein and AA. In addition, egg yolks from hens fed diets containing high levels of cassava are very pale in colour (Udo et al., 1980) and not acceptable in many markets. In this situation the diet requires supplementation with a source of carotenoids (such as maize, lucerne or grass meals, or marigold petals; Karadas et al., 2006) or the hens have to be given access to green crops providing yolk-pigmenting compounds. The dustiness of cassava-based diets can be reduced by adding molasses or suitable oils and by pelleting.

Potatoes (*Solanum tuberosum*)

Potatoes have potential as a poultry feed but are more widely used in pig and ruminant feeding, because they are more difficult to incorporate into poultry feed mixtures than other feed ingredients. On a worldwide basis this crop is superior to any of the major cereal crops in its yield of DM and protein per hectare. The potato is especially susceptible to disease and insect problems and may have received chemical treatment. Therefore, all potatoes used should meet organic guidelines, including those from non-GM varieties.

This tuber crop originated in the Andes but is now cultivated all over the world except in the humid tropics. It is grown in some countries as a feed crop. In others it is available for animal feeding as cull potatoes or as potatoes surplus to the human market. In addition to raw potatoes, the processing of potatoes as human food products has become increasingly common. Potatoes are also used in the industrial production of starch and alcohol. By-products of these industries are potentially useful feedstuffs. The nutritive value of these by-products depends on the industry from which they are derived. Potato protein concentrate (PPC) provides a high-quality protein source, whereas potato pulp, the total residue from the starch-extraction industry or steamed peelings from the human food processing industry provide lower-quality products for poultry feeding because of their higher CF content and lower starch content.

Nutritional features

As with most root crops, the major drawback is the relatively low DM content and consequent low nutrient density. Potatoes are variable in composition, depending on variety, soil type, growing and storage conditions and processing treatment. About 70% of the DM is starch, the CP content is similar to that of maize and the fibre and mineral contents are low (with the exception of potassium). Potatoes are very low in magnesium. The DM concentration of raw potatoes varies from 180 to 250 g/kg. Consequently, when fed raw, the low DM content results in a very low concentration of nutrients per unit of weight. Expressed on a DM basis, whole potatoes contain about 60–120 g CP/kg, 2–6 g fat/kg, 20–50 g CF/kg and 40–70 g ash/kg. Potato protein has a high biological value (BV), among the highest of the plant proteins and similar to that of soybeans. The AA profile is typically 5.3 g lysine, 2.7 g cystine + methionine, 3.2 g threonine and 1.1 g tryptophan per 100 g CP. In PPC, these values are higher (6.8, 3.6, 5.5 and 1.2 g, respectively).

The digestibility of raw potatoes is low in poultry, coefficients as low as 0.22 and 0.36 being recorded for organic matter and N, respectively (Whittemore, 1977). Halnan (1944) reported an ME value of 2.9 MJ/kg (DM basis) for raw potatoes and a comparable value of 13.6 MJ/kg for boiled potatoes. Consequently, potatoes should be cooked before being fed to poultry, otherwise utilization is low and the droppings become very wet.

Anti-nutritional factors

The protein fraction in raw potato is poorly digested due to the presence of a powerful protease inhibitor (Whittemore, 1977). This inhibitor is destroyed by cooking, as it is absent in cooked potato. It has been shown to cause a reduction in N digestibility in pigs, not only in the potato itself but also in other feedstuffs in the diet (Edwards and Livingstone, 1990). Potatoes may contain the glycoside solanin, particularly if the potatoes are green and sprouted, and may result in poisoning, though poultry appear to be less susceptible than other livestock. Consequently, such potatoes should be avoided for feeding. The water used for cooking should be discarded because it may contain solanin.

Poultry diets

There is a lack of recent information on the feeding of potatoes to poultry, the most recent published review being that of Whittemore (1977). No further published reports on investigations involving the feeding of potatoes to poultry appear to be available. Inclusion of cooked potatoes or potato flakes at levels up to 200 g/kg in poultry grower diets has given good results in some investigations but growth appears to be depressed at higher levels (Whittemore, 1977). These investigations showed also that with pelleted diets the pellets became harder, feed intake was reduced and the litter became progressively wetter as the level of potatoes in the diet was increased. The wetter litter might have resulted from an increased content of potassium in the diet, being almost doubled as the level of potatoes increased from 0 to 400 g/kg (Whittemore, 1977). Similar findings have been reported with laying hens (Vogt, 1969; Whittemore, 1977). Egg production was significantly lower when the diet contained 200 g potato meal/kg and the droppings tended to be wetter, even though the dietary ME content increased with the level of potato meal in the diet. A dietary inclusion rate of 150 g potato meal/kg gave satisfactory egg production.

Potato by-products

Several dehydrated processed potato products may be available for feeding to poultry. These include potato meal, potato flakes, potato slices and potato pulp. These products are very variable in their nutritive value, depending on the processing method. This is particularly true of potato pulp, whose protein and fibre content depends on the proportion of potato solubles added back into the material. Therefore, it is necessary to have such materials chemically analysed before using them for feeding to poultry or to purchase them on the basis of a guaranteed analysis.

Dehydrated cooked potato flakes or flour are very palatable and can be used as a cereal replacement (Whittemore, 1977). However, because of the high energy costs involved in their production, they are generally limited to diets for young mammals.

Dehydrated potato waste

Dehydrated potato waste meal, as defined by Association of American Feed Control Officials (AAFCO), consists of the dehydrated ground by-product of whole potatoes (culls), potato peelings, pulp, potato chips and off-colour french fries obtained from the manufacture of processed potato products for human consumption. It may contain calcium carbonate at up to 30 g/kg added as an aid in processing. It is generally marketed with guarantees for minimum CP, minimum CF, maximum CF, maximum ash and maximum moisture. If heated sufficiently during processing, this product can be used successfully in poultry diets.

Potato pulp

This by-product comprises the residue after starch removal. The composition of the dehydrated product can be quite variable (Edwards and Livingstone, 1990), depending on the content of potato solubles. The product has characteristics similar to those of raw potato and is not suitable for feeding to poultry.

Potato protein concentrate

PPC is a high-quality product, widely used in the human food industry because of its high digestibility and high biological value of the protein. It is a high-quality protein source suitable for use in all poultry diets. However, its high cost is a limiting factor and its use in organic diets may not be permitted since it may not be organic in origin.

Basal nutrients are CP 760 g/kg, digestible protein *in vitro* 703 g/kg, ash 44 g/kg, EE 33 g/kg, CF 20 g/kg and non-N-extract 68 g/kg. There are high contents of EAA in PPC, especially of lysine (58.9 g/kg), available lysine (57.2 g/kg), threonine (46.5 g/kg), leucine (77.0 g/kg), phenylalanine (46.7 g/kg), tyrosine (37.9 g/kg) and histidine (17.2 g/kg). EAA index of potato protein was found to range from 86 to 93 (Gelder and van Vonk, 1980). Energy value for poultry expressed as AME_n of PPC evaluated on the basis of chemical analysis was 14.58 MJ/kg (Korol *et al.*, 2001).

PPC is used mainly to replace milk and fish protein in diets for calves and young pigs at inclusion levels of up to 150 g/kg, with no detrimental effects on growth or feed conversion ratio (Seve, 1977). Soybean protein replaced by potato protein in the range of 0–100% gave good results. The best results were obtained when PPC was incorporated into mixed feed at 70 g/kg and replaced 67% of soybean protein, at which level body weight was higher by 5.4% and feed conversion efficiency was improved by 10.3% compared with results of diet containing no PPC. No adverse effects of ANFs on growth and feed conversion were observed. These findings can be used as a guide to the use of PPC in growing meat birds.

Forages and Roughages

Cabbage (*Brassica oleracea*, Capitata group)

Cabbages have a high yield of nutrients per hectare and are of potential interest for organic feeding as a source of roughage. However, little documented research appears to have been conducted on this crop as a feed ingredient for poultry. Livingstone *et al.* (1980) used comminuted cabbage (variety 'Drumhead') in diets for growing–finishing pigs as a partial replacement for barley and soybean meal. The cabbage contained 100 g DM/kg and, per kilogram DM, 18 MJ gross energy, 230 g CP, 79 g true protein, 7.6 g total lysine, 4.7 g methionine + cystine, 142 g ADF and 132 g ash. Replacing a mixture of 805 g barley and 180 g soybean meal with 150 g or 300 g cabbage (DM basis) reduced the rate of carcass weight gain by 12.2% and 18.5%, respectively. These findings suggest that cabbage has a low feed value for poultry.

Grass meal

Poultry with access to outdoors are likely to have access to grass. Some organic producers also wish to include grass meal in diets for growing and laying poultry as a source of roughage and as a natural source of carotenoids and nutrients. Additionally, certain poultry, especially geese, are grazing birds in which intake of grass is a traditional part of the diet. Thus, there is current interest in the utilization of grass and grass meal by poultry.

Nutritional features

Organic grass meal is likely to be a mixture of species, including clover, sainfoin, etc. and true grasses and is therefore unlikely to be a consistent product. This factor has to be taken into account in assessing its suitability for feeding to poultry. Another main factor influencing the nutritive quality of grass meal is the stage of maturity at harvesting. Grass meal derived from the first, second and third cut of a ryegrass–red-clover pasture showed reduced sugar content and an increased content of dietary fibre with stage at maturity (Vestergaard *et al.*, 1995). Nevertheless, this product is an established feed ingredient in Europe, based on research carried out over 50 years ago. Bolton and Blair (1974) recommended that for use in poultry diets, good-quality dried grass should contain at least 170 g CP/kg.

The meal from grass cut at a young stage is a good source of CP, carotene and xanthophylls, riboflavin and minerals.

Poultry diets

Bolton and Blair (1974) suggested that grass meal could be included in poultry diets at levels up to 100 g/kg. Metwally (2003) fed broiler chickens from 4 to 7 weeks of age on a control diet or a diet in which yellow maize at 250 g/kg was replaced by a mixture consisting of poultry by-product meal (100 g/kg), dried lucerne meal (50 g/kg), dried grass meal (50 g/kg) and sorghum (50 g/kg). The grass meal contained 159 g CP/kg. No significant differences were found in growth performance or carcass traits. Rybina and Reshetova (1981) conducted a study in which laying hens were given a 170 g CP/kg diet containing lucerne and grass meals at 50, 75, 100 or 150 g/kg. Egg production was not affected by dietary grass meal level and averaged 60% over the laying year. Egg cholesterol content decreased as lucerne and grass meal in the diet were increased. With increasing age there was a decrease in utilization and digestibility of protein, and, in particular, CF. Dietary levels of grass meal in layer diets were evaluated by Zhavoronkova and D'yakonova (1983). In this study the diet contained grass meal at 30, 100 or 150 g/kg. CP content of the three diets was 137, 134 and 133 g/kg, respectively. Body weight at start of lay was 1633, 1494 and 1374 g, respectively; at 226 days old the hens weighed 1900, 1900 and 1833 g, respectively. Egg production was similar on the three diets. Average weight of eggs was 51.5, 53.0 and 52.5 g, respectively. The diets with grass meal at 100 and 150 g/kg resulted in greater values for blood haemoglobin (Hb), erythrocyte count, alkali reserve, haematocrit and lysozyme. Dietary addition of grass meal was found to have a beneficial effect on fertility in laying hens (Davtyan and Manukyan, 1987). Hens at 21 weeks of age were fed up to 67 weeks of age on diets containing grass meal at 50, 80, 110 or 140 g/kg. Fertility of eggs laid 1–7 days after insemination was 94.3%, 95.7%, 97.7% and 96.0%, respectively; and the hatchability was 81.8%, 86.7%, 87.1% and 88.8%, respectively.

When the hens were inseminated every 10 days, they produced eggs with 77.0%, 82.6%, 84.4% and 86.0% hatchability. As the level of grass meal in the diet was increased there was a linear increase in egg yolk vitamin A and carotene content. Addition of pure carotene to the diet was found to have no effect on egg fertility.

Utilization of grass would be expected to be higher in geese because they are a grazing species. This was confirmed by Nagy *et al.* (2002) who showed that young geese grew as well on a diet containing chopped grass at 250 g/kg as on a grain pellet diet. In a similar study Arslan (2004) fed day-old native Turkish goslings on starter (0–6 weeks) and grower (7–12 weeks) diets containing grass meal or dried sugarbeet pulp meal at 100 g/kg (starter) and 200 g/kg (grower). The feeding regimes did not affect live weight, live weight gain and feed consumption up to 6 or 12 weeks but feed conversion efficiency was higher with a control diet. Length and weight of examined gastrointestinal tract sections were not affected by feeding regimes at any stage.

Lucerne (alfalfa) (*Medicago sativa* L.)

Lucerne is one of the most widely grown forage crops on a worldwide basis and is used commonly in livestock diets. It is a good source of many nutrients and in the past the inclusion of lucerne meal was considered to be essential in diets for a wide range of animals as a source of 'unidentified factors'. Currently lucerne meal is used in poultry diets as a source of pigments to provide a yellow colour in egg yolks, shanks and skin (Karadas *et al.*, 2006).

Lucerne meal can be regarded as being similar to grass meal in its usage potential in poultry diets. One advantage of lucerne meal over grass meal is that it is composed of a single species and is therefore less variable in composition. Its nutritional value is, however, greatly affected by maturity of the plant, nutritional quality being highest in cuts taken at the pre-bloom or early bud stage. Nutrient losses can occur through field damage due to sunlight and rain; therefore, much of the lucerne for animal feeding is dehydrated rapidly after harvesting. A moisture content of 90 g/kg maximum has been suggested for storage of sun-cured (air-dried) lucerne.

Nutritional features

Sun-cured lucerne is the aerial portion of the plant. According to the North American feed regulations the traded product has to be reasonably free of other crop plants, weeds and mould, and is sun-cured and finely ground.

Most of the data refer to the dehydrated product, which is generally of higher quality than the sun-cured product. The CP content of lucerne ranges from 120 to 220 g/kg and the CF ranges from 250 to 300 g/kg. The accessibility of digestive enzymes to the soluble cellular proteins is reduced by the high fibre content. Lucerne contains a relatively high content of lysine and has a good AA balance. Sun-cured lucerne is high in Ca and its bioavailability can be similar to that of calcium carbonate. Lucerne is low in P, but is a good source of other minerals and of most vitamins.

Anti-nutritional factors

Lucerne contains saponins and tannins (Cheeke and Shull, 1985). The saponins are bitter compounds, which can affect feed intake. The discovery of Turkish strains that are essentially devoid of haemolytic saponins may lead to a greater usage of lucerne in poultry diets. Lucerne also contains tannins that can depress protein digestibility and reduce feed intake. In addition, lucerne may contain a trypsin inhibitor and a photosensitizing agent.

Poultry diets

There appears to be a lack of published data on the utilization of sun-cured lucerne meal in poultry diets. Information about the dehydrated product has therefore to be used as a guide. Kuchta et al. (1992) fed laying hens on diets containing high-protein, low-fibre dried lucerne meal at 50, 80, 110 or 140 g/kg. Protein digestibility, feed intake and feed conversion ratio were higher in hens given diets containing lucerne meal. Yolk colour was more intense in hens given diets containing 110 or 140 g lucerne/kg. These researchers suggested that not more than 110 g lucerne/kg should be included in layer diets since it reduced dietary energy value and feed conversion efficiency. In a similar study, Halaj et al. (1998) fed laying hens on diets containing 0, 35 or 70 g lucerne meal/kg. The results showed a positive influence of lucerne meal on egg weight (61.71, 62.2 and 64.82 g, respectively) and egg output per day (51.91, 55.65 and 59.05 g, respectively). Feed intake was higher in the hens fed lucerne meal. Dietary inclusion of 0, 35 and 70 g lucerne meal/kg had a significant effect on egg yolk pigmentation (6.59, 7.60 and 7.92, respectively), the increase appearing after 7–10 days. On the basis of the findings the researchers recommended an inclusion rate of 35 g lucerne meal/kg in layer diets.

Jiang et al. (2012) fed diets containing graded levels of lucerne meal to growing ducks in an attempt to reduce carcass fat content. A total of 240 14-day-old white Muscovy ducks were selected and randomly allocated to diets containing 0, 30, 60, or 90 g lucerne meal/kg for 5 weeks. Results showed no significant effects on growth performance, but a significantly higher dressing percentage and lower abdominal fat content in the birds fed the diets containing lucerne meal. Also, the ducks fed the diet containing 90 g lucerne meal/kg had a higher breast meat percentage than those fed the diet without lucerne meal. In addition the concentrations of triglyceride, total cholesterol, low density lipoprotein (LDL), very low density lipoprotein (VLDL) and free fatty acid in serum of ducks fed on lucerne meal were decreased.

Molasses

Molasses is used in organic diets as a pellet binder. This is a by-product of sugar production, and is produced mainly from sugarcane (Saccharum officinarum) and sugarbeets (Beta vulgaris).

Nutritional features

Molasses generally contains 670–780 g DM/kg and can vary widely in composition due to soil and growing and processing conditions. The carbohydrate content is high, being composed mainly of highly digestible sugars (primarily sucrose, fructose and glucose). The CP content is low (range 30–60 g/kg). Molasses, therefore, should be regarded as a low-energy product. It is usually a rich source of minerals, the Ca content of cane molasses being high (up to 10 g/kg) due to the addition of calcium hydroxide during processing, but the P content is low. Cane molasses is also high in sodium, potassium and magnesium. Beet molasses tends to be higher in both potassium and sodium but lower in Ca content. Molasses also contains significant quantities of copper, zinc, iron and Mn and can be a good source of some B vitamins such as pantothenic acid, choline and niacin.

Poultry diets

Molasses can be used in organic poultry diets at levels around 25–50 g/kg as a pellet-binding agent. This is attributed to its capacity to allow the feed granules to stick together during the pelleting process and produce pellets that are less likely to break down during transportation and passage through feeding equipment. Other benefits of molasses addition are a possible increase in palatability of the diet and a reduction in dustiness of the dietary mixture. Higher dietary levels are known to have a laxative effect.

In tropical countries with a surplus of sugarcane the juice can be used as a feed source. Sugarcane juice has been fed to ducks with resulting rates of growth and feed conversion efficiency being only slightly inferior to those obtained with cereal-based diets (Bui and Vuong, 1992).

Habibu et al. (2014) found that the addition of molasses to the drinking water at a concentration of 5–7 ml/l increased live weight gain, enhanced formation of blood cells and reduced the effect of oxidative stress induced by heat stress during the hot dry season in 7-week-old broiler chickens in Nigeria.

Seaweeds

Seaweeds (kelp) contain significant quantities of minerals but tend to be low in other nutrients. In some regions they are harvested as dried fodder for farm livestock in coastal areas. China is a major seaweed producer, growing over 2.5 million tonnes of the brown alga *Laminaria japonica*. There is considerable potential for increasing kelp production, particularly in regions such as the Pacific coast of North America. Canada has the longest coastline of any nation, suggesting that greater use should be made of this plant resource. In addition, Canada's marine environment is probably less polluted than elsewhere. A recent development in North America is the use of seaweed as a substrate for biogas (methane) production.

The composition of seaweed differs according to species and to naturally occurring changes in the plant. Ventura *et al.* (1994) found that *Ulva rigida* was not a suitable ingredient for poultry diets, at least at inclusion rates of 100 g/kg or higher. However, this seaweed did not have an anti-nutritive effect as has been reported with other species, since the addition of seaweed did not modify the TME_n of the other constituents in the diet. The proximate composition of this seaweed was (per kilogram DM) 206 g CP, 17 g EE, 47 g CF, 312 g NDF, 153 g ADF, 13 g pentosans and 228 g ash. The TME_n value was (per kilogram DM) 5.7 MJ/kg and 4.3 MJ/kg for chicks and cockerels, respectively; and the AME_n value for chicks was 2.9 MJ/kg DM. Feed intake and growth rate decreased as the dietary content of seaweed was increased from 0 to 300 g/kg. Vogt (1967) found a slight benefit from the inclusion of brown seaweed meal (*Macrocystis pyrifera*) at 10 g/kg in broiler diets. Inclusion of the seaweed meal did not affect growth performance but it reduced mortality of the chicks when the diet did not contain a source of animal protein.

El-Deekx and Brikaa (2009) conducted two experiments to assess the potential value of ground, dried seaweed (*Polysiphonis* spp.) as a feedstuff in starter and finisher diets for ducks. In a starter trial, 96 1-day-old commercial ducks were fed diets containing

0, 40, 80 and 120 g seaweed/kg, *ad libitum* in pellet and mash form from 1 day to 5 weeks of age. In a finisher trial 160 commercial ducklings (35 days of age) were fed diets containing 0, 50, 100 and 150 g seaweed/kg, *ad libitum* in pellet and mash form from 35 to 63 days of age. No significant differences in feed intake or feed conversion ratio were found in the starter trial, though ducks given pelleted diets utilized feed more efficiently than those given mash diets. Results of the second trial indicated that seaweed meal could be included in finisher diets at up to 150 g/kg either in pellet or mash form without adversely affecting growth or feed conversion ratio. Carcass dressing percentage and weight of liver, gizzard, thigh muscles and breast muscles were not significantly affected by inclusion of up to 150 g seaweed/kg in finisher duck diets. Seaweed at 50 and 100 g/kg in finisher duck diets significantly increased the relative weight of breast muscles; at up 150 g/kg significantly improved the texture of breast muscles; and at 50 and 100 g/kg improved the texture of thigh muscles. There were no significant differences in the aroma, taste, juiciness and colour of meat as a result of inclusion of seaweed at up to 150 g/kg in duck diets.

Seaweed has been shown to be of value in the control of parasites. Jensen (1972) reported that inclusion of seaweed in the diet markedly reduced the incidence of liver condemnations from ascarid damage in pigs at slaughter. This aspect of seaweed inclusion is of interest to organic farmers but does not appear to have been investigated in poultry.

Milk and Milk Products

As outlined in Chapter 3, milk is not ideally suited for use in poultry diets, because poultry have difficulty in digesting lactose (milk sugar). Generally it is milk by-products that are used in poultry feeding, either in liquid or dried (dehydrated) form. In the past, dried milk was used in poultry diets as a source of riboflavin, a practice no longer used because of the cost involved and the introduction of synthetic riboflavin. However, dried milk may be available in certain countries when the surplus is sold off by government agencies for animal feeding. In these circumstances dried milk powder may be of potential value as a poultry feedstuff. Also, liquid milk by-products may be available locally from milk-processing plants.

Adequate heat treatment (pasteurization) should be applied to all milk products to ensure that any pathogenic organisms have been destroyed. Generally, the AA availability of milk products is considered high, but the quality can be impaired by overheating.

By-products in liquid form

Skimmed (separated) milk

This product consists of milk from which most of the fat has been removed but which contains all the protein. The protein has a high biological value and is very digestible. Skimmed milk is a good source of B vitamins, but the fat-soluble vitamins (A and D) are removed with the fat. Skimmed milk should either be fresh or always at the same degree of sourness. Attention should be paid to the cleanliness of the equipment used for feeding. Normal bacterial acidification can be used as an effective and convenient method of stabilization.

Buttermilk

This is the liquid product remaining after whole milk is churned and the butter removed. It usually contains more fat than skimmed milk. Buttermilk is more acidic than skimmed milk and can have more of a laxative action. It is an excellent source of supplemental protein.

Whey

Whey is the liquid by-product remaining after cheese production. Whey contains about 90% of the lactose, 20% of the protein, 40% of the Ca and 43% of the P originally present in milk. However, its DM content is low, around 70 g/kg. Most of the fat and protein are removed during processing, leaving the whey high in lactose and minerals. The high lactose content can cause problems in poultry, resulting in wet droppings. Usually two types of whey are available commercially, fresh (sweet) and acidified. However, fresh whey deteriorates rapidly and needs to be used within a short time of production. Acid whey is allowed to ferment and become acidic naturally. It is more stable than fresh whey as a result of the acidity. Sometimes acids, such as formic acid and hydrochloric acid, are used to stabilize whey. If used, they should be acceptable for organic poultry production.

Whey is a very dilute feed, consisting of 930 g water/kg and 70 g DM/kg and contains less than 10 g CP/kg on a wet basis (130 g CP/kg on a DM basis). The protein is of excellent quality because of the AA balance.

Poultry can be fed liquid diets, provided that the appropriate feeders are used. Young birds tend to avoid whey, however, and this product is better used with older birds. Shariatmadari and Forbes (2005) found that starting broilers offered water and liquid whey avoided whey completely. When whey was offered from the 4- or 6-week stage it was better accepted. These researchers concluded that whey can be used in diets for broiler chickens by incorporating it in the feed, provided that drinking water is offered *ad libitum*. Whey may be offered as a liquid if the feed is mixed with 1.8 times its weight of water, but it is better to dilute the whey with an equal volume of water whether it is added to feed or given as a liquid. Good results can also be obtained when undiluted whey is offered alternately with water, either in half-day or full-day periods.

One factor related to the low acceptability of whey by young birds is the high lactose content of this ingredient, lactose being difficult for poultry to digest. However, this feature is being investigated as a possible way of reducing salmonella colonization in organic poultry, birds given access to outdoors being more susceptible to salmonella infection than conventional poultry. In one

study (Corrier *et al.*, 1990) day-old chicks were divided into four groups and provided with lactose alone or in combination with a volatile fatty acid-producing anaerobic culture. The birds were then infected with salmonellae. Salmonella growth in the caecal contents was significantly decreased on day 10 in the chicks given lactose from days 1–10. After the removal of lactose from the diet, the chicks were susceptible to salmonella colonization. The number of salmonellae in the caeca was significantly reduced in the chicks given lactose throughout the 40-day growing period. Dietary lactose decreased the pH of the caecal contents and was accompanied by marked increases in the concentrations of undissociated bacteriostatic volatile fatty acids. In another study DeLoach *et al.* (1990) found that dietary supplementation with whey at 5 g/kg for chicks during the first 10 days of life reduced the mean $\log_{10}$ number of *Salmonella typhimurium* from 5.68 in control chickens to 3.38 in whey-fed chickens. Lactose in drinking water or reconstituted dry milk (5% weight/volume) in drinking water reduced the mean $\log_{10}$ number of *S. typhimurium* to 2.60 and 2.11, respectively. Milk (50 g/kg weight basis) in feed was not effective in reducing *S. typhimurium* colonization. The lack of effect of milk in the feed was believed to be because not enough lactose was provided at the 50 g/kg concentration. Kassaify and Mine (2004) tested the effects of supplementation of layer feed with non-immunized egg yolk powder (did not contain anti-*S. enteritidis* antibodies), immunized egg yolk powder (with anti-*S. enteritidis* antibodies), egg yolk proteins, egg white and skim milk powder. The hens were infected orally with *S. enteritidis* then given a supplemented feed of 5%, 10% or 15% (weight/weight) of each of the test samples. Excreta samples tested weekly showed an absence of the organism after the first week of feeding non-immunized egg yolk powder and a gradual decrease with the other samples.

In related work Omara (2012) reported that fresh skimmed milk or whey added to broiler diets at a concentration of 5–10 g/kg improved growth rate, efficiency of feed conversion and nutrient utilization, results similar to those obtained with a commercial probiotic.

An interesting finding with pigs was that whey feeding was found to reduce ascarid egg count in the faeces (Alfredsen, 1980), suggesting that this product could be used as a natural dewormer.

Dried milk products

Dried milk products include dried whole milk, dried skimmed milk and dried whey. The products derived from whole milk and skimmed milk are very palatable and are highly digestible protein supplements with an excellent balance of AA. They are good sources of vitamins and minerals except fat-soluble vitamins, iron and copper. Generally, though, they are too expensive for use as a feed ingredient for poultry.

Fish, Other Marine Animals, Their Products and By-products

Fishmeals

Although not an organic product in the strict sense, fishmeal is approved for use in organic poultry diets. However, as noted above, it has to be from sustainable sources and any antioxidant added to prevent spoilage has to be from the approved list. Oil-extracted fishmeal should be processed by a mechanical method. Fishmeal is defined as clean, dried and ground tissues of non-decomposed whole fish or fish cuttings, with or without extraction of the oil. Some is made from fish waste from fish processed for the human food market, and the rest from whole fish caught specifically to be made into fishmeal. The type of fish used has a major influence on the composition of the meal. White fish have a low oil content, while members of the herring family (e.g. menhaden, anchovy and pilchard) contain large amounts of oil. Major producers of fishmeal are Peru and Chile. Fishmeal used for feed in North America must not contain

more than 100 g moisture/kg or 70 g salt/kg, and the amount of salt must be specified if it is greater than 30 g/kg. Antioxidants are commonly added to the meal to prevent oxidation and spoilage; therefore, this aspect should be checked for acceptability in organic diets.

Nutritional features

Fishmeal is well established internationally as an excellent feedstuff for poultry, though its current availability and cost (because of competing demands from aquaculture and the pet-food industry) tend to limit its use in poultry diets. The nutritional value of fishmeal can vary greatly depending on the quality of fish materials used, processing factors such as overheating and oxidation of the meal (Seerley, 1991; Wiseman et al., 1991). Fishmeals are generally high in protein (500–750 g/kg) and EAA, particularly lysine, which are deficient in many cereal grains and other feedstuffs. Most mineral elements, especially Ca and P, and B vitamins are relatively high when compared with other protein sources and the P is of high availability.

Fishmeal is an important, and sometimes the only, source of animal protein available for poultry feeding in most Asian countries (Ravindran and Blair, 1993). It is either imported or is produced locally. The local fishmeals contain between 400 and 500 g CP/kg, but are generally of low quality due to lack of control over raw fish quality, processing and storage conditions. Also, they are often adulterated with cheap diluents such as sand. Samples containing as much as 150 g salt/kg are not uncommon. This situation underlines the need for strict enforcement of quality control measures.

Poultry diets

Fishmeal is regarded as one of the best sources of protein in diets for young poultry, especially turkeys. For instance, Karimi (2006) studied the effect of diets varying in fishmeal level (0, 25 and 50 g/kg during the starter and 0, 12.5 and 25 g/kg during the grower periods) on the growth performance of broilers. Results showed that the body weight at 32 and 42 days, daily gain during 0–42 days and feed intake during 11–20, 21–32 and 0–42 days increased significantly with fishmeal inclusion. The results indicated that beneficial effects of fishmeal on broiler performance became most evident at higher use levels and during the later growth periods, mainly via a stimulation of feed intake. Generally, however, only low levels of fishmeal are used in poultry diets because of the costs involved.

Other reasons for restricting fishmeal to low levels in poultry diets include the role of fishmeal in the aetiology of gizzard erosion. Gizzerosine, a toxic substance related to gizzard erosion, is produced in fishmeal when the processing temperature exceeds 120°C for 2–4 h. It appears to produce erosion of the gizzard by an increase in gastric secretion in chicks. Another reason is that high levels of dietary fishmeal may result in a fishy flavour in eggs and poultry meat.

Fishmeal prepared from locally caught fish is important in developing regions, in providing protein to balance poultry diets (Ravindran and Blair, 1993).

Mineral Sources

Before outlining the approved ingredients in detail it may be useful to clarify the use of the term 'organic' which can be found in relation to certain classes of ingredients such as mineral sources. Organic minerals are minerals containing carbon, following standard chemical nomenclature. In this context, the term 'organic' does not mean derived from an organic source. Minerals not containing carbon are termed inorganic, following standard chemical nomenclature. Organic minerals such as selenomethionine are used in conventional feed manufacture, but very few appear to be approved for use in organic poultry diets. Some are approved under the US regulations. These organic sources may provide minerals in a more bioavailable form than in inorganic sources (Tables 4.5 and 4.6). Producers wishing to use them should check their acceptability with the local certifying agency.

Other sources may be permitted in organic diets and producers should check with the local certifying agency for details. For instance, in the US regulations (FDA, 2001) the approved list of trace minerals with GRAS (Generally Recognized As Safe) status when added at levels consistent with good feeding practice includes those listed in Table 4.6.

Vitamin Sources

Vitamins from synthetic sources are permitted in organic diets for poultry. A main concern for the nutritionist and feed manufacturer in selecting vitamins for inclusion in diets is their stability. In general, the fat-soluble vitamins are unstable and must be protected from heat, oxygen, metal ions and ultraviolet (UV) light. Antioxidants are frequently used in conventional feeds, to protect these vitamins from breakdown. All of the naturally occurring forms of vitamin A (retinol, retinal and β-carotene), with the exception of retinoic acid, are particularly unstable and sensitive to UV light, heat, oxygen, acids and metal ions. The naturally occurring forms of vitamin E (mainly tocopherols) are readily oxidized and destroyed by peroxides and oxygen in a process accelerated by polyunsaturated fatty acids and metal ions. Because of the instabilities of their naturally occurring vitamers (forms of the vitamin), the concentrations of fat-soluble vitamins in natural foods and feedstuffs are highly variable, as they are greatly affected by the conditions of production, processing and storage. Consequently, the synthetic esterified forms (acetate and palmitate), which are much more stable, are preferred for diet formulation.

Vitamin D is available as D_2 (ergocalciferol) and D_3 (cholecalciferol). Poultry can utilize only the D_3 form; therefore, it is usual to supplement all feeds (where necessary) with the D_3 form. Fish oil is allowed under the Canadian Feeds Regulations as a source of vitamins A and D.

The commonly available source of stable vitamin E used in animal feed is synthetic DL-alpha-tocopheryl acetate. An

Table 4.5. Concentrations of mineral elements in common dietary mineral sources.

Source	IFN	Formula	Mineral	Concentration %
Limestone, ground	6-02-632	$CaCO_3$ (mainly)	Calcium	38
Calcium carbonate	6-01-069	$CaCO_3$	Calcium	40
Oyster shell, ground	6-03-481	$CaCO_3$	Calcium	38
Dicalcium phosphate	6-01-080	$CaHPO_4.2H_2O$	Calcium	23
			Phosphorus	18
Defluorinated phosphate	6-01-780		Calcium	32
			Phosphorus	18
Phosphate, Curaçao	6-05-586		Calcium	36
			Phosphorus	14
Salt, common	6-14-013	NaCl	Sodium	39.3
			Chloride	60.7
Copper sulfate		$CuSO_4.5H_2O$	Copper	25.4
Copper carbonate		$CuCO_3 Cu(OH)_2$	Copper	55
Copper oxide		CuO	Copper	76
Calcium iodate		$Ca(IO_3)_2$	Iodine	62
Potassium iodide		KI	Iodine	70
Ferrous sulfate		$FeSO_4.H_2O$	Iron	31
Ferrous sulfate		$FeSO_4.7H_2O$	Iron	21
Ferrous carbonate		$FeCO_3$	Iron	45
Manganous oxide		MnO	Manganese	60
Manganous sulfate		$MnSO_4.H_2O$	Manganese	27
Sodium selenite		$NaSeO_3$	Selenium	45
Sodium selenate		$NaSeO_4$	Selenium	41.8
Zinc oxide		ZnO	Zinc	79
Zinc sulfate		$ZnSO_4.H_2O$	Zinc	36
Zinc carbonate		$ZnCO_3$	Zinc	52

Notes: 1. The bioavailability of the named minerals in the above forms is high or very high.
2. The exact concentration of minerals will vary, depending on the purity of the source.
3. The above sources may also provide trace amounts of minerals other than those listed, such as sodium, fluoride and selenium.
4. Cobalt-iodized salt is often used as a source of sodium, chloride, iodine and cobalt.

Table 4.6. FDA-approved trace minerals for use in animal feed.

Trace mineral	Approved forms
Cobalt	Cobalt acetate, carbonate, chloride, oxide, sulfate
Copper	Copper carbonate, chloride, gluconate, hydroxide, orthophosphate, pyrophosphate, sulfate
Iodine	Calcium iodate, iodobehenate, cuprous iodide, 3,5-diiodosalicylic acid, ethylenediamine dihydriodide, potassium iodate, potassium iodide, sodium iodate, sodium iodide
Iron	Iron ammonium citrate, iron carbonate, chloride, gluconate, lactate, oxide, phosphate, pyrophosphate, sulfate, reduced iron
Manganese	Manganese acetate, carbonate, citrate (soluble), chloride, gluconate, orthophosphate, phosphate (dibasic), sulfate, manganous oxide
Zinc	Zinc acetate, carbonate, chloride, oxide, sulfate

alternative form of stable vitamin E is D-alpha-tocopheryl acetate, which is derived from plant oils (such as soybean, sunflower and maize oil). This form has a relative biopotency of more than 136% in comparison with DL-alpha-tocopheryl acetate. The potency of the fat-soluble vitamins is expressed in terms of international units (IUs).

The water-soluble vitamins tend to be more stable under most practical conditions, exceptions being riboflavin (which is sensitive to light, heat and metal ions), pyridoxine (pyridoxal, which is sensitive to light and heat), biotin (which is sensitive to oxygen and alkaline conditions), pantothenic acid (which is sensitive to light, oxygen and alkaline conditions) and thiamin (which is sensitive to heat, oxygen, acidic and alkaline conditions, and metal ions). Again, the more stable synthetic forms of these vitamins are used in conventional feed formulation. Choline chloride is very hygroscopic (absorbs water when exposed to air) and the non-hygroscopic choline bitartrate is a preferred source of this vitamin.

Vitamins that are allowed for addition to animal feeds under the Canadian Feeds Regulations (Class 7. Vitamin products) are listed below. All have to be labelled with a guarantee of declared potency.

7.1.1: p-Aminobenzoic acid (IFN 7-03-513).
7.1.2: Ascorbic acid (IFN 7-00-433), vitamin C.
7.1.3: Betaine hydrochloride (IFN 7-00-722), the hydrochloride of betaine.
7.1.4: D-Biotin (or Biotin, D-) (IFN 7-00-723).
7.1.5: Calcium D-pantothenate (IFN 7-01-079).
7.1.6: Calcium DL-pantothenate (IFN 7-17-904).
7.1.7: Choline chloride solution (IFN 7-17-881).
7.1.8: Choline chloride with carrier (IFN 7-17-900).
7.1.9: Fish oil (IFN 7-01-965), oil of fish origin used as a source of vitamins A and D.
7.1.10: Folic acid (or Folacin) (IFN 7-02-066).
7.1.11: Inositol (IFN 7-09-354).
7.1.12, 7.1.13: Menadione and menaphthone in several forms (sources of vitamin K).
7.1.15: Niacin (or Nicotinic acid) (IFN 7-03-219).
7.1.16: Niacinamide (or Nicotinamide) (IFN 7-03-215), the amide of nicotinic acid.
7.1.17: Pyridoxine hydrochloride (IFN 7-03-822).
7.1.18: Riboflavin (IFN 7-03-920).
7.1.19: Riboflavin-5′-phosphate sodium (IFN 7-17-901), the sodium salt of the phosphate ester of riboflavin.
7.1.20: Thiamine hydrochloride (IFN 7-04-828).
7.1.21: Thiamine mononitrate (IFN 7-04-829).
7.1.22: Vitamin B_{12} (IFN 7-05-146), cyanocobalamin.

7.1.23: Sodium ascorbate (IFN 7-00-433), the sodium salt of ascorbic acid.
7.2: Beta-carotene (IFN 7-01-134).
7.3: Vitamin A (IFN 7-05-142), as the acetate ester, palmitate ester, propionate ester or a mixture of these esters.
7.4: Vitamin D_3 (IFN 7-05-699), cholecalciferol.
7.5: Vitamin E (IFN 7-05-150), as the acetate ester, succinate ester or a mixture of these esters.

Enzymes

Certain enzymes are permitted for addition to organic feed, to improve nutrient utilization, but not to stimulate growth unnaturally. Their use has been shown to be generally beneficial, but not in all studies. Their main benefit is in helping to release more of the nutrients in the feed during digestion, resulting in a lower excretion of undigested nutrients and feed components into the environment. Thus, their use helps to reduce environmental pollution and aids environmental sustainability. The main issue is N and P contents of animal manure. Excessive N yields ammonia, which can result in air pollution. Also, soil bacteria can convert N into nitrate, resulting in soil and water contamination. Undigested P in manure contributes to P pollution. A high proportion of undigested fibre in the manure is also undesirable since it increases the bulk of material for land application.

The enzymes permitted for use are usually extracted from edible, non-toxic plants, non-pathogenic fungi or non-pathogenic bacteria and may not be produced by genetic engineering technology. They have to be non-toxigenic. They are termed exogenous enzymes to explain that they do not originate in the gut of animals.

Enzymes permitted in animal feeds in the EC are shown in Table 4.7. Various combinations are allowed, as shown in the table. This list does not include enzymes such as alpha-galactosidase, which are marketed internationally and which may be permitted by other organic agencies. Producers wishing to use enzyme products should check with the local certifying agency for a permitted list.

Table 4.7. Abbreviated list of authorized feed enzymes in the European Community (Directive 70/524/EEC and Annex to Directive 82/471/EEC).

Number	Enzyme (alone or in combination)
15	Beta-glucanase
2	Phytase
8	Beta-glucanase and beta-xylanase
20	Beta-xylanase
21	Beta-xylanase
25	Beta-glucanase and endo-beta-xylanase
25 = E 1601	Beta-glucanase and beta-xylanase
26	Beta-glucanase
27	Beta-xylanase and beta-glucanase
28	Phytase
30	Beta-glucanase and beta-xylanase
31	Beta-xylanase
34	Beta-glucanase and beta-xylanase and alpha-amylase
43	Beta-xylanase and beta-glucanase and alpha-amylase
46	Beta-glucanase and beta-xylanase and polygalacturonase
48	Alpha-amylase and beta-glucanase
52	Beta-glucanase, beta-glucanase *(different source)* and alpha-amylase
53	Beta-glucanase, beta-glucanase *(different source)*, alpha-amylase and bacillolysin
54	Beta-glucanase, beta-glucanase *(different source)*, alpha-amylase and beta-xylanase
55	Beta-glucanase, beta-glucanase *(different source)*, alpha-amylase and bacillolysin
56	Beta-glucanase, beta-glucanase *(different source)*, alpha-amylase and bacillolysin
57	Beta-glucanase, beta-glucanase *(different source)*, alpha-amylase and bacillolysin
58	Beta-glucanase, beta-glucanase *(different source)*, alpha-amylase and bacillolysin
61	Beta-xylanase and beta-glucanase *(different source)*
E 1601	Beta-glucanase and beta-xylanase
E 1602	Beta-glucanase, beta-glucanase *(different source)* and beta-xylanase
E 1603	Beta-glucanase
E 1604	Beta-glucanase and beta-xylanase
E 1605	Beta-xylanase
E 1607	Beta-xylanase
E 1608	Beta-xylanase and beta-glucanase
E 1613	Beta-xylanase

Note: Some are approved in dry and/or liquid form.

Examples of enzymes being used in feeds internationally include those outlined below. Often a mixture of enzymes is used, targeted at dealing with the particular profile of substrates in the diet in question. In addition, the enzymes have to be stable to processing and pelleting of the diet (Inborr and Bedford, 1993).

1. Phytase acts on phytate P in plant materials, releasing more of the contained P. As a result, less P supplementation is required and a lower amount of P is excreted in the manure (possibly 30%).
2. β-Glucanase added to a barley-based diet helps to break down the barley β-glucan, improving the digestibility of carbohydrate, fat and protein.
3. Xylanase added to a wheat-based diet helps to break down arabinoxylan, an NSP, improving the digestibility of carbohydrate and enhancing fat, protein and starch digestion.
4. Alpha-galactosidase is used in feeds that are high in plant protein feedstuffs such as soybean meal. These legumes contain oligosaccharides, which cannot be degraded by the endogenous gut enzymes and are fermented in the large intestine causing flatulence. Nutritional constraints in the use of soy products in animals have been reviewed by Baker (2000).

5. Addition of alpha-amylase can improve the digestion of starch and addition of protease has been shown to improve the digestion of protein.

Microorganisms

Microorganisms approved for feed use under the EU regulations comprise *Enterococcus faecium* (in various forms) and *Saccharomyces cerevisiae*. Their use as probiotics (as an alternative to antibiotics) is based on the principle of promoting the growth of lactobacilli and reducing the numbers of enteropathogenic bacteria in the gut. Sometimes this principle is referred to as competitive exclusion. This aspect will be addressed in more detail in Chapter 7.

Brewer's yeast

Brewer's yeast (*S. cerevisiae*) is permitted as a feed ingredient in organic diets. This by-product was used traditionally in poultry diets in the past as a source of AA and micronutrients (Bolton and Blair, 1974), a practice largely discontinued because of economics. Inactivated yeast should be used for animal feeding since live yeast may grow in the intestinal tract and compete for nutrients. This ingredient can be used only at low levels in poultry diets because of its relatively low palatability (Onifade and Babatunde, 1996). These researchers found that broiler growth was improved when the diet was supplemented with dried yeast, daily gain being highest with a diet containing 3.0 g/kg. Feed intake was similar with all levels of dried yeast from 0 to 4.5 g/kg but was lowest in chicks fed a diet containing 6.0 g/kg.

Other yeasts

Other yeasts have been included in poultry diets and may be acceptable as ingredients for organic production. Some countries producing sugarcane and molasses use these as substrates.

Yeasts have also been used to detoxify the diet of mycotoxins. Modified yeast cell wall mannanoligosaccharide (MOS) has been reported to effectively bind aflatoxin and to bind ochratoxin and the fusariotoxins to a lesser degree. This product has advantages over other binding agents in that it does not bind vitamins or minerals (Devegowda *et al.*, 1998).

References

AAFCO (2005) *Official Publication*. Association of American Feed Control Officials, Oxford, Indiana.

Adamu, M.S., Nafarnda, W.D., Iliya, D.S. and Kubkomawa, I.H. (2006) Replacement value of yellow sorghum (*sorghum bicola*) variety for maize in broiler diets. *Global Journal of Agricultural Sciences* 5, 151–154.

Aherne, F.X. and Kennelly, J.J. (1985) Oilseed meals for livestock feeding. In: Haresign, W. (ed.) *Recent Advances in Animal Nutrition*. Butterworths, London, pp. 39–89.

Aimonen, E.M.J. and Näsi, M. (1991) Replacement of barley by oats and enzyme supplementation in diets for laying hens. 1. Performance and balance trial results. *Acta Agriculturae Scandinavica* 41, 179–192.

Ajuyah, A.O., Hardin, R.T. and Sim, J.S. (1993) Effect of dietary full-fat flax seed with and without antioxidant on the fatty acid composition of major lipid classes of chicken meats. *Poultry Science* 72, 125–136.

Alfredsen, S.A. (1980) The effect of feeding whey on ascarid infection in pigs. *Veterinary Record* 107, 179–180.

Allen, P.C., Danforth, H. and Levander, O.A. (1997) Interaction of dietary flaxseed with coccidia infections in chickens. *Poultry Science* 76, 822–827.

Amaefule, K.U. and Osuagwu, F.M. (2005) Performance of pullet chicks fed graded levels of raw Bambarra groundnut (*Vigna subterranean* (L.) Verdc) offal diets as replacement for soybean meal and maize. *Livestock Research for Rural Development* 17, 55.

Amin, Z., Akram, M., Barque, A. and Rafique, M. (1986) Study on broiler's ration: comparative nutritive value of expeller and solvent-extracted decorticated cottonseed cake and rapeseed cake in broiler's ration. *Pakistan Veterinary Journal* 6, 109–111.

Anon. (2013) Organic Diets for Small Poultry Flocks. Manitoba Agriculture Canada. Available at: https://www.gov.mb.ca/agriculture/livestock/production/poultry/print,organic-diets-for-small-poultry-flocks.html (accessed May 29, 2016).

Arslan, C. (2004) Effects of diets supplemented with grass meal and sugar beet pulp meal on abdominal fat fatty acid profile and ceacal volatile fatty acid composition in geese. *Revue de Médecine Vétérinaire* 155, 619–623.

Ashes, J.R. and Peck, N.J. (1978) A simple device for dehulling seeds and grain. *Animal Feed Science and Technology* 3, 109–116.

Aymond, W.M. and Van Elswyk, M.E. (1995) Yolk thiobarbituric acid reactive substances and n-3 fatty acids in response to whole and ground flaxseed. *Poultry Science* 74, 1388–1394.

Bacon, J.R., Lambert, N., Mathews, P., Arthur, A.E. and Duchene, C. (1991) Variation of trypsin inhibitor levels in peas. *Aspects of Applied Biology* 27, 199–203.

Bai, Y., Sunde, M.L. and Cook, M.E. (1992) Wheat middlings as an alternate feedstuff for laying hens. *Poultry Science* 71, 1007–1014.

Baker, D.H. (2000) Nutritional constraints to use of soy products by animals. In: Drackley, J.K. (ed.) *Soy in Animal Nutrition*. Federation of Animal Science Societies, Savoy, Illinois, pp. 1–12.

Basmacioğlu, H., Cabuk, M., Ünal, K., Özkan, K., Akkan, S. and Yalcin, H. (2003) Effects of dietary fish oil and flax seed on cholesterol and fatty acid composition of egg yolk and blood parameters of laying hens. *South African Journal of Animal Science* 33, 266–273.

Batal, A.B. and Dale, N.M. (2006) True metabolizable energy and amino acid digestibility of distillers dried grains with solubles. *Journal of Applied Poultry Research* 15, 89–93.

Batal, A., and Dale, N. (2010) Feedstuffs Ingredient Analysis Table: 2011 edition. Available at: http://fdsmagissues.feedstuffs.com/fds/Reference_issue_2010/03_Ingredient%20Analysis%20Table%202011%20Edition.pdf (accessed June 14, 2014).

Batal, A., Dale, N.M. and Café, M. (2005) Nutrient composition of peanut meal. *Journal of Applied Poultry Research* 14, 254–257.

Batterham, E.S., Andersen, L.M., Baigent, D.R. and Green, A.G. (1991) Evaluation of meals from Linola(R) low-linolenic acid linseed and conventional linseed as protein sources for growing pigs. *Animal Feed Science and Technology* 35, 181–190.

Begum, A.N., Nicolle, C., Mila, I., Lapierre, C., Nagano, K. *et al.* (2004) Dietary lignins are precursors of mammalian lignans in rats. *Journal of Nutrition* 134, 120–127.

Bekta, M., Fabijan´ka, M. and Smulikowska, S. (2006) The effect of β-glucanase on the nutritive value of hulless barley cv. Rastik for broiler chickens. *Journal of Animal and Feed Sciences* 15, 107–110.

Bell, J.M., Tyler, R.T. and Rakow, G. (1998) Nutritional composition and digestibility by 80-kg to 100-kg pigs of prepress solvent-extracted meals from low glucosinolate *Brassica juncea*, *B. napus* and *B. rapa* seed and of solvent-extracted soybean meal. *Canadian Journal of Animal Science* 78, 199–203.

Beltranena, E. (2016) Triticale performs in pig feeds. Available at: https://www1.agric.gov.ab.ca/$Department/deptdocs.nsf/all/lr14763/$FILE/Triticale-Performs-in-Pig-Feeds.pdf (accessed July 17 2018).

Blair, R. (1984) Nutritional evaluation of ammoniated mustard meal for chicks. *Poultry Science* 63, 754–759.

Blair, R. (2012) *Organic Production and Food Quality: A Down to Earth Analysis*. Wiley-Blackwell, New Jersey, USA.

Blair, R. and Misir, R. (1989) Biotin bioavailability from protein supplements and cereal grains for growing broiler chickens. *International Journal for Vitamin and Nutrition Research* 59, 55–58.

Blair, R. and Reichert, R.D. (1984) Carbohydrate and phenolic constituents in a comprehensive range of rapeseed and canola fractions: nutritional significance for animals. *Journal of the Science of Food and Agriculture* 35, 29–35.

Blair, R., Wilson, B.J. and Bolton, W. (1970) Growth of broilers given diets containing field beans (*Vicia faba* L.) during the 0 to 4 week period. *British Poultry Science* 11, 387–398.

Blair, R., Dewar, W.A. and Downie, J.N. (1973) Egg production responses of hens given a complete mash or unground grain together with concentrate pellets. *British Poultry Science* 14, 373–377.

Bolton, W. and Blair, R. (1974) *Poultry Nutrition*, 4th edn. Bulletin 174, Ministry of Agriculture, Fisheries and Food, HMSO, London, pp. v + 134.

Briggs, K.G. (2002) Western Canadian triticale – reinvented for the forage and feed needs of the 21st century. *Proceedings of the 23rd Western Nutrition Conference*, University of Saskatchewan, Saskatoon, Canada, pp. 65–78.

Bui, X.M. and Vuong, V.S. (1992) Sugar cane juice and 'A' molasses as complete replacement for cereal byproducts in diets for ducks. *Livestock Research for Rural Development On-line Edition* 4, 3.

Burrows, V.D. (2004) Hulless oats. In: Abdel-Aal, E. and Wood, P. (eds) *Specialty Grains for Food and Feed*. American Society of Cereal Chemists, St Paul, Minnesota, pp. 223–251.

Butler, E.J., Pearson, A.W. and Fenwick, G.R. (1982) Problems which limit the use of rapeseed meal as a protein source in poultry diets. *Journal of the Science of Food and Agriculture* 33, 866–875.

Campbell, G.L. and Campbell, L.D. (1989) Rye as a replacement for wheat in laying hen diets. *Canadian Journal of Animal Science* 69, 1041–1047.

Carrouée, B. and Gatel, F. (1995) *Peas: Utilization in Animal Feeding*. UNIP-ITCP, Paris, France.

Casartelli, E.M., Filardi, R.S., Junqueira, O.M., Laurentiz, A.C., Assuena, V. and Duarte, K.F. (2006) Sunflower meal in commercial layer diets formulated on total and digestible amino acids basis. *Brazilian Journal of Poultry Science* 8, 167–171.

Castanon, J.I.R. and Marquardt, R.R. (1989) Effect of enzyme addition, autoclave treatment and fermenting on the nutritive value of field beans (*Vicia faba* L.). *Animal Feed Science and Technology* 26, 71–79.

Castell, A.G. (1990) Lentils. In: Thacker, P.A. and Kirkwood, R.N. (eds) *Nontraditional Feed Sources for Use in Swine Production*. Butterworths, Stoneham, Massachusetts, pp. 265–273.

Castell, A.G., Guenter, W., Igbasan, F.A. and Blair, R. (1996) Nutritive value of peas for nonruminant diets. *Animal Feed Science and Technology* 69, 209–227.

Caston, L. and Leeson, S. (1990) Dietary flax and egg composition. *Poultry Science* 69, 1617–1620.

Castro, L.F.R.V., de Taveira, A.M.C.F. and da Costa, J.S.P. (1992) Faba beans (*Vicia faba* L.) for feeding laying hens and meat chickens. *Avances en Alimentación y Mejora Animal* 32, 3–8.

CFIA (2018) Approved feed ingredients. Canadian Food Inspection Agency, Ottawa. Available at: http://laws-lois.justice.gc.ca/eng/regulations/SOR-83-593/page-11.html (accessed June 20 2018).

Cheeke, P.R. and Shull, L.R. (1985) *Natural Toxicants in Feeds and Poisonous Plants*. AVI Publishing Company, Westport, Connecticut, 492 pp.

Cherian, G. and Sim, J.S. (1991) Effect of feeding full fat flax and canola seeds to laying hens on the fatty acid composition of eggs, embryos and newly hatched chicks. *Poultry Science* 70, 917–922.

Cheva-Isarakul, B., Tangtaweewipat, S. and Sangsrijun, P. (2001) The effect of mustard meal in laying hen diets. *Asian-Australasian Journal of Animal Sciences* 14, 1605–1609.

Cheva-Isarakul, B., Tangtaweewipat, S., Sangsrijun, P. and Yamauchi, K. (2003) Chemical composition and metabolizable energy of mustard meal. *Journal of Poultry Science* 40, 221–225.

Chiang, C.C., Yu, B. and Chiou, W.S.P. (2005) Effects of xylanase supplementation to wheat-based diet on the performance and nutrient availability of broiler chickens. *Asian-Australasian Journal of Animal Sciences* 18, 1141–1146.

Chiba, L.I. (2001) Protein supplements. In: Lewis, A.J. and Southern, L.L. (eds) *Swine Nutrition*. CRC Press Boca Raton, Florida, pp. 803–837.

Choct, M., Hughes, R., Trimble, R.P., Angkanaporn, K. and Annison, G. (1995) Non-starch polysaccharide-degrading enzymes increase the performance of broiler chickens fed wheat of low apparent metabolizable energy. *Journal of Nutrition* 125, 485–492.

Classen, H.L., Campbell, G.L. and Groot Wassink, J.W.D. (1988a) Improved feeding value of Saskatchewan-grown barley for broiler chickens with dietary enzyme supplementation. *Canadian Journal of Animal Science* 68, 1253–1259.

Classen, H.L., Campbell, G.L., Rossnagel, B.G. and Bhatty, R.S. (1988b) Evaluation of hulless barley as replacement for wheat or conventional barley in laying hen diets. *Canadian Journal of Animal Science* 68, 1261–1266.

Conners, W.E. (2000) Importance of n-3 fatty acids in health and disease. *American Journal of Clinical Nutrition* 71, 171S–175S.

Corrier, D.E., Hinton, A. Jr., Ziprin, R.L. and DeLoach, J.R. (1990) Effect of dietary lactose on salmonella colonization of market-age broiler chickens. *Avian Diseases* 34, 668–676.

Cromwell, G.L., Herkelmad, K.L. and Stahly, T.S. (1993) Physical, chemical, and nutritional characteristics of distillers dried grains with solubles for chicks and pigs. *Journal of Animal Science* 71, 679–686.

Cromwell, G.L., Cline, T.R., Crenshaw, J.D., Crenshaw, T.D., Easter, R.A. *et al.* (2000) Variability among sources and laboratories in analyses of wheat middlings. *Journal of Animal Science* 78, 2652–2658.

Cuadrado, C., Ayet, G., Burbano, C., Muzquiz, M., Camacho, L. *et al.* (1995) Occurrence of saponins and sapogenols in Andean crops. *Journal of the Science of Food and Agriculture* 67, 169–172.

Cumming, R.B. (1992) The biological control of coccidiosis by choice feeding. *Proceedings of 19th World's Poultry Congress*, Vol 2. Amsterdam, The Netherlands, 20–24 September 1992, 425–428.

Dänner, E.E. (2003) Use of low-vicin/convicin faba beans (*Vicia faba*) in laying hens. *Archiv für Geflügelkunde* 67, 249–252.

Darroch, C.S. (1990) Safflower meal. In: Thacker, P.A. and Kirkwood, R.N. (eds) *Nontraditional Feed Sources for Use in Swine Production.* Butterworths, Stoneham, Massachusetts, pp. 373–382.

Daun, J.K. and Pryzbylski, R. (2000) Environmental effects on the composition of four Canadian flax cultivars. *Proceedings of 58th Flax Institute*, Fargo, North Dakota, 23–25 March, 2000, pp. 80–91.

Daveby, Y.D., Razdan, A. and Aman, P. (1998) Effect of particle size and enzyme supplementation of diets based on dehulled peas on the nutritive value for broiler chickens. *Animal Feed Science and Technology* 74, 229–239.

Davtyan, A. and Manukyan, V. (1987) Effect of grass meal on fertility of hens. *Ptitsevodstvo* 6, 28–29.

DeClercq, D.R. (2006) *Quality of Canadian Flax.* Canadian Grain Commission, Winnipeg, Manitoba, Canada.

DeLoach, J.R., Oyofo, B.A., Corrier, D.E., Kubena, L.F., Ziprin, R.L. and Norman, J.O. (1990) Reduction of *Salmonella typhimurium* concentration in broiler chickens by milk or whey. *Avian Diseases* 34, 389–392.

Devegowda, G., Raju, M.V.L.N., Afzali, N. and Swamy, H.V.L.N. (1998) Mycotoxin picture worldwide: novel solutions for their counteraction. In: Lyons, T.P. and Jacques, K.A. (eds) *Biotechnology in the Feed Industry, Proceedings of the 14th Annual Alltech Symposium*, Nottingham University Press, Nottingham, UK, pp. 241–255.

Draganov, I.F. (1986) Brewer's grains in the feeding of farm animals – a review. *Zhivotnovodstvo* 11, 61–63.

Edwards, S.A. and Livingstone, R.M. (1990) Potato and potato products. In: Thacker, P.A. and Kirkwood, R.N. (eds) *Nontraditional Feed Sources for Use in Swine Production.* Butterworth, Stoneham, Massachusetts, pp. 305–314.

El-Boushy, A.R. and Raterink, R. (1989) Replacement of soybean meal by cottonseed meal and peanut meal or both in low energy diets for broilers. *Poultry Science* 68, 799–804.

El-Deekx, A.A. and Brikaa, A.M. (2009) Effect of different levels of seaweed in starter and finisher diets in pellet and mash form on performance and carcass quality of ducks. *International Journal of Poultry Science* 8, 1014-1021.

Eldred, A.R., Damron, B.L. and Harms, R.H. (1975) Evaluation of dried brewers grains and yeast in laying hen diets containing various sulfur amino acid levels. *Poultry Science* 54, 856–860.

Elfverson, C., Andersson, A.A.M., Åman, P. and Regnér, S. (1999) Chemical composition of barley cultivars fractionated by weighing, pneumatic classification, sieving, and sorting on a specific gravity table. *Cereal Chemistry* 76, 434–438.

Elzubeir, E.A. and Ibrahim, M.A. (1991) Effect of dietary full-fat raw sunflower seed on performance and carcass skin colour of broilers. *Journal of the Science of Food and Agriculture* 55, 479–481.

Emmert, J.L. and Baker, D.H. (1997) A chick bioassay approach for determining the bioavailable choline concentration in normal and overheated soybean meal, canola meal and peanut meal. *Journal of Nutrition* 127, 745–752.

Ernst, R.A., Vohra, P., Kratzer, F.H. and Ibanga, O. (1994) A comparison of feeding corn, oats, and barley on the growth of White Leghorn chickens, gastrointestinal weights of males, and sexual maturity of females. *Journal of Applied Poultry Research* 3, 253–260.

Etuk, E.B., Ifeduba, A.V., Okata U.E., Chiaka I., Okoli, I.C. *et al.* (2012) Nutrient composition and feeding value of sorghum for livestock and poultry : a review. *Journal of Animal Science Advances* 2, 510-524.

European Commission (2007) Council Regulation EC No 834/2007 on organic production and labelling of organic and repealing regulation (EEC) No 2092/91. *Official Journal of the European Communities* L 189205, 1–23.

Evans, R.J. and Bandemer, S.L. (1967) Nutritive values of some oilseed proteins. *Cereal Chemistry* 44, 417–426.

Farrell, D.J. (1978) A nutritional evaluation of buckwheat (*Fagopyrum esculentum*). *Animal Feed Science and Technology* 3, 95–108.

Farrell, D.J. (1994) Utilization of rice bran in diets for domestic fowl and ducklings. *World's Poultry Science Journal* 50, 115–131.

Farrell, D.J. (1995) Effects of consuming seven omega-3 fatty acid enriched eggs per week on blood profiles of human volunteers. *Poultry Science* 74(supplement), 148.

Farrell, D.J. and Hutton, K. (1990) Rice and rice milling by-products. In: Thacker, P.A. and Kirkwood, R.N. (eds) *Nontraditional Feed Sources for Use in Swine Production*. Butterworths, Massachusetts, pp. 339–354.

Farrell, D.J., Takhar, B.S., Barr, A.R. and Pell, A.S. (1991) Naked oats: their potential as a complete feed for poultry. In: Farrell, D.J. (ed.) *Recent Advances in Animal Nutrition in Australia 1991*. University of New England, Armidale, New South Wales, Australia, pp. 312–325.

Farrell, D.J., Perez-Malondano, R.A. and Mannion, P.F. (1999) Optimum inclusion of field peas, faba beans, chick peas and sweet lupins in poultry diets. II. Broiler experiments. *British Poultry Science* 40, 674–680.

FDA (2001) *Food and Drug Administration Code of Federal Regulations*, Title 21, Vol 6, Revised as of 1 April, 2001, p. 515. From the US Government Printing Office via GPO Access [CITE: 21CFR582.80].

Fru-Nji, F., Niess, E. and Pfeffer, E. (2007) Effect of graded replacement of soyabean meal by faba beans (*Vicia faba* L.) or field peas (*Pisum sativum* L.) in rations for laying hens on egg production and quality. *Journal of Poultry Science* 44, 34–41.

Garcia, M. and Dale, N. (1999) Cassava root meal for poultry. *Journal of Applied Poultry Research* 8, 132–137.

Gatel, F. (1993) Protein quality of legume seeds for non-ruminant animals: a literature review. *Animal Feed Science and Technology* 45, 317–348.

Gdala, J., Jansman, A.J.M., van Leeuwen, P., Huisman, J. and Verstegen, M.W.A. (1996) Lupins (*L. luteus, L. albus, L. angustifolius*) as a protein source for young pigs. *Animal Feed Science and Technology* 62, 239–249.

Gelder, W.M.J. and van Vonk, C.R. (1980) Amino acid composition of coagulable protein from tubers of 34 potato varieties and its relationship with protein content. *Potato Research* 23, 427–434.

Grashorn, M. and Ritteser, C. (2016) Potential of native organic feeding stuffs in poultry production. *Veterinarija Ir Zootechnika*, 73, Supplement.

Grela, E.R. and Semeniuk, V. (2008) Chemical composition and nutritional value of feeding-stuffs from organic and conventional farms. In *BIOACADEMY 2008 – PROCEEDINGS, New Developments in Science and Research on Organic Agriculture*, 3–5/9/2008. Lednice na Moravě, Czech Republic. Available at: http://orgprints.org/20541/1/konference_abstrakta.pdf (accessed 8 July 2017).

Gupta, J.J., Yadav, B.P.S. and Hore, D.K. (2002) Production potential of buckwheat grain and its feeding value for poultry in Northeast India. *Fagopyrum* 19, 101–104.

Habibu, B., Ikira, N.M., Buhari, H.U., Aluwong, T., Kawu, M.U. et al. (2014) Effect of molasses supplementation on live weight gain, haematologic parameters and erythrocyte osmotic fragility of broiler chickens in the hot-dry season. *International Journal of Veterinary Science* 3, 181–188.

Halaj, M., Halaj, P., Najdúch, L. and Arpášová, H. (1998) Effect of alfalfa meal contained in hen feeding diet on egg yolk pigmentation. *Acta Fytotechnica et Zootechnica* 1, 80–83.

Halnan, E.T. (1944) Digestibility trials with poultry. II. The digestibility and metabolizable energy of raw and cooked potatoes, potato flakes, dried potato slices and dried potato shreds. *Journal of Agricultural Science* 34, 139–154.

Hamid, K. and Jalaludin, S. (1972) The utilization of tapioca in rations for laying poultry. *Malaysian Journal of Agricultural Research* 1, 48–53.

Harris, L.E. (1980) *International Feed Descriptions, International Feed Names, and Country Feed Names*. International Network of Feed Information Centers, Logan, Utah.

Harris, R.K. and Haggerty, W.J. (1993) Assays for potentially anti-carcinogenic phytochemicals in flax-seed. *Cereal Foods World* 38, 147–151.

Harrold, R.L. (2002) *Field Pea in Poultry Diets*. Extension bulletin EB-76. North Dakota State University, Extension Service, Fargo, North Dakota.

Heartland Lysine (1998) *Digestibility of Essential Amino Acids for Poultry and Swine*. Version 3.51. Heartland Lysine, Chicago, Illinois.

Hede, A.R. (2001) A new approach to triticale improvement. *Research Highlights of the CIMMYT Wheat Program, 1999–2000*. International Maize and Wheat Improvement Center, Oaxaca, Mexico, pp. 21–26.

Helm, C.V. and de Francisco, A. (2004) Chemical characterization of Brazilian hulless barley varieties, flour fractionation, and protein concentration. *Scientia Agricola* 61, 593–597.

Henry, M.H., Pesti, G.M. and Brown, T.P. (2001) Pathology and histopathology of gossypol toxicity in broiler chicks. *Avian Diseases* 45, 598–604.

Hermes, J.C. and Johnson, R.C. (2004) Effects of feeding various levels of triticale var. Bogo in the diet of broiler and layer chickens. *Journal of Applied Poultry Research* 13, 667–672.

Hsun, C.L. and Maurice, D.V. (1992) Nutritional value of naked oats (*Avena nuda*) in laying hen diets. *British Poultry Science* 33, 355–361.

Huthail, N. and Al-Khateeb, S.A. (2004) The effect of incorporating different levels of locally produced canola seeds (*Brassica napus, L.*) in the diet of laying hens. *International Journal of Poultry Science* 3, 490–496.

Igbassen, F.A. and Guenter, W. (1997a) The influence of feeding yellow-, green-, and brown-seeded peas on production performance of laying hens. *Journal of the Science of Food and Agriculture* 73, 120–128.

Igbassen, F.A. and Guenter, W. (1997b) The influence of micronization, dehulling, and enzyme supplementation on the nutritive value of peas for laying hens. *Poultry Science* 76, 331–337.

Igbasan, F.A., Guenter, W. and Slominski, B.A. (1997) The effect of pectinase and alpha-galactosidase supplementation on the nutritive value of peas for broiler chickens. *Canadian Journal of Animal Science* 77, 537–539.

Im, H.L., Ravindran, V., Ravindran, G., Pittolo, P.H. and Bryden, W.L. (1999) The apparent metabolisable energy and amino acid digestibility of wheat, triticale and wheat middlings for broiler chickens as affected by exogenous xylanase supplementation. *Journal of the Science of Food and Agriculture* 79, 1727–1732.

Inborr, J. and Bedford, M.R. (1993) Stability of feed enzymes to steam pelleting during feed processing. *Animal Feed Science and Technology* 46, 179–196.

Jacob, J.P. (2007) Nutrient content of organically grown feedstuffs. *Journal of Applied Poultry Research* 16, 642–651.

Jacob, J.P. (2013) Including Cottonseed Meal in Organic Poultry Diets. eXtension.org. Available at: http://articles.extension.org/pages/69975/including-cottonseed-meal-in-organic-poultry-diets (accessed 21 May 2017).

Jacob, J.P. (2015) Feeding Triticale to Poultry. eXtension.org. Available at: http://articles.extension.org:80/pages/69295/feeding-triticale-to-poultry (accessed 29 May, 2016).

Jacob, J.P. and Pescatore, A.J. (2012) Use of barley in poultry diets: A review. *Journal of Applied Poultry Research* 21, 915–940.

Jacob, J.P., Mitaru, B.N., Mbugua, P.N. and Blair, R. (1996) The feeding value of Kenyan sorghum, sunflower seed cake and sesame seed cake for broilers and layers. *Animal Feed Science and Technology* 61, 41–56.

Jamroz, D., Wiliczkiewicz, A. and Skorupinska, J. (1992) The effect of diets containing different levels of structural substances on morphological changes in the intestinal walls and the digestibility of the crude fibre fractions in geese (Part 3). *Journal of Animal Feed Science* 1, 37–50.

Jaroni, D., Scheideler, S.E., Beck, M. and Wyatt, C. (1999) The effect of dietary wheat middlings and enzyme supplementation. 1. Late egg production efficiency, egg yields, and egg composition in two strains of Leghorn hens. *Poultry Science* 78, 841–847.

Jensen, A. (1972) The nutritive value of seaweed meal for domestic animals. In: *Proceedings of the 7th International Seaweed Symposium*. University of Tokyo Press, Tokyo, pp. 7–14.

Jeroch, H. and Dänicke, S. (1995) Barley in poultry feeding: a review. *World's Poultry Science Journal* 51, 271–291.

Jiang, J.F., Song, X.M., Huang, X., Wu, J.L., Zhou, W.D. *et al.* (2012). Effects of alfalfa meal on carcase quality and fat metabolism of Muscovy ducks. *British Poultry Science* 53, 681–688.

Jiang, Z., Ahn, D.U., Ladner, L. and Sim, J.S. (1992) Influence of feeding full-fat flax and sunflower seeds on internal and sensory qualities of eggs. *Poultry Science* 71, 378–382.

Józefiak, D., Rutkowski, A., Jensen, B.B. and Engberg, R.M. (2007) Effects of dietary inclusion of triticale, rye and wheat and xylanase supplementation on growth performance of broiler chickens and fermentation in the gastrointestinal tract. *Animal Feed Science and Technology* 132, 79–93.

Karadas, F., Grammenidis, E., Surai, P.F., Acamovic, T. and Sparks, N.H.C. (2006) Effects of carotenoids from lucerne, marigold and tomato on egg yolk pigmentation and carotenoid composition. *British Poultry Science* 47, 561–566.

Karimi, A. (2006) The effects of varying fish meal inclusion levels (%) on performance of broiler chicks. *International Journal of Poultry Science* 5, 255–258.

Karunajeewa, H., Tham, S.H. and Abu-Serewa, S. (1989) Sunflower seed meal, sunflower oil and full-fat sunflower seeds, hulls and kernels for laying hens. *Animal Feed Science and Technology* 26, 45–54.

Kassaify, Z.G. and Mine, Y. (2004) Effect of food protein supplements on *Salmonella enteritidis* infection and prevention in laying hens. *Poultry Science* 83, 753–760.

Kiiskinen, T. (1989) Effect of long-term use of rapeseed meal on egg production. *Annales Agriculturae Fenniae* 28, 385–396.

Kim, E.M., Choi, J.H. and Chee, K.M. (1997) Effects of dietary safflower and perilla oils on fatty acid composition in egg yolk. *Korean Journal of Animal Science* 39, 135–144.

Kluth, H. and Rodehutscord, M. (2006) Comparison of amino acid digestibility in broiler chickens, turkeys and Pekin ducks. *Poultry Science* 85, 1953–1960.

Korol, W., Adamczyk, M. and Niedzwiadek, T. (2001) Evaluation of chemical composition and utility of potato protein concentrate in broiler chickens feeding. *Biuletyn Naukowy Przemyslu Paszowego* 40, 25–35.

Korver, D.R., Zuidhof, M.J. and Lawes, K.R. (2004) Performance characteristics and economic comparison of broiler chickens fed wheat and triticale-based diets. *Poultry Science* 83, 716–725.

Kratzer, F.H. and Vohra, P. (1996) *The Use of Flaxseed as a Poultry Feedstuff.* Poultry Fact Sheet No. 21, Cooperative Extension, University of California, Davis, California.

Kuchta, M., Koreleski, J. and Zegarek, Z. (1992) High level of fractional dried lucerne in the diet for laying hens. *Roczniki Naukowe Zootechniki* 19, 119–129.

Kyntäjä, S., Partanen, K., Siljander-Rasi, H. and Jalava, T. (2014) Tables of composition and nutritional values of organically produced feed materials for pigs and poultry. *MTT REPORT 164*, Finland. Available at: http://www.mtt.fi/mttraportti/pdf/mttraportti164.pdf (accessed 21 February 2016).

Lázaro, R., García, M., Aranibar, M.J. and Mateos, G.G. (2003) Effect of enzyme addition to wheat-, barley- and rye-based diets on nutrient digestibility and performance of laying hens. *British Poultry Science* 44, 256–265.

Lee, M.R.F., Parkinson, S., Fleming, H.R., Theobald, V.J., Leemans, D.J. and Burgess, T. (2016) The potential of blue lupins as a protein source in the diets of laying hens. *Veterinary and Animal Science* 1, 29–35.

Leeson, S., Hussar, N. and Summers, J.D. (1988) Feeding and nutritive value of hominy and corn grits for poultry. *Animal Feed Science and Technology* 19, 313–325.

Li, J.H., Vasanthan, T., Rossnagel, B. and Hoover, R. (2001) Starch from hull-less barley: I. Granule morphology, composition and amylopectin structure. *Food Chemistry* 74, 395–405.

Liener, I.E. (1994) Implications of antinutritional components in soybean foods. *CRC Critical Reviews in Food Science and Nutrition* 34, 31–67.

Livingstone, R.M., Baird, B.A. and Atkinson, T. (1980) Cabbage (*Brassica oleracea*) in the diet of growing–finishing pigs. *Animal Feed Science and Technology* 5, 69–75.

Lu, J., Wang, K.H., Tong, H.B., Shi, S.R. and Wang, Q. (2013) Effects of graded replacement of soybean meal by peanut meal on performance, egg quality, egg fatty acid composition and cholesterol content in laying hens. *Archiv für Geflugelkunde* 77, 43–50.

Lumpkins, B.S., Batal, A.B. and Dale, N.M. (2004) Evaluation of distillers dried grains with solubles as a feed ingredient for broilers. *Poultry Science* 83, 1891–1896.

Maddock, T.D., Anderson, V.L. and Lardy, G.P. (2005) *Using Flax in Livestock Diets.* Extension Report AS-1283, Department of Animal and Range Sciences, North Dakota State University, Fargo and Carrington Research Extension Center, Carrington, ND.

Madhusudhan, K.T., Ramesh, H.P., Ogawa, T., Sasaoka, K. and Singh, N. (1986) Detoxification of commercial linseed meal for use in broiler rations. *Poultry Science* 65, 164–171.

Marquardt, R.R. and Bell, J.M. (1988) Future potential of pulses for use in animal feeds. In: Summerfield, R.J. (ed.) *World Crops: Cool Season Food Legumes.* Kluwer, Dordrecht, pp. 421–444.

Marquardt, R.R., Boros, D., Guenter, W. and Crow, G. (1994) The nutritive value of barley, rye, wheat and corn for young chicks as affected by use of a trichoderma reesei enzyme preparation. *Animal Feed Science and Technology* 45, 363–378.

Marshall, A.C., Kubena, K.S., Hinton, K.R., Hargis, P.S. and Van Elswyk, M.E. (1994) n-3 Fatty acid enriched table eggs: a survey of consumer acceptability. *Poultry Science* 73, 1334–1340.

Maurice, D.V., Jones, J.E., Hall, M.A., Castaldo, D.J., Whisenhunt, J.E. and McConnell, J.C. (1985) Chemical composition and nutritive value of naked oats (*Avena nuda L.*) in broiler diets. *Poultry Science* 64, 529–535.

Meng, X., Slominski, B.A., Campbell, L.D., Guenter, W. and Jones, O. (2006) The use of enzyme technology for improved energy utilization from full-fat oilseeds. Part I: canola seed. *Poultry Science* 85, 1025–1030.

Metwally, M.A. (2003) Evaluation of slaughterhouse poultry byproduct meal, dried alfalfa meal, sorghum grains and grass meal as non-conventional feedstuffs for poultry diets. *Egyptian Poultry Science Journal* 23, 875–892.

Mikulski, D., Faruga, A., Kriz, L. and Klecker, D. (1997) The effect of thermal processing of faba beans, peas, and shelled grains on the results of raising turkeys. *Zivocisna-Vyroba* 42, 72–81.

Mitaru, B.N., Blair, R., Bell, J.M. and Reichert, R.D. (1982) Tannin and fibre contents of rapeseed and canola hulls. *Canadian Journal of Animal Science* 62, 661–663.

Mitaru, B.N., Reichert, R.D. and Blair, R. (1983) Improvement of the nutritive value of high tannin sorghums for broiler chickens by high moisture storage (reconstitution). *Poultry Science* 62, 2065–2072.

Mitaru, B.N., Reichert, R.D. and Blair, R. (1985) Protein and amino acid digestibilities for chickens of reconstituted and boiled sorghum grains varying in tannin contents. *Poultry Science* 64, 101–106.

Mohan, L., Reddy, C.V., Rao, P.V. and Siddiqui, S.M. (1984) Safflower (*Carthamus tinctorius* Linn.) oilcake as a source of protein for broilers. *Indian Journal of Animal Science* 54, 870–873.

Morgan, K.N. and Choct, M. (2016) Cassava: Nutrient composition and nutritive value in poultry diets. *Animal Nutrition* 2, 253–261.

Moschini, M., Masoero, F., Prandini, A., Fusconi, G., Morlacchini, M. and Piva, G. (2005) Raw pea (*Pisum sativum*), raw faba bean (*Vicia faba* var. minor) and raw lupin (*Lupinus albus* var. multitalia) as alternative protein sources in broiler diets. *Italian Journal of Animal Science* 4, 59–69.

Mullan, B.P., van Barneveld, R.J. and Cowling, W.A. (1997) Yellow lupins (*Lupinus luteus*): a new feed grain for the Australian pig industry. In: Cranwell, P. (ed.) *Manipulating Pig Production VI*. Australasian Pig Science Association Conference, Canberra, p. 237.

Mullan, B.P., Pluske, J.R., Allen, J. and Harris, D.J. (2000) Evaluation of Western Australian canola meal for growing pigs. *Australian Journal of Agricultural Research* 51, 547–553.

Mulyantini, N.G.A., Choct, M., Li, X. and Lole, U.R. (2005) The effect of xylanase, phytase and lipase supplementation on the performance of broiler chickens fed a diet with a high level of rice bran. In: Scott, T.A. (ed.) *Proceedings of the 17th Australian Poultry Science Symposium*. Sydney, New South Wales, Australia, 7–9 February 2005, pp. 305–307.

Nagy, G., Gyüre, P. and Mihók, S. (2002) Goose production responses to grass based diets. Multifunction grasslands: quality forages, animal products and landscapes. In: Durand, J.L., Emile, J.C., Huyghe, C. and Lemaire, G. (eds) *Proceedings of the 19th General Meeting of the European Grassland Federation*. La Rochelle, France, 27–30 May 2002, pp. 1060–1061.

Najib, H. and Y. M. Al-Yousef. (2011) Performance and essential fatty acids content of dark meat as affected by supplementing the broiler diet with different levels of flax seed. *Annual Review and Research in Biology* 1, 22–32.

Nalle, C.L., Ravindran, V., and Ravindran, G. (2011) Nutritional value of peas (*Pisum sativum* L.) for broilers: apparent metabolisable energy, apparent ileal amino acid digestibility and production performance. *Animal Production Science* 51, 150–155.

Naseem, M.Z., Khan, S.H. and Yousaf, M. (2006) Effect of feeding various levels of canola meal on the performance of broiler chicks. *Journal of Animal and Plant Sciences* 16, 75–78.

Newkirk, R.W., Classen, H.L. and Tyler, R.T. (1997) Nutritional evaluation of low glucosinolate mustard meals (*Brassica juncea*) in broiler diets. *Poultry Science* 76, 1272–1277.

Nilipour, A.H., Savage, T.F. and Nakaue, H.S. (1987) The influence of feeding triticale (Var: Flora) and varied crude protein diets on the seminal production of medium white turkey breeder toms. *Nutrition Reports International* 36, 151–160.

Nimruzi, R. (1998) The value of field peas and fig powder. *World Poultry* 14, 20.

NRC (1994) *Nutrient Requirements of Poultry*, 9th rev. edn. National Research Council, National Academy of Sciences, Washington, DC.

Nwokolo, E. and Sim, J. (1989a) Barley and full-fat canola seed in broiler diets. *Poultry Science* 68, 1374–1380.

Nwokolo, E. and Sim, J. (1989b) Barley and full-fat canola seed in layer diets. *Poultry Science* 68, 1485–1489.

Nzekwe, N.M. and Olomu, J.M. (1984) Cottonseed meal as a substitute for groundnut meal in the rations of laying chickens and growing turkeys. *Journal of Animal Production Research* 4, 57–71.

NZFSA (2011) *NZFSA Technical Rules for Organic Production, Version 7*. New Zealand Food Safety Authority, Wellington.

Ochetim, S. and Solomona, S.L. (1994) The feeding value of dried brewers spent grains for laying chickens. In: Djajanegara, A. and Sukmawati, A. (eds) *Sustainable animal production and the environment, Proceedings of the 7th AAAP Animal Science Congress*. Bali, Indonesia, 11–16 July, 1994. Vol 2, pp. 283–284.

Offiong, S.A., Flegal, C.J. and Sheppard, C.C. (1974) The use of raw peanuts and peanut meal in practical chick diets. *East African Agricultural and Forestry Journal* 39, 344–348.

Ojewola, G.S., Ukachukwu, S.N. and Okulonye, E.I. (2006) Cottonseed meal as substitute for soyabean meal in broiler ration. *International Journal of Poultry Science* 5, 360–364.

Oke, O.L. (1990) Cassava. In: Thacker, P.A. and Kirkwood, R.N. (eds) *Nontraditional Feed Sources for Use in Swine Production*. Butterworth, Stoneham, Massachusetts, pp. 103–112.

Omara, I.I. (2012). Nutritive value of skimmed milk and whey, added as natural probiotics in broiler diets. *Egyptian Journal of Animal Production* 49, 207–217.

Onifade, A.A. and Babatunde, G.M. (1996) Supplemental value of dried yeast in a high-fibre diet for broiler chicks. *Animal Feed Science and Technology* 62, 91–96.

Onifade, A.A. and Babatunde, G.M. (1998) Comparison of the utilisation of palm kernel meal, brewers' dried grains and maize offal by broiler chicks. *British Poultry Science* 39, 245–250.

Opapeju, F.O., Golian, A., Nyachoti, C.M. and Campbell, L.D. (2006) Amino acid digestibility in dry extruded-expelled soyabean meal fed to pigs and poultry. *Journal of Animal Science* 84, 1130–1137.

Ortiz, L.T., Centeno, C. and Treviño, J. (1993) Tannins in faba bean seeds: effects on the digestion of protein and amino acids in growing chicks. *Animal Feed Science and Technology* 41, 271–278.

Panigrahi, S. (1996) A review of the potential for using cassava root meal in poultry diets. In: Kurup, G.T., Palaniswami, M.S., Potty, V.P., Padmaja, G., Kabeerathumma, S. and Pillai, S.V. (eds) *Tropical Tuber Crops: Problems, Prospects and Future Strategies*. Science Publishers, Lebanon, Indiana, pp. 416–428.

Panigrahi, S., Plumb, V.E. and Machin, D.H. (1989) Effects of dietary cottonseed meal, with and without iron treatment, on laying hens. *British Poultry Science* 30(3), 641–651, 19 ref.

Parviz, F. (2006) Performance and carcass traits of lentil seed fed broilers. *Indian Veterinary Journal* 83, 187–190.

Pârvu, M., Iofciu, A., Grossu, D. and Iliescu, M. (2001) Efficiency of toasted full fat soyabean utilization in broiler feeding. *Archiva Zootechnica* 6, 121–124.

Patterson, P.H., Sunde, M.L., Schieber, E.M. and White, W.B. (1988) Wheat middlings as an alternate feedstuff for laying hens. *Poultry Science* 67, 1329–1337.

Perez-Maldonado, R.A. and Barram, K.M. (2004) Evaluation of Australian canola meal for production and egg quality in two layer strains. *Proceedings of the 16th Australian Poultry Science Symposium*. New South Wales, Sydney, pp. 171–174.

Perez-Maldonado, R.A., Mannion, P.F. and Farrell, D.J. (1999) Optimum inclusion of field peas, faba beans, chick peas and sweet lupins in poultry diets. I. Chemical composition and layer experiments. *British Poultry Science* 40, 667–673.

Pesti, G.M., Bakalli, R.I., Driver, J.P., Sterling, K.G., Hall, L.E. and Bell, E.M. (2003) Comparison of peanut meal and soybean meal as protein supplements for laying hens. *Poultry Science* 82, 1274–1280.

Petterson, D.S. (1998) Composition and food uses of legumes. In: Gladstones, J.S., Atkins, C.A. and Hamblin, J. (eds) *Lupins as Crop Plants. Biology, Production and Utilization*. CAB International, Wallingford, UK.

Petterson, D.S., Sipsas, S. and Mackintosh, J.B. (1997) *The Chemical Composition and Nutritive Value of Australian Pulses*, 2nd edn. Grains Research and Development Corporation, Canberra, 65 pp.

Purushothaman, M.R., Vasan, P., Mohan, B. and Ravi, R. (2005) Utilization of tallow and rice bran oil in feeding broilers. *Indian Journal of Poultry Science* 40, 175–178.

Ranhotra, G.S., Gelroth, J.A., Glaser, B.K. and Lorenz, K.J. (1996a) Nutrient composition of spelt wheat. *Journal of Food Composition and Analysis* 9, 81–84.

Ranhotra, G.S., Gelroth, J.A., Glaser, B.K. and Stallknecht, G.F. (1996b) Nutritional profile of three spelt wheat cultivars grown at five different locations. *Cereal Chemistry* 73, 533–535.

Rao, S.V.R., Raju, M.V.L.N., Panda, A.K. and Shashibindu, M.S. (2005) Utilization of low glucosinalate and conventional mustard oilseed cakes in commercial broiler chicken diets. *Asian-Australasian Journal of Animal Sciences* 18, 1157–1163.

Rao, S.V.R., Raju, M.V.L.N., Panda, A.K. and Reddy, M.R. (2006) Sunflower seed meal as a substitute for soybean meal in commercial broiler chicken diets. *British Poultry Science* 47, 592–598.

Ravindran, V. and Blair, R. (1991) Feed resources for poultry production in Asia and the Pacific region. I. Energy sources. *World's Poultry Science Journal* 47, 213–262.

Ravindran, V. and Blair, R. (1992) Feed resources for poultry production in Asia and the Pacific. II. Plant protein sources. *World's Poultry Science Journal* 48, 205–231.

Ravindran, V. and Blair, R. (1993) Feed resources for poultry production in Asia and the Pacific. III. Animal protein sources. *World's Poultry Science Journal* 49, 219–235.

Ravindran, V., Bryden, W.L. and Kornegay, E.T. (1995) Phytates: occurrence, bioavailability and implications in poultry nutrition. *Avian Biology and Poultry Science Reviews* 6, 125–143.

Ravindran, V., Tilman, Z.V., Morel, P.C.H., Ravindran, G. and Coles, G.D. (2007) Influence of β-glucanase supplementation on the metabolisable energy and ileal nutrient digestibility of normal starch and waxy barleys for broiler chickens. *Animal Feed Science and Technology* 134, 45–55.

Rebolé, A., Rodríguez, M.L., Ortiz, L.T., Alzueta, C., Centeno, C. *et al.* (2006) Effect of dietary high-oleic acid sunflower seed, palm oil and vitamin E supplementation on broiler performance, fatty acid composition and oxidation susceptibility of meat. *British Poultry Science* 47, 581–591.

Reddy, N.R., Sathe, S.K. and Salunkhe, D.K. (1982) Phytates in legumes and cereals. *Advances in Food Research* 28, 1–92.

Richter, G. and Lenser A. (1993). The use of native triticale in poultry. 3. Use in laying hens. *Archives of Animal Nutrition* 43, 237–244

Richter, G., Schurz, M., Ochrimenko, W.I., Kohler, H., Schubert, R. *et al.* (1999) The effect of NSP-hydrolysing enzymes in diets of laying hens and broilers. *Vitamine und Zusatzstoffe in der Ernahrung von Mensch und Tier: 7. Symposium.* Jena-Thuringen, Germany, pp. 519–522.

Roberson, K.D. (2003) Use of dried distillers grains with solubles in growing-finishing diets of turkey hens. *International Journal of Poultry Science* 2, 389–393.

Rodríguez, M.L., Ortiz, L.T., Treviño, J., Rebolé, A., Alzueta, C. and Centeno, C. (1997) Studies on the nutritive value of full-fat sunflower seed in broiler chick diets. *Animal Feed Science and Technology* 71, 341–349.

Roth-Maier, D.A. (1999) Investigations on feeding full-fat canola seed and canola meal to poultry. In: Santen, E., van Wink, M. and Weissmann, S. (eds) *Proceedings of the 10th International Rapeseed Congress.* Canberra, Australia.

Roth-Maier, D.A. and Paulicks, B.R. (2004) Blue and yellow lupin seed in the feeding of broiler chicks. In: Santen, E. and van Hill, G.D. (eds) *Wild and Cultivated Lupins from the Tropics to the Poles. Proceedings of the 10th International Lupin Conference.* Laugarvatn, Iceland, 19–24 June 2002, pp. 333–335.

Rubio, L.A., Brenes, A. and Centeno, C. (2003) Effects of feeding growing broiler chickens with practical diets containing sweet lupin (*Lupinus angustifolius*) seed meal. *British Poultry Science* 44, 391–397.

Rybina, E.A. and Reshetova, T.A. (1981) Digestibility of nutrients and biochemical values of eggs in relation to the amount of lucerne and grass meal and the quality of supplementary fat in the diet of laying hens. *Zhivotnovodstva* 35, 148–152.

Sakomura, N.K., da Silva, R., Basaglia, R., Malheiros, E.B. and Junqueira, O.M. (1998) Whole steam-toasted and extruded soyabean in diets of laying hens. *Revista Brasileira de Zootecnia* 27, 754–765.

Salih, M.E., Classen, H.L. and Campbell, G.L. (1991) Response of chickens fed on hull-less barley to dietary β-glucanase at different ages. *Animal Feed Science and Technology* 33, 139–149.

Salmon, R.E., Stevens, V.I. and Ladbrooke, B.D. (1988) Full-fat canola seed as a feedstuff for turkeys. *Poultry Science* 67, 1731–1742.

Savage, T.F., Holmes, Z.A., Nilipour, A.H. and Nakaue, H.S. (1987) Evaluation of cooked breast meat from male breeder turkeys fed diets containing varying amounts of triticale, variety Flora. *Poultry Science* 66, 450–452.

Scheideler, S.E., Cuppett, S. and Froning, G. (1994) Dietary flaxseed for poultry: production effects, dietary vitamin levels, fatty acid incorporation into eggs and sensory analysis. In: *Proceedings of the 55th Flax Institute.* 26–28 January, 1994. Fargo, North Dakota, pp. 86–95.

Seerley, R.W. (1991) Major feedstuffs used in swine diets. In: Miller, E.R., Ullrey, D.E. and Lewis, A.J. (eds) *Swine Nutrition*. Butterworth-Heinemann, Boston, Massachusetts, pp. 451–481.

Senkoylu, N. and Dale, N. (1999) Sunflower meal in poultry diets: a review. *World's Poultry Science Journal* 55, 153–174.

Senkoylu, N. and Dale, N. (2006) Nutritional evaluation of a high-oil sunflower meal in broiler starter diets. *Journal of Applied Poultry Research* 15, 40–47.

Šerman, V., Mas, N., Melenjuk, V., Dumanovski, F. and Mikulec, Ž. (1997) Use of sunflower meal in feed mixtures for laying hens. *Acta Veterinaria Brno* 66, 219–227.

Seve, B. (1977) Utilisation d'un concentre de proteine de pommes de terre dans l'aliment de sevrage du porcelet a 10 jours et a 21 jours. *Journees de la Recherche Porcine en France* 205–210.

Shariatmadari, F. and Forbes, J.M. (2005) Performance of broiler chickens given whey in the food and/ or drinking water. *British Poultry Science* 46, 498–505.

Shim, K.F., Chen, T.W., Teo, L.H. and Khin, M.W. (1989) Utilization of wet spent brewer's grains by ducks. *Nutrition Reports International* 40, 261–270.

Singh, K.K., Mridula, D., Rehal, J. and Barnwal, P. (2011) Flaxseed: A potential source of food, feed and fiber. *Critical Reviews in Food Science and Nutrition* 51, 210–222.

Smithard, R. (1993) Full-fat rapeseed for pig and poultry diets. *Feed Compounder* 13, 35–38.

Sokól, J.L., Niemiec, J. and Fabijan´ska, M. (2004) Effect of naked oats and enzyme supplementation on egg yolk fatty acid composition and performance of hens. *Journal of Animal and Feed Sciences* 13(supplement 2), 109–112.

Stacey, P., O'Kiely, P., Rice, B., Hackett, R. and O'Mara, F.P. (2003) Changes in yield and composition of barley, wheat and triticale grains with advancing maturity. In: Gechie, L.M. and Thomas, C. (eds) *Proceedings of the XIIIth International Silage Conference*. Ayr, UK, 11–13 September, 2002, p. 222.

Steenfeldt, S. (2001) The dietary effect of different wheat cultivars for broiler chickens. *British Poultry Science* 42, 595–609.

Svihus, B. and Gullord, M. (2002) Effect of chemical content and physical characteristics on nutritional value of wheat, barley and oats for poultry. *Animal Feed Science and Technology* 102, 71–92.

Światkiewicz, S. and Koreleski, J. (2006) Effect of maize distillers dried grains with solubles and dietary enzyme supplementation on the performance of laying hens. *Journal of Animal and Feed Sciences* 15, 253–260.

Talebali, H. and Farzinpour, A. (2005) Effect of different levels of full-fat canola seed as a replacement for soyabean meal on the performance of broiler chickens. *International Journal of Poultry Science* 12, 982–985.

Tanksley, T.D. Jr. (1990) Cottonseed meal. In: Thacker, P.A. and Kirkwood, R.N. (eds) *Nontraditional Feed Sources for Use in Swine Production*. Butterworth Publishers, Stoneham, Massachusetts, pp. 139–152.

Thacker, P.A., Willing, B.P. and Racz, V.J. (2005) Performance of broiler chicks fed wheat-based diets supplemented with combinations of non-extruded or extruded canola, flax and peas. *Journal of Animal and Veterinary Advances* 4, 902–907.

Thomas, V.M., Katz, R.J., Auld, D.A., Petersen, C.F., Sauter, E.A. and Steele, E.E. (1983) Nutritional value of expeller extracted rape and safflower oilseed meals for poultry. *Poultry Science* 62, 882–886.

Tsuzuki, E.T., de Garcia, E.R.M., Murakami, A.E., Sakamoto, M.I. and Galli, J.R. (2003) Utilization of sunflower seed in laying hen rations. *Revista Brasileira de Ciência Avícola* 5, 179–182.

Udo, H., Foulds, J. and Tauo, A. (1980) Comparison of cassava and maize in commercially formulated poultry diets for Western Samoa. *Alafua Agricultural Bulletin* 5, 18–26.

Valdebouze, P., Bergeron, E., Gaborit, T. and Delort-Laval, J. (1980) Content and distribution of trypsin inhibitors and hemagglutinins in some legume seeds. *Canadian Journal of Plant Science* 60, 695–701.

van Barnveld, R.J. (1999) Understanding the nutritional chemistry of lupin (*Lupinus* spp.) seed to improve livestock production efficiency. *Nutrition Research Reviews* 12, 203–230.

Ventura, M.R., Castañon, J.I.R. and McNab, J.M. (1994) Nutritional value of seaweed (*Ulva rigida*) for poultry. *Animal Feed Science and Technology* 49, 87–92.

Vestergaard, E.M., Danielsen, V. and Larsen, A.E. (1995) Utilisation of dried grass meal by young growing pigs and sows. *Proceedings of the 45th Annual Meeting of the European Association for Animal Production*, Prague, paper N2b.

Vieira, S.L., Penz, A.M. Jr., Kessler, A.M. and Catellan, E.V. Jr. (1995) A nutritional evaluation of triticale in broiler diets. *Journal of Applied Poultry Research* 4, 352–355.

Vogt, H. (1967) Brown seaweed meal (*Macrocystis pyrifera*) in feed for fattening poultry. *Archiv für Geflugelkunde* 31, 145–149.

Vogt, H. (1969) Potato meal for laying hens. *Archiv für Geflugelkunde* 33, 439–443.

Waibel, P.E., Noll, S.L., Hoffbeck, S., Vickers, Z.M. and Salmon, R.E. (1992) Canola meal in diets for market turkeys. *Poultry Science* 71, 1059–1066.

Weder, J.K.P. (1981) Protease inhibitors in the leguminosae. In: Pothill, R.M. and Raven, P.H. (eds) *Advances in Legume Systematics*. British Museum of Natural History, London, pp. 533–560.

Westendorf, M.L. and Wohlt, J.E. (2002) Brewing by-products: their use as animal feeds. *Journal of Animal Science* 42, 871–875.

Whittemore, C.T. (1977) The potato (*Solanum tuberosum*) as a source of nutrients for pigs, calves and fowl – a review. *Animal Feed Science and Technology* 2, 171–190.

Wilson, A.S. (1876) On wheat and rye hybrids. *Transactions and Proceedings of the Botanical Society of Edinburgh* 12, 286–288.

Wiseman, J., Jagger, S., Cole, D.J.A. and Haresign, W. (1991) The digestion and utilization of amino acids of heat-treated fish meal by growing/finishing pigs. *Animal Production* 53, 215–225.

Woodworth, J.C., Tokach, M.D., Goodband, R.D., Nelsen, J.L., O'Quinn, P.R. *et al.* (2001) Apparent ileal digestibility of amino acids and the digestible and metabolizable energy content of dry extruded-expelled soybean meal and its effect on growth performance of pigs. *Journal of Animal Science* 79, 1280–1287.

Wylie, P.W., Talley, S.M. and Freeeman, J.N. (1972) Substitution of linseed and safflower meal for soybean meal in diets of growing pullets. *Poultry Science* 51, 1695–1701.

Yamasaki, S., Manh, L.H., Takada, R., Men, L.T., Dung, N.N.X. *et al.* (2003) Admixing synthetic antioxidants and sesame to rice bran for increasing pig performance in Mekong Delta, Vietnam. *Japan International Research Center for Agricultural Science, Research Highlights* 38–39.

Zdunczyk, Z., Jankowski, J., Juskiewicz, J., Mikulski, D. and Slominski, B.A. (2013). Effect of different dietary levels of low-glucosinolate rapeseed (canola) meal and non-starch polysaccharide-degrading enzymes on growth performance and gut physiology of growing turkeys. *Canadian Journal of Animal Science* 93, 353–362.

Zettl, A., Lettner, F. and Wetscherek, W. (1993) Home-produced sunflower oilmeal for pig feeding. *Förderungsdienst* 41, 362–365.

Zhavoronkova, L.D. and D'yakonova, E.V. (1983) Productivity and resistance [to infection] of hens in relation to different levels of grass meal in their diet. *Sbornik Nauchnykh Trudov Moskovskoi Veterinarnoi Akademii* 26–28.

Zijlstra, R.T., Ekpe, M.N., Casano, E.D. and Patience, J.F. (2001) Variation in nutritional value of western Canadian feed ingredients for pigs. In: *Proceedings of 22nd Western Nutrition Conference*. University of Saskatchewan, Saskatoon, Canada, pp. 12–24.

Further Reading

Ajinomoto (2018) Feedstuffs amino acids database. http://ajinomoto-eurolysine.com/feedstuffs-amino-acid-database.html (accessed 17 July 2018).

Queensland Department of Agriculture and Fisheries (2018) Nutrient composition. Available at: https://www.daf.qld.gov.au/business-priorities/animal-industries/pigs/feed-nutrition/ingredients-contaminants/nutrient-composition (accessed 17 July 2018).

NRC (1971) *Atlas of Nutritional Data on United States and Canadian Feeds*. National Research Council, National Academy of Sciences, Washington, DC.

NRC (1982) *United States – Canadian Tables of Feed Composition*. National Research Council, National Academy of Sciences, Washington, DC.

NRC (1988) *Nutrient Requirements of Dairy Cattle*. National Research Council, National Academy of Sciences, Washington, DC.

NRC (1998) *Nutrient Requirements of Swine*, 10th rev. edn. National Research Council, National Academy of Sciences, Washington, DC.

Sauvent, D., Perez, J.-M. and Tran, G. (2004) *Tables of Composition and Nutritional Value of Feed Materials*. Translated by Andrew Potter. Wageningen Academic Publishers, The Netherlands and INRA, Paris, France, 304 pp.

UBC (1997) *Tables of Analysed Composition of Feedstuffs*. Department of Animal Science, University of British Columbia, Vancouver, Canada (unpublished).

US Department of Agriculture, Agricultural Research Service, Nutrient Data Laboratory. USDA National Nutrient Database for Standard Reference, Legacy. Version Current: April 2018. Internet: /nea/bhnrc/ndl (accessed 17 July 2018).

Wiseman, J. (1987) *Feeding of Non-Ruminant Livestock*: collective edited work by the research staff of the Departement de l'elevage des monogastriques, INRA, under the responsibility of Jean-Claude Blum; translated and edited by Julian Wiseman. Butterworths, London.

Zduńczyk, Z., Juíkiewicz, J., Flis, M. and Frejnagel, S. (1996) The chemical composition and nutritive value of low-alkaloid varieties of white lupin. 2. Oligosaccharides, phytates, fatty acids and biological value of protein. *Journal of Animal and Feed Science* 5, 73–82.

Appendix 4.1

Tables of nutrient content of feedstuffs

The following tables present data on average values of energy and nutrients for a range of feedstuffs likely to be used in organic poultry feeding (as-fed basis). Each feedstuff is listed under its International Feed Number (IFN) (Harris, 1980), and definitions of the feedstuffs taken from AAFCO (2005) or CFIA (2005, 2007).

Feedstuffs can be highly variable in composition, especially organic feedstuffs which are grown on land fertilized with manures; therefore, the values should only be used as a guide in diet formulation. In addition, some of the data quoted are old, particularly the vitamin values, since more recent data are not available. Although old, the vitamin data should be of interest since organic producers wish to maximize the use of natural ingredients. The vitamin A values assume 1667 IU vitamin A per 1 mg β-carotene (NRC, 1994). The abbreviation NA in the tables indicates that no data are available.

Most of the AME (apparent ME) values quoted in the tables have been derived using growing chickens and may not apply precisely to adult birds and other species.

Table 4A.1. Barley (IFN 4-00-549). The entire seed of the barley plant. (From CFIA, 2007.)

Component	Amount	Component	Amount
Dry matter (g/kg)	890	Trace minerals (mg/kg)	
AME (kcal/kg)	2640	Copper	7.0
AME (MJ/kg)	11.5	Iodine	0.35
Crude fibre (g/kg)	50.0	Iron	78.0
Neutral detergent fibre (g/kg)	180.0	Manganese	18.0
Acid detergent fibre (g/kg)	62.0	Selenium	0.19
Crude fat (g/kg)	19.0	Zinc	25.0
Linoleic acid (g/kg)	8.8	Vitamins (IU/kg)	
Crude protein (g/kg)	113.0	Beta-carotene (mg/kg)	4.1
Amino acids (g/kg)		Vitamin A (IU/kg)	6835
Arginine	5.4	Vitamin E (IU/kg)	7.4
Glycine + serine	9.2	Vitamins (mg/kg)	
Histidine	2.5	Biotin	0.14
Isoleucine	3.9	Choline	1034
Leucine	7.7	Folacin	0.31
Lysine	4.1	Niacin	55.0
Methionine (Met)	2.0	Pantothenic acid	8.0
Met + cystine	4.8	Pyridoxine	5.0
Phenylalanine (Phe)	5.5	Riboflavin	1.8
Phe + tyrosine	8.4	Thiamin	4.5
Threonine	3.5	Vitamins (µg/kg)	
Tryptophan	1.1	Cobalamin (vitamin B_{12})	0
Valine	5.2		
Major minerals (g/kg)			
Calcium	0.6		
Chloride	1.4		
Magnesium	1.2		
Phosphorus (total)	3.5		
Phosphorus (non-phytate)	1.7		
Potassium	4.5		
Sodium	0.4		

Table 4A.2. Buckwheat (IFN 4-00-994). The entire seed of the buckwheat (not defined by AAFCO or CFIA).

Component	Amount	Component	Amount
Dry matter (g/kg)	880	Trace minerals (mg/kg)	
AME (kcal/kg)	2670	Copper	10.0
AME (MJ/kg)	11.5	Iodine	NA
Crude fibre (g/kg)	105	Iron	44.0
Neutral detergent fibre (g/kg)	NA	Manganese	34.0
Acid detergent fibre (g/kg)	NA	Selenium	0.08
Crude fat (g/kg)	25	Zinc	24.0
Linoleic acid (g/kg)	NA	Vitamins (IU/kg)	
Crude protein (g/kg)	108	Beta-carotene (mg/kg)	NA
Amino acids (g/kg)		Vitamin A (IU/kg)	NA
Arginine	10.2	Vitamin E (IU/kg)	NA
Glycine + serine	11.2	Vitamins (mg/kg)	
Histidine	2.6	Biotin	NA
Isoleucine	3.7	Choline	440
Leucine	5.6	Folacin	0.03
Lysine	6.1	Niacin	70.0
Methionine (Met)	2.0	Pantothenic acid	12.3
Met + cystine	4.0	Pyridoxine	NA
Phenylalanine (Phe)	4.4	Riboflavin	10.6
Phe + tyrosine	6.5	Thiamin	4
Threonine	4.6	Vitamins (µg/kg)	
Tryptophan	1.9	Cobalamin (vitamin B_{12})	0
Valine	5.4		
Major minerals (g/kg)			
Calcium	0.9		
Chloride	0.4		
Magnesium	0.9		
Phosphorus (total)	3.2		
Phosphorus (non-phytate)	1.2		
Potassium	4.0		
Sodium	0.5		

Table 4A.3. Cabbage raw (IFN 2-01-046). The aerial part of the *Brassica oleracea* (Capitata group) plant (not defined by AAFCO or CFIA).

Component	Amount	Component	Amount
Dry matter (g/kg)	81.0	Trace minerals (mg/kg)	
AME (kcal/kg)	NA	Copper	0.23
AME (MJ/kg)	NA	Iodine	0.06
Crude fibre (g/kg)	11.6	Iron	5.9
Neutral detergent fibre (g/kg)	14.1	Manganese	1.59
Acid detergent fibre (g/kg)	11.5	Selenium	0.01
Crude fat (g/kg)	1.2	Zinc	1.8
Linoleic acid (g/kg)	0.26	Vitamins (IU/kg)	
Crude protein (g/kg)	18.6	Beta-carotene (mg/kg)	0.9
Amino acids (g/kg)		Vitamin A (IU/kg)	1500
Arginine	0.75	Vitamin E (IU/kg)	1.5
Glycine + serine	NA	Vitamins (mg/kg)	
Histidine	0.34	Biotin	NA
Isoleucine	0.43	Choline	NA
Leucine	0.60	Folacin	0.43
Lysine	0.67	Niacin	3
Methionine (Met)	0.14	Pantothenic acid	1.4
Met + cystine	0.28	Pyridoxine	0.96
Phenylalanine (Phe)	0.45	Riboflavin	0.4
Phe + tyrosine	0.69	Thiamin	0.61
Threonine	0.44	Vitamins (µg/kg)	
Tryptophan	0.15	Cobalamin (vitamin B_{12})	0
Valine	0.65		
Major minerals (g/kg)			
Calcium	0.67		
Chloride	0.43		
Magnesium	0.12		
Phosphorus (total)	0.15		
Phosphorus (non-phytate)	0.05		
Potassium	2.88		
Sodium	0.12		

Table 4A.4. Canola seed cooked (IFN 5-04-597). The entire seed of the species *Brassica napus* or *B. campestris*, the oil component of which contains less than 2% erucic acid and the solid component of which contains less than 30 µmol of any one or any mixture of 3-butenyl glucosinolate, 4-pentenyl glucosinolate, 2-hydroxy-3-butenyl glucosinolate and 2-hydroxy-4-pentenyl glucosinolate per g of air-dry, oil-free solid (GLC method of the Canadian Grain Commission). (From CFIA, 2007.)

Component	Amount	Component	Amount
Dry matter (g/kg)	940	Trace minerals (mg/kg)	
AME (kcal/kg)	4640	Copper	6.0
AME (MJ/kg)	19.4	Iodine	NA
Crude fibre (g/kg)	71.3	Iron	88.0
Neutral detergent fibre (g/kg)	248.0	Manganese	33.0
Acid detergent fibre (g/kg)	122.0	Selenium	0.7
Crude fat (g/kg)	397.4	Zinc	42.0
Linoleic acid (g/kg)	109.9	Vitamins (IU/kg)	
Crude protein (g/kg)	242.0	Beta-carotene (mg/kg)	NA
Amino acids (g/kg)		Vitamin A (IU/kg)	NA
Arginine	13.6	Vitamin E (IU/kg)	115
Glycine + serine	NA	Vitamins (mg/kg)	
Histidine	6.9	Biotin	0.67
Isoleucine	9.6	Choline	4185
Leucine	16.0	Folacin	1.4
Lysine	14.4	Niacin	100
Methionine (Met)	4.3	Pantothenic acid	5.9
Met + cystine	8.6	Pyridoxine	4.5
Phenylalanine (Phe)	10.2	Riboflavin	3.6
Phe + tyrosine	16.4	Thiamin	3.2
Threonine	9.9	Vitamins ((µg/kg)	
Tryptophan	2.7	Cobalamin (vitamin B_{12})	0
Valine	11.4		
Major minerals (g/kg)			
Calcium	3.9		
Chloride	0.52		
Magnesium	3.0		
Phosphorus (total)	6.4		
Phosphorus (non-phytate)	2.0		
Potassium	5.0		
Sodium	0.12		

Table 4A.5. Canola meal expeller (IFN 5-06-870). The residual product obtained after extraction of most of the oil from canola seeds by a mechanical extraction process (not defined by AAFCO or CFIA).

Component	Amount	Component	Amount
Dry matter (g/kg)	940	Trace minerals (mg/kg)	
AME (kcal/kg)	2500	Copper	6.8
AME (MJ/kg)	10.5	Iodine	0.6
Crude fibre (g/kg)	120	Iron	180.0
Neutral detergent fibre (g/kg)	244	Manganese	55.3
Acid detergent fibre (g/kg)	180	Selenium	1.0
Crude fat (g/kg)	107.0	Zinc	43.2
Linoleic acid (g/kg)	12.8	Vitamins (IU/kg)	
Crude protein (g/kg)	352	Beta-carotene (mg/kg)	NA
Amino acids (g/kg)		Vitamin A (IU/kg)	NA
Arginine	28.0	Vitamin E (IU/kg)	18.8
Glycine + serine	33.5	Vitamins (mg/kg)	
Histidine	14.0	Biotin	0.9
Isoleucine	18.0	Choline	6532
Leucine	34.0	Folacin	2.2
Lysine	20.0	Niacin	155.0
Methionine (Met)	10.0	Pantothenic acid	8.0
Met + cystine	16.0	Pyridoxine	NA
Phenylalanine (Phe)	19.0	Riboflavin	3.0
Phe + tyrosine	30.0	Thiamin	1.8
Threonine	21.0	Vitamins (µg/kg)	
Tryptophan	6.0	Cobalamin (vitamin B_{12})	0
Valine	19.0		
Major minerals (g/kg)			
Calcium	6.8		
Chloride	5.0		
Magnesium	0.5		
Phosphorus (total)	11.5		
Phosphorus (non-phytate)	4.0		
Potassium	8.3		
Sodium	0.7		

Table 4A.6. Cassava (manioc) (IFN 4-01-152). The whole root of cassava (or tapioca) chipped mechanically into small pieces and sun-dried. It must be free of sand and other debris except for that which occurs unavoidably as a result of good harvesting practices. The levels of HCN equivalent (HCN, linamarin and cyanohydrins combined) must not exceed 50 mg/kg in the complete feed and the maximum level of cassava permitted in poultry in Canada is 200 g/kg. (From CFIA, 2005.)

Component	Amount	Component	Amount
Dry matter (g/kg)	880	Trace minerals (mg/kg)	
AME (kcal/kg)	3450	Copper	NA
AME (MJ/kg)	14.4	Iodine	NA
Crude fibre (g/kg)	44.0	Iron	NA
Neutral detergent fibre (g/kg)	85.0	Manganese	NA
Acid detergent fibre (g/kg)	61.0	Selenium	NA
Crude fat (g/kg)	5.0	Zinc	NA
Linoleic acid (g/kg)	0.9	Vitamins (IU/kg)	
Crude protein (g/kg)	33.0	Beta-carotene (mg/kg)	NA
Amino acids (g/kg)		Vitamin A (IU/kg)	NA
Arginine	1.2	Vitamin E (IU/kg)	NA
Glycine + serine	NA	Vitamins (mg/kg)	
Histidine	1.2	Biotin	NA
Isoleucine	0.8	Choline	NA
Leucine	0.7	Folacin	NA
Lysine	0.2	Niacin	3.0
Methionine (Met)	0.4	Pantothenic acid	1.0
Met + cystine	0.9	Pyridoxine	1.0
Phenylalanine (Phe)	1.0	Riboflavin	0.8
Phe + tyrosine	1.7	Thiamin	1.7
Threonine	0.8	Vitamins (µg/kg)	
Tryptophan	0.2	Cobalamin (vitamin B_{12})	0
Valine	1.6		
Major minerals (g/kg)			
Calcium	2.2		
Chloride	0.2		
Magnesium	1.1		
Phosphorus (total)	1.5		
Phosphorus (non-phytate)	0.7		
Potassium	7.8		
Sodium	0.3		

Table 4A.7. Cottonseed meal expeller (IFN 5-01-609). The residual product obtained after extraction of most of the oil from cottonseeds by a mechanical extraction process. (From CFIA, 2007.)

Component	Amount	Component	Amount
Dry matter (g/kg)	920	Trace minerals (mg/kg)	
AME (kcal/kg)	2450	Copper	19.0
AME (MJ/kg)	10.3	Iodine	0.1
Crude fibre (g/kg)	119	Iron	160
Neutral detergent fibre (g/kg)	257	Manganese	23.0
Acid detergent fibre (g/kg)	160	Selenium	0.9
Crude fat (g/kg)	61.0	Zinc	64.0
Linoleic acid (g/kg)	31.5	Vitamins (IU/kg)	
Crude protein (g/kg)	424	Beta-carotene (mg/kg)	NA
Amino acids (g/kg)		Vitamin A (IU/kg)	NA
Arginine	42.6	Vitamin E (IU/kg)	35.0
Glycine + serine	40.3	Vitamins (mg/kg)	
Histidine	11.1	Biotin	0.30
Isoleucine	12.9	Choline	2753
Leucine	24.5	Folacin	1.65
Lysine	15.5	Niacin	38.0
Methionine (Met)	5.5	Pantothenic acid	10.0
Met + cystine	11.7	Pyridoxine	5.3
Phenylalanine (Phe)	19.7	Riboflavin	5.1
Phe + tyrosine	32.0	Thiamin	6.4
Threonine	13.4	Vitamins (µg/kg)	
Tryptophan	5.4	Cobalamin (vitamin B_{12})	57
Valine	17.6		
Major minerals (g/kg)			
Calcium	2.3		
Chloride	0.4		
Magnesium	5.2		
Phosphorus (total)	10.3		
Phosphorus (non-phytate)	3.2		
Potassium	13.4		
Sodium	0.4		

Table 4A.8. Distiller's grains with solubles, dried (IFN 5-02-843).The product obtained after the removal of ethyl alcohol by distillation from the yeast fermentation of a grain or a grain mixture by condensing and drying at least 75% of the solids of the resultant whole stillage by methods employed in the grain-distilling industry. The predominating grain shall be declared as the first word in the name (AAFCO, 2005) and that determines the exact IFN number (maize product in table below).

Component	Amount	Component	Amount
Dry matter (g/kg)	930	Trace minerals (mg/kg)	
AME (kcal/kg)	2480	Copper	57
AME (MJ/kg)	10.3	Iodine	NA
Crude fibre (g/kg)	78.0	Iron	257
Neutral detergent fibre (g/kg)	346	Manganese	24
Acid detergent fibre (g/kg)	163	Selenium	0.39
Crude fat (g/kg)	84	Zinc	80
Linoleic acid (g/kg)	21.5	Vitamins (IU/kg)	
Crude protein (g/kg)	277	Beta-carotene (mg/kg)	3.5
Amino acids (g/kg)		Vitamin A (IU/kg)	5835
Arginine	11.3	Vitamin E (IU/kg)	38
Glycine + serine	31.8	Vitamins (mg/kg)	
Histidine	6.9	Biotin	0.78
Isoleucine	10.3	Choline	2637
Leucine	25.7	Folacin	0.9
Lysine	6.2	Niacin	75
Methionine (Met)	5.0	Pantothenic acid	14.0
Met + cystine	10.2	Pyridoxine	8.0
Phenylalanine (Phe)	13.4	Riboflavin	8.6
Phe + tyrosine	21.7	Thiamin	2.9
Threonine	9.4	Vitamins (µg/kg)	
Tryptophan	2.5	Cobalamin (vitamin B_{12})	57
Valine	1.3		
Major minerals (g/kg)			
Calcium	2.0		
Chloride	2.0		
Magnesium	1.9		
Phosphorus (total)	8.4		
Phosphorus (non-phytate)	6.2		
Potassium	8.4		
Sodium	2.5		

Table 4A.9. Faba beans (IFN 5-09-262). The entire seed of the faba bean plant *Vicia faba*. (From CFIA, 2007.)

Component	Amount	Component	Amount
Dry matter (g/kg)	870	Trace minerals (mg/kg)	
AME (kcal/kg)	2430	Copper	11.0
AME (MJ/kg)	10.3	Iodine	NA
Crude fibre (g/kg)	73.0	Iron	75.0
Neutral detergent fibre (g/kg)	137.0	Manganese	15.0
Acid detergent fibre (g/kg)	97.0	Selenium	0.02
Crude fat (g/kg)	14.0	Zinc	42.0
Linoleic acid (g/kg)	5.8	Vitamins (IU/kg)	
Crude protein (g/kg)	254	Beta-carotene (mg/kg)	NA
Amino acids (g/kg)		Vitamin A (IU/kg)	NA
Arginine	22.8	Vitamin E (IU/kg)	5.0
Glycine + serine	21.7	Vitamins (mg/kg)	
Histidine	6.7	Biotin	0.09
Isoleucine	10.3	Choline	1670
Leucine	18.9	Folacin	4.23
Lysine	16.2	Niacin	26.0
Methionine (Met)	2.0	Pantothenic acid	3.0
Met + cystine	5.2	Pyridoxine	3.66
Phenylalanine (Phe)	10.3	Riboflavin	2.9
Phe + tyrosine	19.0	Thiamin	5.5
Threonine	8.9	Vitamins (μg/kg)	
Tryptophan	2.2	Cobalamin (vitamin B_{12})	0
Valine	11.4		
Major minerals (g/kg)			
Calcium	1.1		
Chloride	0.7		
Magnesium	1.5		
Phosphorus (total)	4.8		
Phosphorus (non-phytate)	1.9		
Potassium	12.0		
Sodium	0.3		

Table 4A.10. Flaxseed (linseed) expeller (IFN 5-02-045). The product obtained by grinding the cake or chips which remain after removal of most of the oil from flaxseed by a mechanical extraction process. It must contain no more than 100 g fibre/kg. (From AAFCO, 2005.)

Component	Amount	Component	Amount
Dry matter (g/kg)	910	Trace minerals (mg/kg)	
AME (kcal/kg)	2350	Copper	26.0
AME (MJ/kg)	9.7	Iodine	0.07
Crude fibre (g/kg)	88.0	Iron	176
Neutral detergent fibre (g/kg)	239	Manganese	38.0
Acid detergent fibre (g/kg)	138	Selenium	0.81
Crude fat (g/kg)	54.0	Zinc	33.0
Linoleic acid (g/kg)	11.0	Vitamins (IU/kg)	
Crude protein (g/kg)	343	Beta-carotene (mg/kg)	0
Amino acids (g/kg)		Vitamin A (IU/kg)	0
Arginine	28.1	Vitamin E (IU/kg)	8.0
Glycine + serine	19.9	Vitamins (mg/kg)	
Histidine	6.5	Biotin	0.33
Isoleucine	16.9	Choline	1780
Leucine	19.2	Folacin	2.80
Lysine	11.8	Niacin	37.0
Methionine (Met)	5.8	Pantothenic acid	14.3
Met + cystine	11.9	Pyridoxine	5.5
Phenylalanine (Phe)	13.5	Riboflavin	3.2
Phe + tyrosine	23.4	Thiamin	4.2
Threonine	11.4	Vitamins (µg/kg)	
Tryptophan	5.0	Cobalamin (vitamin B_{12})	0
Valine	16.1		
Major minerals (g/kg)			
Calcium	4.1		
Chloride	0.4		
Magnesium	5.8		
Phosphorus (total)	8.7		
Phosphorus (non-phytate)	2.7		
Potassium	12.2		
Sodium	1.1		

Table 4A.11. Grass meal dehydrated (IFN 1-02-211). The product of drying and grinding grass which has been cut before formation of the seed. If a species name is used, the produce must correspond thereto. (From AAFCO, 2005.)

Component	Amount	Component	Amount
Dry matter (g/kg)	917	Trace minerals (mg/kg)	
AME (kcal/kg)	1,390	Copper	6.7
AME (MJ/kg)	5.8	Iodine	0.14
Crude fibre (g/kg)	209	Iron	525
Neutral detergent fibre (g/kg)	525	Manganese	53.0
Acid detergent fibre (g/kg)	254	Selenium	0.05
Crude fat (g/kg)	34.2	Zinc	19.0
Linoleic acid (g/kg)	2.9	Vitamins (IU/kg)	
Crude protein (g/kg)	182.2	Beta-carotene (mg/kg)	35.8
Amino acids (g/kg)		Vitamin A (IU/kg)	59,678
Arginine	7.5	Vitamin E (IU/kg)	150
Glycine + serine	13.7	Vitamins (mg/kg)	
Histidine	2.7	Biotin	0.22
Isoleucine	5.6	Choline	1,470
Leucine	12.1	Folacin	NA
Lysine	7.1	Niacin	74.0
Methionine (Met)	3.1	Pantothenic acid	15.4
Met + cystine	5.0	Pyridoxine	11.7
Phenylalanine (Phe)	7.1	Riboflavin	15.5
Phe + tyrosine	11.6	Thiamin	12.6
Threonine	6.2	Vitamins (µg/kg)	
Tryptophan	3.1	Cobalamin (vitamin B_{12})	0
Valine	7.0		
Major minerals (g/kg)			
Calcium	6.3		
Chloride	8.0		
Magnesium	1.74		
Phosphorus (total)	3.48		
Phosphorus (non-phytate)	3.2		
Potassium	25.1		
Sodium	2.84		

Table 4A.12. Herring meal (IFN 5-02-000). Fishmeal is the clean, dried, ground tissue of undecomposed whole fish or fish cuttings, either or both, with or without the extraction of part of the oil. If it bears a name descriptive of its type, it must correspond thereto. (From AAFCO, 2005.)

Component	Amount	Component	Amount
Dry matter (g/kg)	925	Trace minerals (mg/kg)	
AME (kcal/kg)	3190	Copper	6.0
AME (MJ/kg)	13.4	Iodine	2.0
Crude fibre (g/kg)	0	Iron	181.0
Neutral detergent fibre (g/kg)	0	Manganese	8.0
Acid detergent fibre (g/kg)	0	Selenium	1.93
Crude fat (g/kg)	92.0	Zinc	132.0
Linoleic acid (g/kg)	1.5	Vitamins (IU/kg)	
Crude protein (g/kg)	681	Beta-carotene (mg/kg)	0
Amino acids (g/kg)		Vitamin A (IU/kg)	0
Arginine	40.1	Vitamin E (IU/kg)	15.0
Glycine + serine	70.5	Vitamins (mg/kg)	
Histidine	15.2	Biotin	0.13
Isoleucine	29.1	Choline	5306
Leucine	52.0	Folacin	0.37
Lysine	54.6	Niacin	93.0
Methionine (Met)	20.4	Pantothenic acid	17.0
Met + cystine	28.2	Pyridoxine	4.8
Phenylalanine (Phe)	27.5	Riboflavin	9.9
Phe + tyrosine	49.3	Thiamin	0.4
Threonine	30.2	Vitamins (µg/kg)	
Tryptophan	7.4	Cobalamin (vitamin B_{12})	403.0
Valine	34.6		
Major minerals (g/kg)			
Calcium	24.0		
Chloride	11.2		
Magnesium	1.8		
Phosphorus (total)	17.6		
Phosphorus (non-phytate)	17.0		
Potassium	10.1		
Sodium	6.1		

Table 4A.13. Hominy feed (IFN 4-03-011). A mixture of maize bran, maize germ and part of the starchy portion of either white or yellow maize kernels or mixture thereof; as produced in the manufacture of pearl hominy, hominy grits or table meal and must contain not less than 40 g crude fat/kg. If prefixed with the words 'white' or 'yellow', the product must correspond thereto. (From AAFCO, 2005.)

Component	Amount	Component	Amount
Dry matter (g/kg)	900	Trace minerals (mg/kg)	
AME (kcal/kg)	2,950	Copper	13
AME (MJ/kg)	12.1	Iodine	NA
Crude fibre (g/kg)	49.0	Iron	67
Neutral detergent fibre (g/kg)	285	Manganese	15
Acid detergent fibre (g/kg)	81.0	Selenium	0.1
Crude fat (g/kg)	67.0	Zinc	45
Linoleic acid (g/kg)	29.7	Vitamins (IU/kg)	
Crude protein (g/kg)	103	Beta-carotene (mg/kg)	9.0
Amino acids (g/kg)		Vitamin A (IU/kg)	15,003
Arginine	5.6	Vitamin E (IU/kg)	6.5
Glycine + serine	9.0	Vitamins (mg/kg)	
Histidine	2.8	Biotin	0.13
Isoleucine	3.6	Choline	1,155
Leucine	9.8	Folacin	0.21
Lysine	3.8	Niacin	47
Methionine (Met)	1.8	Pantothenic acid	8.2
Met + cystine	3.6	Pyridoxine	11.0
Phenylalanine (Phe)	4.3	Riboflavin	2.1
Phe + tyrosine	8.3	Thiamin	8.1
Threonine	4.0	Vitamins (µg/kg)	
Tryptophan	1.0	Cobalamin (vitamin B_{12})	0
Valine	5.2		
Major minerals (g/kg)			
Calcium	0.5		
Chloride	0.7		
Magnesium	2.4		
Phosphorus (total)	4.3		
Phosphorus (non-phytate)	0.6		
Potassium	6.1		
Sodium	0.8		

Table 4A.14. Kelp meal (seaweed) dehydrated or dried (IFN 1-08-073). The product resulting from drying and grinding non-toxic macroscopic marine algae (marine plants) of the families *Gelidiaceae*, *Gigartinaceae*, *Gracilariaceae*, *Solieriaceae*, *Palmariaceae*, *Bangiaceae*, *Laminariaceae*, *Lessoniaceae*, *Alariaceae*, *Fucaceae*, *Sargassaceae*, *Monostromataceae* and *Ulvaceae*. (From CFIA, 2007.)

Component	Amount	Component	Amount
Dry matter (g/kg)	930	Trace minerals (mg/kg)	
AME (kcal/kg)	700	Copper	45.0
AME (MJ/kg)	2.9	Iodine	3500
Crude fibre (g/kg)	239	Iron	444
Neutral detergent fibre (g/kg)	NA	Manganese	2.0
Acid detergent fibre (g/kg)	100	Selenium	0.4
Crude fat (g/kg)	30.0	Zinc	12.3
Linoleic acid (g/kg)	1.01	Vitamins (IU/kg)	
Crude protein (g/kg)	16.8	Beta-carotene (mg/kg)	0.7
Amino acids (g/kg)		Vitamin A (IU/kg)	1167
Arginine	0.65	Vitamin E (IU/kg)	8.7
Glycine + serine	NA	Vitamins (mg/kg)	
Histidine	0.24	Biotin	0.09
Isoleucine	0.76	Choline	1670
Leucine	0.83	Folacin	1.65
Lysine	0.82	Niacin	26.0
Methionine (Met)	0.25	Pantothenic acid	3.0
Met + cystine	1.23	Pyridoxine	0.02
Phenylalanine (Phe)	0.43	Riboflavin	2.9
Phe + tyrosine	0.69	Thiamin	5.5
Threonine	0.55	Vitamins (µg/kg)	
Tryptophan	0.48	Cobalamin (vitamin B_{12})	50.0
Valine	0.72		
Major minerals (g/kg)			
Calcium	1.68		
Chloride	1.2		
Magnesium	1.21		
Phosphorus (total)	0.42		
Phosphorus (non-phytate)	NA		
Potassium	0.9		
Sodium	0.8		

Table 4A.15. Lucerne (alfalfa) dehydrated (IFN 1-00-023). The aerial portion of the lucerne plant, which has been dried by thermal means and finely ground. (From AAFCO, 2005.)

Component	Amount	Component	Amount
Dry matter (g/kg)	920	Trace minerals (mg/kg)	
AME (kcal/kg)	1,450	Copper	10.0
AME (MJ/kg)	6.1	Iodine	NA
Crude fibre (g/kg)	240	Iron	333
Neutral detergent fibre (g/kg)	412	Manganese	32.0
Acid detergent fibre (g/kg)	302	Selenium	0.34
Crude fat (g/kg)	26.0	Zinc	24.0
Linoleic acid (g/kg)	3.5	Vitamins (IU/kg)	
Crude protein (g/kg)	170	Beta-carotene (mg/kg)	170
Amino acids (g/kg)		Vitamin A (IU/kg)	283,390
Arginine	7.1	Vitamin E (IU/kg)	49.8
Glycine + serine	15.4	Vitamins (mg/kg)	
Histidine	3.7	Biotin	0.54
Isoleucine	6.8	Choline	1,401
Leucine	12.1	Folacin	4.36
Lysine	7.4	Niacin	38.0
Methionine (Met)	2.5	Pantothenic acid	29.0
Met + cystine	4.3	Pyridoxine	6.5
Phenylalanine (Phe)	8.4	Riboflavin	13.6
Phe + tyrosine	13.9	Thiamin	3.4
Threonine	7.0	Vitamins (µg/kg)	
Tryptophan	2.4	Cobalamin (vitamin B_{12})	0
Valine	8.6		
Major minerals (g/kg)			
Calcium	13.3		
Chloride	4.7		
Magnesium	2.3		
Phosphorus (total)	2.3		
Phosphorus (non-phytate)	2.3		
Potassium	23.0		
Sodium	0.9		

Table 4A.16. Lucerne (alfalfa) sun-cured (IFN 1-00-059). The aerial portion of the lucerne plant, which has been dried by solar means and finely or coarsely ground. (From AAFCO, 2005.)

Component	Amount	Component	Amount
Dry matter (g/kg)	907	Trace minerals (mg/kg)	
AME (kcal/kg)	770	Copper	10.0
AME (MJ/kg)	3.2	Iodine	NA
Crude fibre (g/kg)	207	Iron	173
Neutral detergent fibre (g/kg)	368	Manganese	27.0
Acid detergent fibre (g/kg)	290	Selenium	0.49
Crude fat (g/kg)	32.0	Zinc	22.0
Linoleic acid (g/kg)	4.3	Vitamins (IU/kg)	
Crude protein (g/kg)	162	Beta-carotene (mg/kg)	145.8
Amino acids (g/kg)		Vitamin A (IU/kg)	243,049
Arginine	7.3	Vitamin E (IU/kg)	200
Glycine + serine	NA	Vitamins (mg/kg)	
Histidine	3.4	Biotin	0.005
Isoleucine	6.0	Choline	919
Leucine	10.7	Folacin	0.24
Lysine	8.1	Niacin	8.2
Methionine (Met)	1.9	Pantothenic acid	28.0
Met + cystine	5.0	Pyridoxine	4.4
Phenylalanine (Phe)	7.1	Riboflavin	15.0
Phe + tyrosine	11.9	Thiamin	4.2
Threonine	6.0	Vitamins (µg/kg)	
Tryptophan	1.8	Cobalamin (vitamin B_{12})	0
Valine	7.9		
Major minerals (g/kg)			
Calcium	13.7		
Chloride	3.4		
Magnesium	2.9		
Phosphorus (total)	2.0		
Phosphorus (non-phytate)	2.0		
Potassium	22.7		
Sodium	1.3		

Table 4A.17. Lupinseed meal sweet white (IFN 5-27-717). The ground whole seed of the species *Lupinus albus, L. angustifolius* or *L. luteus*. It has to contain less than 0.3 g total alkaloids/kg. The species of seed must be listed after the name 'Sweet lupin seeds, ground'. (From CFIA, 2007.)

Component	Amount	Component	Amount
Dry matter (g/kg)	890	Trace minerals (mg/kg)	
AME (kcal/kg)	2950	Copper	10.3
AME (MJ/kg)	12.2	Iodine	NA
Crude fibre (g/kg)	110	Iron	54.0
Neutral detergent fibre (g/kg)	203	Manganese	23.8
Acid detergent fibre (g/kg)	167	Selenium	0.08
Crude fat (g/kg)	97.5	Zinc	47.5
Linoleic acid (g/kg)	35.6	Vitamins (IU/kg)	
Crude protein (g/kg)	349	Beta-carotene (mg/kg)	NA
Amino acids (g/kg)		Vitamin A (IU/kg)	NA
Arginine	33.8	Vitamin E (IU/kg)	8.0
Glycine + serine	NA	Vitamins (mg/kg)	
Histidine	7.7	Biotin	0.05
Isoleucine	14.0	Choline	NA
Leucine	24.3	Folacin	3.6
Lysine	15.4	Niacin	21.9
Methionine (Met)	2.7	Pantothenic acid	7.5
Met + cystine	7.8	Pyridoxine	3.6
Phenylalanine (Phe)	12.2	Riboflavin	2.2
Phe + tyrosine	25.7	Thiamin	6.4
Threonine	12.0	Vitamins (µg/kg)	
Tryptophan	2.6	Cobalamin (vitamin B_{12})	0
Valine	10.5		
Major minerals (g/kg)			
Calcium	3.4		
Chloride	0.3		
Magnesium	1.9		
Phosphorus (total)	3.8		
Phosphorus (non-phytate)	1.6		
Potassium	11.0		
Sodium	0.2		

Table 4A.18. Maize yellow (IFN 4-02-935). The whole yellow maize kernel. (From CFIA, 2007.)

Component	Amount	Component	Amount
Dry matter (g/kg)	890	Trace minerals (mg/kg)	
AME (kcal/kg)	3,350	Copper	3.0
AME (MJ/kg)	14.2	Iodine	0.05
Crude fibre (g/kg)	26.0	Iron	29.0
Neutral detergent fibre (g/kg)	96.0	Manganese	7.0
Acid detergent fibre (g/kg)	28.0	Selenium	0.07
Crude fat (g/kg)	39.0	Zinc	18.0
Linoleic acid (g/kg)	19.2	Vitamins (IU/kg)	
Crude protein (g/kg)	83.0	Beta-carotene (mg/kg)	17.0
Amino acids (g/kg)		Vitamin A (IU/kg)	28,339
Arginine	3.7	Vitamin E (IU/kg)	17
Glycine + serine	7.0	Vitamins (mg/kg)	
Histidine	2.3	Biotin	0.06
Isoleucine	2.8	Choline	620
Leucine	9.9	Folacin	0.15
Lysine	2.6	Niacin	24.0
Methionine (Met)	1.7	Pantothenic acid	6.0
Met + cystine	3.9	Pyridoxine	5.0
Phenylalanine (Phe)	3.9	Riboflavin	1.2
Phe + tyrosine	6.4	Thiamin	3.5
Threonine	2.9	Vitamins (μg/kg)	
Tryptophan	0.6	Cobalamin (vitamin B_{12})	0
Valine	3.9		
Major minerals (g/kg)			
Calcium	0.3		
Chloride	0.5		
Magnesium	1.2		
Phosphorus (total)	2.8		
Phosphorus (non-phytate)	0.4		
Potassium	3.3		
Sodium	0.2		

Table 4A.19. Menhaden fishmeal (IFN 5-02-009). Fishmeal is the clean, dried, ground tissue of undecomposed whole fish or fish cuttings, either or both, with or without the extraction of part of the oil. If it bears a name descriptive of its type, it must correspond thereto. (From AAFCO, 2005.)

Component	Amount	Component	Amount
Dry matter (g/kg)	920	Trace minerals (mg/kg)	
AME (kcal/kg)	2820	Copper	11.0
AME (MJ/kg)	11.8	Iodine	2.0
Crude fibre (g/kg)	0	Iron	440
Neutral detergent fibre (g/kg)	0	Manganese	37.0
Acid detergent fibre (g/kg)	0	Selenium	2.1
Crude fat (g/kg)	94.0	Zinc	147
Linoleic acid (g/kg)	1.2	Vitamins (IU/kg)	
Crude protein (g/kg)	623	Beta-carotene (mg/kg)	0
Amino acids (g/kg)		Vitamin A (IU/kg)	0
Arginine	36.6	Vitamin E (IU/kg)	5.0
Glycine + serine	68.3	Vitamins (mg/kg)	
Histidine	17.8	Biotin	0.13
Isoleucine	25.7	Choline	3056
Leucine	45.4	Folacin	0.37
Lysine	48.1	Niacin	55.0
Methionine (Met)	17.7	Pantothenic acid	9.0
Met + cystine	23.4	Pyridoxine	4.0
Phenylalanine (Phe)	25.1	Riboflavin	4.9
Phe + tyrosine	45.5	Thiamin	0.5
Threonine	26.4	Vitamins (µg/kg)	
Tryptophan	6.6	Cobalamin (vitamin B_{12})	143
Valine	30.3		
Major minerals (g/kg)			
Calcium	52.1		
Chloride	5.5		
Magnesium	1.6		
Phosphorus (total)	30.4		
Phosphorus (non-phytate)	30.4		
Potassium	7.0		
Sodium	4.0		

Table 4A.20. Milk skimmed fluid (IFN 5-01-170). Defatted milk of cattle (not defined by AAFCO or CFIA).

Component	Amount	Component	Amount
Dry matter (g/kg)	91.0	Trace minerals (mg/kg)	
AME (kcal/kg)	286	Copper	0.04
AME (MJ/kg)	1.2	Iodine	0.03
Crude fibre (g/kg)	0	Iron	1.0
Neutral detergent fibre (g/kg)	0	Manganese	0.1
Acid detergent fibre (g/kg)	0	Selenium	0.05
Crude fat (g/kg)	0.9	Zinc	4.0
Linoleic acid (g/kg)	0.3	Vitamins (IU/kg)	
Crude protein (g/kg)	3.5	Beta-carotene (mg/kg)	0
Amino acids (g/kg)		Vitamin A (IU/kg)	29
Arginine	1.2	Vitamin E (IU/kg)	0.4
Glycine + serine	NA	Vitamins (mg/kg)	
Histidine	0.9	Biotin	NA
Isoleucine	1.9	Choline	135.0
Leucine	3.4	Folacin	0.05
Lysine	2.7	Niacin	0.88
Methionine (Met)	0.7	Pantothenic acid	3.29
Met + cystine	1.0	Pyridoxine	0.4
Phenylalanine (Phe)	1.8	Riboflavin	1.4
Phe + tyrosine	3.3	Thiamin	0.36
Threonine	1.5	Vitamins (µg/kg)	
Tryptophan	0.5	Cobalamin (vitamin B_{12})	3.8
Valine	2.3		
Major minerals (g/kg)			
Calcium	1.2		
Chloride	1.0		
Magnesium	0.1		
Phosphorus (total)	1.0		
Phosphorus (non-phytate)	1.0		
Potassium	1.5		
Sodium	0.6		

Table 4A.21. Milk skimmed dehydrated (IFN 5-01-175). Cattle skimmed milk dehydrated (or dried skimmed milk) is composed of the residue obtained by drying defatted milk by thermal means. (From CFIA, 2007.)

Component	Amount	Component	Amount
Dry matter (g/kg)	960	Trace minerals (mg/kg)	
AME (kcal/kg)	2950	Copper	0.1
AME (MJ/kg)	12.2	Iodine	NA
Crude fibre (g/kg)	0	Iron	0.9
Neutral detergent fibre (g/kg)	0	Manganese	0.2
Acid detergent fibre (g/kg)	0	Selenium	0.14
Crude fat (g/kg)	9.0	Zinc	4.0
Linoleic acid (g/kg)	0.1	Vitamins (IU/kg)	
Crude protein (g/kg)	346	Beta-carotene (mg/kg)	0
Amino acids (g/kg)		Vitamin A (IU/kg)	290
Arginine	12.0	Vitamin E (IU/kg)	4.1
Glycine + serine	NA	Vitamins (mg/kg)	
Histidine	8.4	Biotin	0.33
Isoleucine	13.5	Choline	1408
Leucine	34.0	Folacin	0.62
Lysine	28.0	Niacin	11.0
Methionine (Met)	8.4	Pantothenic acid	33.0
Met + cystine	11.7	Pyridoxine	4.0
Phenylalanine (Phe)	16.0	Riboflavin	19.8
Phe + tyrosine	29.0	Thiamin	3.52
Threonine	15.1	Vitamins (µg/kg)	
Tryptophan	4.4	Cobalamin (vitamin B_{12})	40.3
Valine	23.0		
Major minerals (g/kg)			
Calcium	13.1		
Chloride	10.0		
Magnesium	1.2		
Phosphorus (total)	10.0		
Phosphorus (non-phytate)	10.0		
Potassium	16.0		
Sodium	4.4		

Table 4A.22. Molasses (cane) (IFN 4-04-696). A by-product of the manufacture of sucrose from sugarcane. It must contain not less than 430 g/kg total sugars expressed as invert. (From AAFCO, 2005.)

Component	Amount	Component	Amount
Dry matter (g/kg)	710	Trace minerals (mg/kg)	
AME (kcal/kg)	2150	Copper	59.6
AME (MJ/kg)	9.6	Iodine	NA
Crude fibre (g/kg)	0	Iron	175
Neutral detergent fibre (g/kg)	0	Manganese	42.2
Acid detergent fibre (g/kg)	0	Selenium	NA
Crude fat (g/kg)	1.1	Zinc	13.0
Linoleic acid (g/kg)	0	Vitamins (IU/kg)	
Crude protein (g/kg)	40.0	Beta-carotene (mg/kg)	0
Amino acids (g/kg)		Vitamin A (IU/kg)	0
Arginine	0.2	Vitamin E (IU/kg)	4.4
Glycine + serine	NA	Vitamins (mg/kg)	
Histidine	0.1	Biotin	0.7
Isoleucine	0.3	Choline	660
Leucine	0.5	Folacin	0.1
Lysine	0.1	Niacin	45
Methionine (Met)	0.2	Pantothenic acid	39
Met + cystine	0.5	Pyridoxine	7.0
Phenylalanine (Phe)	0.2	Riboflavin	2.3
Phe + tyrosine	0.8	Thiamin	0.9
Threonine	0.6	Vitamins (µg/kg)	
Tryptophan	0.1	Cobalamin (vitamin B_{12})	0
Valine	1.2		
Major minerals (g/kg)			
Calcium	8.2		
Chloride	15.9		
Magnesium	3.5		
Phosphorus (total)	0.8		
Phosphorus (non-phytate)	0.72		
Potassium	23.8		
Sodium	9.0		

Table 4A.23. Molasses (sugarbeet) (IFN 4-00-669). A by-product of the manufacture or refining of sucrose from sugarbeets. (From CFIA, 2007.)

Component	Amount	Component	Amount
Dry matter (g/kg)	770	Trace minerals (mg/kg)	
AME (kcal/kg)	2100	Copper	13.0
AME (MJ/kg)	9.5	Iodine	NA
Crude fibre (g/kg)	0	Iron	117
Neutral detergent fibre (g/kg)	0	Manganese	10.0
Acid detergent fibre (g/kg)	0	Selenium	NA
Crude fat (g/kg)	0	Zinc	40.0
Linoleic acid (g/kg)	0	Vitamins (IU/kg)	
Crude protein (g/kg)	60.0	Beta-carotene (mg/kg)	0
Amino acids (g/kg)		Vitamin A (IU/kg)	0
Arginine	NA	Vitamin E (IU/kg)	4.0
Glycine + serine	NA	Vitamins (mg/kg)	
Histidine	0.4	Biotin	0.46
Isoleucine	NA	Choline	716.0
Leucine	NA	Folacin	NA
Lysine	0.1	Niacin	41.0
Methionine (Met)	NA	Pantothenic acid	7.0
Met + cystine	0.1	Pyridoxine	NA
Phenylalanine (Phe)	NA	Riboflavin	2.3
Phe + tyrosine	NA	Thiamin	NA
Threonine	0.4	Vitamins (µg/kg)	
Tryptophan	NA	Cobalamin (vitamin B_{12})	0
Valine	NA		
Major minerals (g/kg)			
Calcium	2.0		
Chloride	9.0		
Magnesium	2.3		
Phosphorus (total)	0.3		
Phosphorus (non-phytate)	0.2		
Potassium	47.0		
Sodium	10.0		

Table 4A.24. Oats (IFN 4-03-309). The whole seed of the oat plant. (From CFIA, 2007.)

Component	Amount	Component	Amount
Dry matter (g/kg)	890	Trace minerals (mg/kg)	
AME (kcal/kg)	2610	Copper	6.0
AME (MJ/kg)	10.9	Iodine	0.1
Crude fibre (g/kg)	108	Iron	85.0
Neutral detergent fibre (g/kg)	270	Manganese	43.0
Acid detergent fibre (g/kg)	135	Selenium	0.3
Crude fat (g/kg)	47.0	Zinc	38.0
Linoleic acid (g/kg)	26.8	Vitamins (IU/kg)	
Crude protein (g/kg)	115	Beta-carotene (mg/kg)	3.7
Amino acids (g/kg)		Vitamin A (IU/kg)	6168
Arginine	8.7	Vitamin E (IU/kg)	12.0
Glycine + serine	7.0	Vitamins (mg/kg)	
Histidine	3.1	Biotin	0.24
Isoleucine	4.8	Choline	946
Leucine	9.2	Folacin	0.30
Lysine	4.0	Niacin	19.0
Methionine (Met)	2.2	Pantothenic acid	13.0
Met + cystine	5.8	Pyridoxine	2.0
Phenylalanine (Phe)	6.5	Riboflavin	1.7
Phe + tyrosine	10.6	Thiamin	6.0
Threonine	4.4	Vitamins (µg/kg)	
Tryptophan	1.4	Cobalamin (vitamin B_{12})	0
Valine	6.60		
Major minerals (g/kg)			
Calcium	0.7		
Chloride	1.0		
Magnesium	1.6		
Phosphorus (total)	3.1		
Phosphorus (non-phytate)	0.68		
Potassium	4.2		
Sodium	0.8		

Table 4A.25. Peanut meal expeller (IFN 5-03-649). The ground residual product obtained after extraction of most of the oil from groundnut kernels by a mechanical extraction process. (From CFIA, 2007.)

Component	Amount	Component	Amount
Dry matter (g/kg)	930	Trace minerals (mg/kg)	
AME (kcal/kg)	2300	Copper	15.0
AME (MJ/kg)	9.6	Iodine	0.4
Crude fibre (g/kg)	69.0	Iron	285
Neutral detergent fibre (g/kg)	146	Manganese	39.0
Acid detergent fibre (g/kg)	91.0	Selenium	0.28
Crude fat (g/kg)	65.0	Zinc	47.0
Linoleic acid (g/kg)	17.3	Vitamins (IU/kg)	
Crude protein (g/kg)	432	Beta-carotene (mg/kg)	0
Amino acids (g/kg)		Vitamin A (IU/kg)	0
Arginine	47.9	Vitamin E (IU/kg)	3.0
Glycine + serine	48.0	Vitamins (mg/kg)	
Histidine	10.1	Biotin	0.35
Isoleucine	14.1	Choline	1848
Leucine	27.7	Folacin	0.7
Lysine	14.8	Niacin	166
Methionine (Met)	5.0	Pantothenic acid	47.0
Met + cystine	11.0	Pyridoxine	7.4
Phenylalanine (Phe)	20.2	Riboflavin	5.2
Phe + tyrosine	37.6	Thiamin	7.1
Threonine	11.6	Vitamins (µg/kg)	
Tryptophan	4.1	Cobalamin (vitamin B_{12})	0
Valine	17.0		
Major minerals (g/kg)			
Calcium	1.7		
Chloride	0.3		
Magnesium	3.3		
Phosphorus (total)	5.9		
Phosphorus (non-phytate)	2.7		
Potassium	12.0		
Sodium	0.6		

Table 4A.26. Peas, field (IFN 5-03-600). The entire seed from the field pea plant *Pisum sativum*. (From CFIA, 2007.)

Component	Amount	Component	Amount
Dry matter (g/kg)	890	Trace minerals (mg/kg)	
AME (kcal/kg)	2570	Copper	9.0
AME (MJ/kg)	10.7	Iodine	0.26
Crude fibre (g/kg)	61.0	Iron	65.0
Neutral detergent fibre (g/kg)	127	Manganese	23.0
Acid detergent fibre (g/kg)	72.0	Selenium	0.38
Crude fat (g/kg)	12.0	Zinc	23.0
Linoleic acid (g/kg)	4.7	Vitamins (IU/kg)	
Crude protein (g/kg)	228	Beta-carotene (mg/kg)	1.0
Amino acids (g/kg)		Vitamin A (IU/kg)	1667
Arginine	18.7	Vitamin E (IU/kg)	4.0
Glycine + serine	20.8	Vitamins (mg/kg)	
Histidine	5.4	Biotin	0.15
Isoleucine	8.6	Choline	547
Leucine	15.1	Folacin	0.2
Lysine	15.0	Niacin	31.0
Methionine (Met)	2.1	Pantothenic acid	18.7
Met + cystine	5.2	Pyridoxine	1.0
Phenylalanine (Phe)	9.8	Riboflavin	1.8
Phe + tyrosine	16.9	Thiamin	4.6
Threonine	7.8	Vitamins (µg/kg)	
Tryptophan	1.9	Cobalamin (vitamin B_{12})	0
Valine	9.8		
Major minerals (g/kg)			
Calcium	1.1		
Chloride	0.5		
Magnesium	1.2		
Phosphorus (total)	3.9		
Phosphorus (non-phytate)	1.7		
Potassium	10.2		
Sodium	0.4		

Table 4A.27. Potatoes cooked (IFN 4-03-787). The tuberous portion of *Solanum tuberosum*, following heating (not defined by AAFCO or CFIA).

Component	Amount	Component	Amount
Dry matter (g/kg)	222	Trace minerals (mg/kg)	
AME (kcal/kg)	798	Copper	1.75
AME (MJ/kg)	3.36	Iodine	0.05
Crude fibre (g/kg)	7.5	Iron	13.75
Neutral detergent fibre (g/kg)	NA	Manganese	1.75
Acid detergent fibre (g/kg)	26.7	Selenium	0.02
Crude fat (g/kg)	1.0	Zinc	3.75
Linoleic acid (g/kg)	0.01	Vitamins (IU/kg)	
Crude protein (g/kg)	24.5	Beta-carotene (mg/kg)	0.01
Amino acids (g/kg)		Vitamin A (IU/kg)	16.7
Arginine	1.1	Vitamin E (IU/kg)	0.13
Glycine + serine	NA	Vitamins (mg/kg)	
Histidine	0.5	Biotin	0.01
Isoleucine	4.1	Choline	206
Leucine	1.4	Folacin	0.18
Lysine	1.43	Niacin	11.9
Methionine (Met)	0.4	Pantothenic acid	3.33
Met + cystine	0.7	Pyridoxine	3.33
Phenylalanine (Phe)	1.0	Riboflavin	0.35
Phe + tyrosine	1.9	Thiamin	0.9
Threonine	0.9	Vitamins (µg/kg)	
Tryptophan	0.4	Cobalamin (vitamin B_{12})	0
Valine	1.3		
Major minerals (g/kg)			
Calcium	0.14		
Chloride	NA		
Magnesium	0.25		
Phosphorus (total)	0.7		
Phosphorus (non-phytate)	0.6		
Potassium	5.5		
Sodium	0.07		

Table 4A.28. Potato protein concentrate (IFN 5-25-392). Derived from de-starched potato juice from which the proteinaceous fraction has been precipitated by thermal coagulation followed by dehydration. (From AAFCO, 2005.)

Component	Amount	Component	Amount
Dry matter (g/kg)	910	Trace minerals (mg/kg)	
AME (kcal/kg)	2840	Copper	29.0
AME (MJ/kg)	11.7	Iodine	0.3
Crude fibre (g/kg)	5.5	Iron	450
Neutral detergent fibre (g/kg)	20.0	Manganese	5.0
Acid detergent fibre (g/kg)	NA	Selenium	1.0
Crude fat (g/kg)	20.0	Zinc	21.0
Linoleic acid (g/kg)	2.2	Vitamins, IU/kg	
Crude protein (g/kg)	755	Beta-carotene (mg/kg)	NA
Amino acids (g/kg)		Vitamin A (IU/kg)	NA
Arginine	42.7	Vitamin E (IU/kg)	35.0
Glycine + serine	NA	Vitamins (mg/kg)	
Histidine	18.2	Biotin	NA
Isoleucine	45.2	Choline	NA
Leucine	86.4	Folacin	NA
Lysine	64.0	Niacin	NA
Methionine (Met)	19.2	Pantothenic acid	NA
Met + cystine	32.0	Pyridoxine	NA
Phenylalanine (Phe)	55.2	Riboflavin	NA
Phe + tyrosine	106.6	Thiamin	NA
Threonine	48.0	Vitamins (µg/kg)	
Tryptophan	10.6	Cobalamin (vitamin B_{12})	0
Valine	56.8		
Major minerals (g/kg)			
Calcium	2.5		
Chloride	3.0		
Magnesium	0.25		
Phosphorus (total)	3.8		
Phosphorus (non-phytate)	2.8		
Potassium	0.05		
Sodium	0.05		

Table 4A.29. Rice bran (IFN 4-03-928). The pericarp or bran layer and germ of the rice, with only such quantity of hull fragments, chipped, broken or brewers' rice and calcium carbonate as is unavoidable in the regular milling of edible rice. It must contain not more than 139 g fibre/kg. (From AAFCO, 2005.)

Component	Amount	Component	Amount
Dry matter (g/kg)	874	Trace minerals (mg/kg)	
AME (kcal/kg)	2980	Copper	9.0
AME (MJ/kg)	12.5	Iodine	0.1
Crude fibre (g/kg)	130	Iron	190
Neutral detergent fibre (g/kg)	237	Manganese	228
Acid detergent fibre (g/kg)	139	Selenium	0.4
Crude fat (g/kg)	130	Zinc	30
Linoleic acid (g/kg)	41.2	Vitamins (IU/kg)	
Crude protein (g/kg)	133	Beta-carotene (mg/kg)	0
Amino acids (g/kg)		Vitamin A (IU/kg)	0
Arginine	10.0	Vitamin E (IU/kg)	35.0
Glycine + serine	12.9	Vitamins (mg/kg)	
Histidine	3.4	Biotin	0.35
Isoleucine	4.4	Choline	1135
Leucine	9.2	Folacin	220
Lysine	5.7	Niacin	293
Methionine (Met)	2.6	Pantothenic acid	23
Met + cystine	5.3	Pyridoxine	26
Phenylalanine (Phe)	5.6	Riboflavin	2.5
Phe + tyrosine	9.6	Thiamin	22.5
Threonine	4.8	Vitamins (μg/kg)	
Tryptophan	1.4	Cobalamin (vitamin B_{12})	0
Valine	6.8		
Major minerals (g/kg)			
Calcium	0.7		
Chloride	0.7		
Magnesium	9.0		
Phosphorus (total)	13.4		
Phosphorus (non-phytate)	3.1		
Potassium	15.6		
Sodium	0.3		

Table 4A.30. Rye (IFN 4-04-047). The whole grain of the rye plant. (From CFIA, 2007.)

Component	Amount	Component	Amount
Dry matter (g/kg)	880	Trace minerals (mg/kg)	
AME (kcal/kg)	2630	Copper	7.0
AME (MJ/kg)	11.0	Iodine	0.07
Crude fibre (g/kg)	22.0	Iron	60.0
Neutral detergent fibre (g/kg)	123.0	Manganese	58.0
Acid detergent fibre (g/kg)	46.0	Selenium	0.38
Crude fat (g/kg)	16.0	Zinc	31.0
Linoleic acid (g/kg)	7.6	Vitamins (IU/kg)	
Crude protein (g/kg)	118	Beta-carotene (mg/kg)	0
Amino acids (g/kg)		Vitamin A (IU/kg)	0
Arginine	5.0	Vitamin E (IU/kg)	15.0
Glycine + serine	10.1	Vitamins (mg/kg)	
Histidine	2.4	Biotin	0.08
Isoleucine	3.7	Choline	419
Leucine	6.4	Folacin	1.0
Lysine	3.8	Niacin	19.0
Methionine (Met)	1.7	Pantothenic acid	8.0
Met + cystine	3.6	Pyridoxine	2.6
Phenylalanine (Phe)	5.0	Riboflavin	1.6
Phe + tyrosine	7.6	Thiamin	3.6
Threonine	3.2	Vitamins (µg/kg)	
Tryptophan	1.2	Cobalamin (vitamin B_{12})	0
Valine	5.1		
Major minerals (g/kg)			
Calcium	0.6		
Chloride	0.3		
Magnesium	1.2		
Phosphorus (total)	3.3		
Phosphorus (non-phytate)	1.1		
Potassium	4.8		
Sodium	0.2		

Table 4A.31. Safflower meal expeller (IFN 5-04-109), partly decorticated. The ground residue obtained after extraction of most of the oil from whole safflower seeds by a mechanical extraction process. (From AAFCO, 2005.)

Component	Amount	Component	Amount
Dry matter (g/kg)	920	Trace minerals (mg/kg)	
AME (kcal/kg)	2770	Copper	9.0
AME (MJ/kg)	11.59	Iodine	NA
Crude fibre (g/kg)	135	Iron	484
Neutral detergent fibre (g/kg)	259	Manganese	40
Acid detergent fibre (g/kg)	180	Selenium	NA
Crude fat (g/kg)	76.0	Zinc	42.0
Linoleic acid (g/kg)	11.0	Vitamins (IU/kg)	
Crude protein (g/kg)	390	Beta-carotene (mg/kg)	NA
Amino acids (g/kg)		Vitamin A(IU/kg)	NA
Arginine	27.5	Vitamin E (IU/kg)	16.0
Glycine + serine	46.4	Vitamins (mg/kg)	
Histidine	9.1	Biotin	1.4
Isoleucine	16.0	Choline	2570
Leucine	24.4	Folacin	0.4
Lysine	10.4	Niacin	22
Methionine (Met)	7.9	Pantothenic acid	39.1
Met + cystine	14.9	Pyridoxine	11.3
Phenylalanine (Phe)	17.5	Riboflavin	4.0
Phe + tyrosine	27.0	Thiamin	4.5
Threonine	13.6	Vitamins (µg/kg)	
Tryptophan	6.8	Cobalamin (Vitamin B_{12})	0
Valine	21.5		
Major minerals (g/kg)			
Calcium	2.6		
Chloride	1.6		
Magnesium	10.2		
Phosphorus (total)	18.0		
Phosphorus (non-phytate)	5.0		
Potassium	7.0		
Sodium	0.5		

Table 4A.32. Sesame meal expeller (IFN 5-04-220). The ground residual product obtained after extraction of most of the oil from *Sesamum indicum* seeds by a mechanical extraction process (not defined by AAFCO or CFIA).

Component	Amount	Component	Amount
Dry matter (g/kg)	930	Trace minerals (mg/kg)	
AME (kcal/kg)	2210	Copper	34.0
AME (MJ/kg)	9.25	Iodine	0.17
Crude fibre (g/kg)	57.0	Iron	93.0
Neutral detergent fibre (g/kg)	180	Manganese	53.0
Acid detergent fibre (g/kg)	132	Selenium	0.21
Crude fat (g/kg)	75.0	Zinc	100
Linoleic acid (g/kg)	30.7	Vitamins (IU/kg)	
Crude protein (g/kg)	426	Beta-carotene (mg/kg)	0.2
Amino acids (g/kg)		Vitamin A (IU/kg)	333
Arginine	48.6	Vitamin E (IU/kg)	3.0
Glycine + serine	30.6	Vitamins (mg/kg)	
Histidine	9.8	Biotin	0.24
Isoleucine	14.7	Choline	1536
Leucine	27.4	Folacin	0.3
Lysine	10.1	Niacin	30.0
Methionine (Met)	11.5	Pantothenic acid	6.0
Met + cystine	19.7	Pyridoxine	12.5
Phenylalanine (Phe)	17.7	Riboflavin	3.6
Phe + tyrosine	32.9	Thiamin	2.8
Threonine	14.4	Vitamins (µg/kg)	
Tryptophan	5.4	Cobalamin (vitamin B_{12})	0
Valine	18.5		
Major minerals (g/kg)			
Calcium	19.0		
Chloride	0.7		
Magnesium	5.4		
Phosphorus (total)	12.2		
Phosphorus (non-phytate)	2.7		
Potassium	11.0		
Sodium	0.4		

Table 4A.33. Sorghum (milo) (IFN 4-04-444). The whole seed of the sorghum plant of the variety milo. (From CFIA, 2007.)

Component	Amount	Component	Amount
Dry matter (g/kg)	880	Trace minerals (mg/kg)	
AME (kcal/kg)	3290	Copper	5.0
AME (MJ/kg)	13.7	Iodine	0.4
Crude fibre (g/kg)	27.0	Iron	45.0
Neutral detergent fibre (g/kg)	180.0	Manganese	15.0
Acid detergent fibre (g/kg)	83.0	Selenium	0.2
Crude fat (g/kg)	29.0	Zinc	15.0
Linoleic acid (g/kg)	13.5	Vitamins (IU/kg)	
Crude protein (g/kg)	92.0	Beta-carotene (mg/kg)	NA
Amino acids (g/kg)		Vitamin A (IU/kg)	110
Arginine	3.8	Vitamin E (IU/kg)	12.1
Glycine + serine	7.1	Vitamins (mg/kg)	
Histidine	2.3	Biotin	0.26
Isoleucine	3.7	Choline	668
Leucine	12.1	Folacin	0.17
Lysine	2.2	Niacin	41.0
Methionine (Met)	1.7	Pantothenic acid	12.4
Met + cystine	3.4	Pyridoxine	5.2
Phenylalanine (Phe)	4.9	Riboflavin	1.3
Phe + tyrosine	8.4	Thiamin	3.0
Threonine	3.1	Vitamins (µg/kg)	
Tryptophan	1.0	Cobalamin (vitamin B_{12})	0
Valine	4.6		
Major minerals (g/kg)			
Calcium	0.3		
Chloride	0.9		
Magnesium	1.5		
Phosphorus (total)	2.9		
Phosphorus (non-phytate)	0.6		
Potassium	3.5		
Sodium	0.1		

Table 4A.34. Soybeans cooked (IFN 5-04-597). The product resulting from heating whole soybean seeds without removing any of the component parts. (From CFIA, 2007.)

Component	Amount	Component	Amount
Dry matter (g/kg)	900	Trace minerals (mg/kg)	
AME (kcal/kg)	3850	Copper	16.0
AME (MJ/kg)	16.1	Iodine	0.05
Crude fibre (g/kg)	43.0	Iron	80.0
Neutral detergent fibre (g/kg)	139	Manganese	30.0
Acid detergent fibre (g/kg)	80.0	Selenium	0.11
Crude fat (g/kg)	180	Zinc	39.0
Linoleic acid (g/kg)	104	Vitamins (IU/kg)	
Crude protein (g/kg)	352	Beta-carotene (mg/kg)	1.0
Amino acids (g/kg)		Vitamin A (IU/kg)	1667
Arginine	26.0	Vitamin E (IU/kg)	18.1
Glycine + serine	34.2	Vitamins (mg/kg)	
Histidine	9.6	Biotin	0.24
Isoleucine	16.1	Choline	2307
Leucine	27.5	Folacin	3.6
Lysine	22.2	Niacin	22.0
Methionine (Met)	5.3	Pantothenic acid	15.0
Met + cystine	10.8	Pyridoxine	10.8
Phenylalanine (Phe)	18.3	Riboflavin	2.6
Phe + tyrosine	31.5	Thiamin	11.0
Threonine	14.1	Vitamins (µg/kg)	
Tryptophan	4.8	Cobalamin (vitamin B_{12})	0
Valine	16.8		
Major minerals (g/kg)			
Calcium	2.5		
Chloride	0.3		
Magnesium	2.8		
Phosphorus (total)	5.9		
Phosphorus (non-phytate)	2.3		
Potassium	17.0		
Sodium	0.3		

Table 4A.35. Soybean meal expeller (IFN 5-04-600). The product obtained by grinding the cake or chips which remain after removal of most of the oil from soybeans by a mechanical extraction process. (From AAFCO, 2005.)

Component	Amount	Component	Amount
Dry matter (g/kg)	900	Trace minerals (mg/kg)	
AME (kcal/kg)	2750	Copper	22.0
AME (MJ/kg)	11.5	Iodine	0.15
Crude fibre (g/kg)	65.0	Iron	157
Neutral detergent fibre (g/kg)	150	Manganese	31.0
Acid detergent fibre (g/kg)	100	Selenium	0.10
Crude fat (g/kg)	81.0	Zinc	60.0
Linoleic acid (g/kg)	27.9	Vitamins (IU/kg)	
Crude protein (g/kg)	420	Beta-carotene (mg/kg)	NA
Amino acids (g/kg)		Vitamin A (IU/kg)	NA
Arginine	30.7	Vitamin E (IU/kg)	7.0
Glycine + serine	NA	Vitamins (mg/kg)	
Histidine	11.4	Biotin	0.33
Isoleucine	26.3	Choline	2623
Leucine	36.2	Folacin	6.4
Lysine	27.9	Niacin	31.0
Methionine (Met)	6.5	Pantothenic acid	14.3
Met + cystine	12.2	Pyridoxine	5.5
Phenylalanine (Phe)	22.0	Riboflavin	3.4
Phe + tyrosine	37.5	Thiamin	3.9
Threonine	17.2	Vitamins (µg/kg)	
Tryptophan	6.1	Cobalamin (vitamin B_{12})	0
Valine	22.8		
Major minerals (g/kg)			
Calcium	2.0		
Chloride	0.7		
Magnesium	2.5		
Phosphorus (total)	6.1		
Phosphorus (non-phytate)	2.3		
Potassium	17.9		
Sodium	0.3		

Table 4A.36. Sunflower meal dehulled expeller (IFN 5-30-033). The meal obtained after the removal of most of the oil from sunflower seeds without hulls by a mechanical extraction process. (From CFIA, 2007.)

Component	Amount	Component	Amount
Dry matter (g/kg)	930	Trace minerals (mg/kg)	
AME (kcal/kg)	2350	Copper	4.0
AME (MJ/kg)	9.8	Iodine	0.6
Crude fibre (g/kg)	122	Iron	31.0
Neutral detergent fibre (g/kg)	263	Manganese	21.0
Acid detergent fibre (g/kg)	174	Selenium	0.6
Crude fat (g/kg)	80.0	Zinc	50.6
Linoleic acid (g/kg)	19.0	Vitamins (IU/kg)	
Crude protein (g/kg)	414	Beta-carotene (mg/kg)	0.3
Amino acids (g/kg)		Vitamin A (IU/kg)	500
Arginine	34.5	Vitamin E (IU/kg)	12.0
Glycine + serine	NA	Vitamins (mg/kg)	
Histidine	9.0	Biotin	1.4
Isoleucine	17.6	Choline	2500
Leucine	24.7	Folacin	2.3
Lysine	16.1	Niacin	200
Methionine (Met)	9.4	Pantothenic acid	5.9
Met + cystine	16.3	Pyridoxine	12.5
Phenylalanine (Phe)	18.0	Riboflavin	3.4
Phe + tyrosine	28.0	Thiamin	33.9
Threonine	13.7	Vitamins (µg/kg)	
Tryptophan	5.0	Cobalamin (vitamin B_{12})	0
Valine	20.1		
Major minerals (g/kg)			
Calcium	3.9		
Chloride	1.9		
Magnesium	7.2		
Phosphorus (total)	10.6		
Phosphorus (non-phytate)	2.7		
Potassium	10.6		
Sodium	2.2		

Table 4A.37. Swedes (IFN 4-04-001). Root crop of *Brassica napus* (*Napobrassica* group) (not defined by AAFCO or CFIA).

Component	Amount	Component	Amount
Dry matter (g/kg)	103.4	Trace minerals (mg/kg)	
AME (kcal/kg)	NA	Copper	0.47
AME (MJ/kg)	NA	Iodine	NA
Crude fibre (g/kg)	12.0	Iron	6.1
Neutral detergent fibre (g/kg)	NA	Manganese	2.0
Acid detergent fibre (g/kg)	NA	Selenium	0.1
Crude fat (g/kg)	NA	Zinc	3.5
Linoleic acid (g/kg)	0.35	Vitamins (IU/kg)	
Crude protein (g/kg)	12.0	Beta-carotene (mg/kg)	0.01
Amino acids (g/kg)		Vitamin A (IU/kg)	17
Arginine	1.74	Vitamin E (IU/kg)	3.0
Glycine + serine	NA	Vitamins (mg/kg)	
Histidine	0.35	Biotin	NA
Isoleucine	0.59	Choline	NA
Leucine	0.45	Folacin	0.25
Lysine	0.46	Niacin	8.24
Methionine (Met)	0.12	Pantothenic acid	1.88
Met + cystine	0.27	Pyridoxine	1.18
Phenylalanine (Phe)	0.36	Riboflavin	0.47
Phe + tyrosine	0.64	Thiamin	1.06
Threonine	0.54	Vitamins (μg/kg)	
Tryptophan	0.15	Cobalamin (vitamin B_{12})	0
Valine	0.56		
Major minerals (g/kg)			
Calcium	0.55		
Chloride	NA		
Magnesium	0.27		
Phosphorus (total)	0.68		
Phosphorus (non-phytate)	NA		
Potassium	3.96		
Sodium	0.24		

Table 4A.38. Triticale (IFN 4-20-362). The entire seed of the triticale plant *Triticale hexaploide*. (From CFIA, 2007.)

Component	Amount	Component	Amount
Dry matter (g/kg)	900	Trace minerals (mg/kg)	
AME (kcal/kg)	3160	Copper	8.0
AME (MJ/kg)	12.7	Iodine	NA
Crude fibre (g/kg)	21.6	Iron	31.0
Neutral detergent fibre (g/kg)	127	Manganese	43.0
Acid detergent fibre (g/kg)	38.0	Selenium	0.5
Crude fat (g/kg)	18.0	Zinc	32.0
Linoleic acid (g/kg)	7.1	Vitamins (IU/kg)	
Crude protein (g/kg)	125	Beta-carotene (mg/kg)	NA
Amino acids (g/kg)		Vitamin A (IU/kg)	NA
Arginine	5.7	Vitamin E (IU/kg)	9.0
Glycine + serine	10.0	Vitamins (mg/kg)	
Histidine	2.6	Biotin	1.0
Isoleucine	3.9	Choline	462
Leucine	7.6	Folacin	0.73
Lysine	3.9	Niacin	14.3
Methionine (Met)	2.0	Pantothenic acid	13.23
Met + cystine	4.6	Pyridoxine	1.38
Phenylalanine (Phe)	4.9	Riboflavin	1.34
Phe + tyrosine	8.1	Thiamin	4.16
Threonine	3.6	Vitamins (µg/kg)	
Tryptophan	1.4	Cobalamin (vitamin B_{12})	0
Valine	5.1		
Major minerals (g/kg)			
Calcium	0.7		
Chloride	0.3		
Magnesium	1.0		
Phosphorus (total)	3.3		
Phosphorus (non-phytate)	1.4		
Potassium	4.6		
Sodium	0.3		

Table 4A.39. Wheat (IFN 4-050-211). The whole seed of the wheat plant. (From CFIA, 2007.)

Component	Amount	Component	Amount
Dry matter (g/kg)	880	Trace minerals (mg/kg)	
AME (kcal/kg)	3160	Copper	6.0
AME (MJ/kg)	12.9	Iodine	0.04
Crude fibre (g/kg)	26.0	Iron	39.0
Neutral detergent fibre (g/kg)	135	Manganese	34.0
Acid detergent fibre (g/kg)	40.0	Selenium	0.33
Crude fat (g/kg)	20.0	Zinc	40.0
Linoleic acid (g/kg)	9.3	Vitamins (IU/kg)	
Crude protein (g/kg)	135	Beta-carotene (mg/kg)	0.4
Amino acids (g/kg)		Vitamin A (IU/kg)	667
Arginine	6.0	Vitamin E (IU/kg)	15.0
Glycine + serine	10.6	Vitamins (mg/kg)	
Histidine	3.2	Biotin	0.11
Isoleucine	4.1	Choline	778
Leucine	8.6	Folacin	0.22
Lysine	3.4	Niacin	48.0
Methionine (Met)	2.0	Pantothenic acid	9.9
Met + cystine	4.9	Pyridoxine	3.4
Phenylalanine (Phe)	6.0	Riboflavin	1.4
Phe + tyrosine	9.8	Thiamin	4.5
Threonine	3.7	Vitamins (µg/kg)	
Tryptophan	1.5	Cobalamin (vitamin B_{12})	0
Valine	5.4		
Major minerals (g/kg)			
Calcium	0.6		
Chloride	0.6		
Magnesium	1.3		
Phosphorus (total)	3.7		
Phosphorus (non-phytate)	1.8		
Potassium	4.9		
Sodium	0.1		

Table 4A.40. Wheat middlings (IFN 4-05-205). Consists of the fine bran particles, germ and a small proportion of floury endosperm particles as separated in the usual processes of commercial flour milling (From CFIA, 2007). It has to contain less than 95 g crude fibre/kg.

Component	Amount	Component	Amount
Dry matter (g/kg)	890	Trace minerals (mg/kg)	
AME (kcal/kg)	2200	Copper	10.0
AME (MJ/kg)	9.0	Iodine	NA
Crude fibre (g/kg)	73.0	Iron	84.0
Neutral detergent fibre (g/kg)	356	Manganese	100.0
Acid detergent fibre (g/kg)	107	Selenium	0.72
Crude fat (g/kg)	42.0	Zinc	92.0
Linoleic acid (g/kg)	17.4	Vitamins (IU/kg)	
Crude protein (g/kg)	159	Beta-carotene (mg/kg)	3.0
Amino acids (g/kg)		Vitamin A (IU/kg)	5000
Arginine	9.7	Vitamin E (IU/kg)	20.1
Glycine + serine	13.8	Vitamins (mg/kg)	
Histidine	4.4	Biotin	0.33
Isoleucine	5.3	Choline	1187
Leucine	10.6	Folacin	0.76
Lysine	5.7	Niacin	72.0
Methionine (Met)	2.6	Pantothenic acid	15.6
Met + cystine	5.8	Pyridoxine	9.0
Phenylalanine (Phe)	7.0	Riboflavin	1.8
Phe + tyrosine	9.9	Thiamin	16.5
Threonine	5.1	Vitamins (µg/kg)	
Tryptophan	2.0	Cobalamin (vitamin B_{12})	0
Valine	7.5		
Major minerals (g/kg)			
Calcium	1.2		
Chloride	0.4		
Magnesium	4.1		
Phosphorus (total)	9.3		
Phosphorus (non-phytate)	3.8		
Potassium	10.6		
Sodium	0.5		

Table 4A.41. Wheat shorts (IFN 4-05-201). The fine particles of wheat bran, wheat germ, wheat flour and the offal from the 'tail of the mill'. This product must be obtained in the usual process of commercial milling and must contain not more than 70 g fibre/kg. (From AAFCO, 2005.)

Component	Amount	Component	Amount
Dry matter (g/kg)	880	Trace minerals (mg/kg)	
AME (kcal/kg)	2750	Copper	12
AME (MJ/kg)	11.3	Iodine	NA
Crude fibre (g/kg)	69.0	Iron	100
Neutral detergent fibre (g/kg)	284	Manganese	89.0
Acid detergent fibre (g/kg)	86.0	Selenium	0.75
Crude fat (g/kg)	46.0	Zinc	100
Linoleic acid (g/kg)	19.0	Vitamins (IU/kg)	
Crude protein (g/kg)	165	Beta-carotene (mg/kg)	NA
Amino acids (g/kg)		Vitamin A (IU/kg)	NA
Arginine	10.7	Vitamin E (IU/kg)	54.0
Glycine + serine	17.3	Vitamins (mg/kg)	
Histidine	4.3	Biotin	0.24
Isoleucine	5.8	Choline	1170
Leucine	10.2	Folacin	1.4
Lysine	7.0	Niacin	107
Methionine (Met)	2.5	Pantothenic acid	22.3
Met + cystine	5.3	Pyridoxine	7.2
Phenylalanine (Phe)	7.0	Riboflavin	3.3
Phe + tyrosine	12.1	Thiamin	18.1
Threonine	5.7	Vitamins (µg/kg)	
Tryptophan	2.2	Cobalamin (vitamin B_{12})	0
Valine	8.7		
Major minerals (g/kg)			
Calcium	1.0		
Chloride	0.4		
Magnesium	2.5		
Phosphorus (total)	8.6		
Phosphorus (non-phytate)	1.7		
Potassium	10.6		
Sodium	0.2		

Table 4A.42. Whey sweet fluid (IFN 4-01-134). Cattle whey fresh (or whey or liquid whey) is the product obtained as a fluid by separating the coagulum from milk, cream, skimmed milk or cheese. It has to be labelled with the following statement in English or French: 'This product is free of antimicrobial activity and is not a source of viable microbial cells'. (From CFIA, 2007.)

Component	Amount	Component	Amount
Dry matter (g/kg)	69.0	Trace minerals (mg/kg)	
AME (kcal/kg)	NA	Copper	0.04
AME (MJ/kg)	NA	Iodine	NA
Crude fibre (g/kg)	0	Iron	0.6
Neutral detergent fibre (g/kg)	0	Manganese	0.01
Acid detergent fibre (g/kg)	0	Selenium	0.02
Crude fat (g/kg)	3.6	Zinc	1.3
Linoleic acid (g/kg)	0.08	Vitamins (IU/kg)	
Crude protein (g/kg)	15.0	Beta-carotene (mg/kg)	0
Amino acids (g/kg)		Vitamin A (IU/kg)	120.0
Arginine	0.23	Vitamin E (IU/kg)	0
Glycine + serine	NA	Vitamins (mg/kg)	
Histidine	0.18	Biotin	0.04
Isoleucine	0.52	Choline	127.7
Leucine	0.90	Folacin	0.01
Lysine	1.1	Niacin	0.74
Methionine (Met)	0.15	Pantothenic acid	3.83
Met + cystine	0.5	Pyridoxine	0.31
Phenylalanine (Phe)	0.3	Riboflavin	1.58
Phe + tyrosine	0.54	Thiamin	0.36
Threonine	0.8	Vitamins (μg/kg)	
Tryptophan	0.21	Cobalamin (vitamin B_{12})	2.8
Valine	0.46		
Major minerals (g/kg)			
Calcium	0.47		
Chloride	NA		
Magnesium	0.08		
Phosphorus (total)	0.46		
Phosphorus (non-phytate)	0.45		
Potassium	1.61		
Sodium	0.54		

Table 4A.43. Whey sweet dehydrated (IFN 4-01-182). Cattle whey dehydrated (or dried whey) consists of the residue remaining after the drying of or evaporating of whey by thermal means. (From CFIA, 2007.)

Component	Amount	Component	Amount
Dry matter (g/kg)	960	Trace minerals (mg/kg)	
AME (kcal/kg)	1900	Copper	13.0
AME (MJ/kg)	8.0	Iodine	NA
Crude fibre (g/kg)	0	Iron	130
Neutral detergent fibre (g/kg)	0	Manganese	3.0
Acid detergent fibre (g/kg)	0	Selenium	0.12
Crude fat (g/kg)	9.0	Zinc	10.0
Linoleic acid (g/kg)	0.1	Vitamins (IU/kg)	
Crude protein (g/kg)	121	Beta-carotene (mg/kg)	NA
Amino acids (g/kg)		Vitamin A (IU/kg)	NA
Arginine	2.6	Vitamin E (IU/kg)	0.3
Glycine + serine	6.4	Vitamins (mg/kg)	
Histidine	2.3	Biotin	0.27
Isoleucine	6.2	Choline	1820
Leucine	10.8	Folacin	0.85
Lysine	9.0	Niacin	10.0
Methionine (Met)	1.7	Pantothenic acid	47.0
Met + cystine	4.2	Pyridoxine	4.0
Phenylalanine (Phe)	3.6	Riboflavin	27.1
Phe + tyrosine	6.1	Thiamin	4.1
Threonine	7.2	Vitamins (µg/kg)	
Tryptophan	1.8	Cobalamin (vitamin B_{12})	23.0
Valine	6.0		
Major minerals (g/kg)			
Calcium	7.5		
Chloride	14.0		
Magnesium	1.3		
Phosphorus (total)	7.2		
Phosphorus (non-phytate)	7.0		
Potassium	19.6		
Sodium	9.4		

Table 4A.44. White fishmeal (IFN 5-00-025). Fishmeal is the clean, dried, ground tissue of undecomposed whole fish or fish cuttings, either or both, with or without the extraction of part of the oil (AAFCO, 2005). White fishmeal is a product containing not more than 60 g fat/kg and not more than 40 g salt/kg, obtained from white fish or white fish waste such as filleting offal. For organic feeding the fat should be extracted by mechanical means.

Component	Amount	Component	Amount
Dry matter (g/kg)	910	Trace minerals (mg/kg)	
AME (kcal/kg)	2820	Copper	8.0
AME (MJ/kg)	11.8	Iodine	2.0
Crude fibre (g/kg)	0	Iron	80.0
Neutral detergent fibre (g/kg)	0	Manganese	10.0
Acid detergent fibre (g/kg)	0	Selenium	1.5
Crude fat (g/kg)	48.0	Zinc	80.0
Linoleic acid (g/kg)	0.8	Vitamins (IU/kg)	
Crude protein (g/kg)	633	Beta-carotene (mg/kg)	0
Amino acids (g/kg)		Vitamin A (IU/kg)	0
Arginine	42.0	Vitamin E (IU/kg)	5.6
Glycine + serine	68.3	Vitamins (mg/kg)	
Histidine	19.3	Biotin	0.14
Isoleucine	31.0	Choline	4050
Leucine	45.0	Folacin	0.3
Lysine	43.0	Niacin	38.0
Methionine (Met)	16.5	Pantothenic acid	4.7
Met + cystine	24.0	Pyridoxine	5.9
Phenylalanine (Phe)	28.0	Riboflavin	4.6
Phe + tyrosine	45.5	Thiamin	1.5
Threonine	26.0	Vitamins (µg/kg)	
Tryptophan	7.0	Cobalamin (vitamin B_{12})	71.0
Valine	32.5		
Major minerals (g/kg)			
Calcium	70.0		
Chloride	5.0		
Magnesium	2.2		
Phosphorus (total)	35.0		
Phosphorus (non-phytate)	35.0		
Potassium	11.0		
Sodium	9.7		

Table 4A.45. Yeast brewer's dehydrated (IFN 7-05-527). The dried, non-fermentative, non-extracted yeast produced from an unmodified strain of the botanical classification *Saccharomyces* resulting as a by-product from the brewing of beer and ale. It has to be labelled with the following statement in English or French: 'This product is not a source of viable *Saccharomyces* cells'. (From CFIA, 2007.)

Component	Amount	Component	Amount
Dry matter (g/kg)	930	Trace minerals (mg/kg)	
AME (kcal/kg)	1990	Copper	33.0
AME (MJ/kg)	8.32	Iodine	0.02
Crude fibre (g/kg)	29.0	Iron	215
Neutral detergent fibre (g/kg)	40.0	Manganese	8.0
Acid detergent fibre (g/kg)	30.0	Selenium	1.0
Crude fat (g/kg)	17.0	Zinc	49.0
Linoleic acid (g/kg)	0.4	Vitamins (IU/kg)	
Crude protein (g/kg)	459	Beta-carotene (mg/kg)	NA
Amino acids (g/kg)		Vitamin A (IU/kg)	NA
Arginine	22.0	Vitamin E (IU/kg)	2.0
Glycine + serine	NA	Vitamins (mg/kg)	
Histidine	10.9	Biotin	0.63
Isoleucine	21.5	Choline	3984
Leucine	31.3	Folacin	9.9
Lysine	32.2	Niacin	448
Methionine (Met)	7.4	Pantothenic acid	109
Met + cystine	12.4	Pyridoxine	42.8
Phenylalanine (Phe)	18.3	Riboflavin	37.0
Phe + tyrosine	33.8	Thiamin	91.8
Threonine	22.0	Vitamins (µg/kg)	
Tryptophan	5.6	Cobalamin (vitamin B_{12})	1.0
Valine	23.9		
Major minerals (g/kg)			
Calcium	1.6		
Chloride	1.2		
Magnesium	2.3		
Phosphorus (total)	14.4		
Phosphorus (non-phytate)	14.0		
Potassium	18.0		
Sodium	1.0		

5

Diets for Organic Poultry Production

In general, the following is suggested as the minimum range of diets necessary for feeding flocks of organic poultry. Additional diets will be required for breeding flocks of chickens, turkeys and waterfowl (See table at bottom of the page).

A key aim of organic farming is environmental sustainability. Consequently organic producers wish to provide most or all of their required inputs, including feed. However, this is not possible on small farms, and even larger farms that may produce some of the feedstuffs required may not have the necessary mixing equipment to allow adequate diets to be prepared on-site. Farms with a land base sufficient for the growing of a variety of crops may be able to mix diets on-site or in a cooperative mill.

Farms Producing No Feed Ingredients

These farms have to rely on purchased complete feeds. The feeds should be obtained from reputable manufacturers, and producers can use the information provided in this book to help them set the dietary specifications with the feed manufacturer. The purchased feed can be manufactured according to specifications set by the feed manufacturer, by the customer or by a consultant. Most feed manufacturers are willing to provide diets tailored to the wishes of customers and may even prepare mixtures according to formulas supplied by the customer. The feeds should be ordered regularly and not stored on the farm for extended periods.

One of the advantages of purchasing a complete feed is that the label provides useful information, including a list of ingredients (in some countries a complete formula) and a guaranteed analysis. The following information is required by law on commercial poultry feed labels (complete feeds and supplements) in North America:

1. The net weight of the batch.
2. The product name and the brand name under which the commercial feed is distributed.

Laying hens	Meat chickens	Meat turkeys	Ducks/Geese	Quail	Ostriches	Emus
Chick starter	Starter	Starter	Starter/grower	Starter/grower	Starter	Starter
Pullet grower	Grower	Grower			Grower	Grower
Layer	Finisher	Finisher			Finisher	Finisher

3. A guaranteed analysis relating to certain nutrients: crude protein (CP), lysine, methionine, crude fat, crude fibre, calcium, phosphorus, sodium (or salt), selenium and zinc). A changeover from crude fibre to the more informative neutral-detergent fibre and acid-detergent fibre is planned once the analytical methods have been approved by the regulatory authorities for universal application in the feed industry.

4. The common or usual name of each ingredient used in the manufacture of the feed. In some jurisdictions the use of a collective term for a group of ingredients that perform similar functions is permitted or the regulations may allow a statement that the feedstuffs used are from an approved list. This is a main difference in labelling from that in some European countries where the exact formula used is provided.

5. The name and principal mailing address of the manufacturer or the person responsible for distributing the feed.

6. Directions for use.

7. Precautionary statements required for the safe and effective use of the feed.

The commercial feed label provides some useful information to the poultry producer but its limitations should be recognized. The metabolizable energy (ME) value is not always stated and cannot be calculated from the label, except that the crude fibre content provides some indication on whether low-energy ingredients may have been used, and the crude fat content provides some information on whether the feed is likely to be low or high in energy. The recent addition of lysine and methionine to the label is a welcome move. The explanation for the apparent omission of other information on the label is that the guarantees must be supported by laboratory methods approved by the regulatory authorities. To date these methods have not yet been agreed for universal application in the feed industry in Canada and the USA. One difference between the European and North American feed labels is that the label in Europe is required to list the contents of feedstuffs included, in descending order of weight.

Producers who purchase complete feeds from a feed manufacturer will avoid the need to mix their own feeds but should adopt the relevant sections of the quality control programme outlined below.

Farms with Grain Available for Use in Poultry Feeding

Farms with an adequate supply of grains but no protein crops can purchase a 'supplement' (sometimes called a 'concentrate') which provides all of the nutrients lacking in the grains. It should be purchased with an accompanying mixing guide. In North America the label information provided with the purchased supplement is similar to that provided above for complete feed. Various supplements are sold by feed manufacturers and compounders, though perhaps only a single poultry supplement may be available and is expected to be used with all classes of birds. Use of supplement is therefore a compromise between convenience and accuracy in formulation. Any supplement purchased would have to comply with organic standards.

Table 5.1 provides an example formula for a supplement (320 g CP/kg) that is designed for use with laying hens.

One of the benefits of using a supplement is that it can be incorporated easily into a feeding programme based on the use of whole (unground) grain. This avoids the need for the farm to purchase mixing equipment. More information on whole grain feeding is given in Chapter 7.

Table 5.1. Example composition of a layer supplement (concentrate) (from Blair *et al.*, 1973).

Ingredients	g/kg (90% DM basis)
Fishmeal	170
Soybean meal	350
Grass meal	177
Ground limestone	230
Dicalcium phosphate	47
Salt (NaCl)	6.0
Trace mineral premix	10.0
Vitamin premix	10.0

Farms with Grain and Protein Feedstuffs Available for Farm Use

Many producers wish to mix their own diets on-farm so that they have complete control over the formula (Fig. 5.1). This is particularly relevant when one or more protein feedstuffs is available on-farm in addition to a supply of suitable grain. In this case it is necessary only to purchase a 'premix', examples of which are shown later in this chapter.

The section below explains the principles of feed formulation and manufacture for those organic producers who choose to mix their feeds on-farm.

Steps in Feed Manufacture On-farm

Feed manufacturing (or compounding) is the process of converting ingredients into balanced diets. It is carried out with equipment to process feedstuffs, making them suitable for mixing in the correct proportions to produce a finished dietary mixture. Often the mixed feed is then pelleted, or it may be fed to the birds as a meal-type (mash) feed.

Fig. 5.1. A small mix-mill for on-farm use.

Feed formulas: chickens

Some producers wish to use formulas devised by others. Unfortunately there is no readily available source of satisfactory feed formulas that can be recommended to organic producers. However, some formulas are available in the published literature from several countries, examples of which are shown below. In some cases these provide details of the production results when the diets were fed to poultry. Some of the formulas provide information on analysed as well as calculated nutrient data, but only the calculated values have been included below since these values are of main interest for formulation purposes.

Laying chickens

Lampkin (1997) described a typical feeding programme in the UK for laying stock as being a chick starter for 8–10 weeks, followed by a grower diet up to 10 days before the onset of lay. For the first 40 weeks of lay a high-protein diet (180 g CP/kg) is fed, followed by a lower-protein diet (160 g CP/kg) for the remainder of the laying period. Examples of dietary formulations for these periods designed to meet prevailing standards for UKROFS, for EU and for EU with no supplemental amino acids are shown in Tables 5.2 and 5.3.

These formulations illustrate an important point: the exclusion of pure amino acids from organic diets usually requires an excessive level of protein to be included in the diet (in the starter diet example, 250 g/kg instead of 201 g or 211 g when pure amino acids can be used). This has a marked effect on the cost of the diet. In addition, it results in an inefficient use of scarce protein supplies as well as contributing to an increased output of nitrogen in manure, a potential environmental hazard.

Bennett (2013) also published examples of organic diets designed for small-scale producers, with and without supplementation with pure amino acids (Fig. 5.2; Table 5.4). Such formulas are very useful in allowing producers to select the appropriate type of formula to adopt in order to comply with

Table 5.2. Example diets and nutrient specifications for organic pullet feeding in the UK (from Lampkin, 1997).

Ingredients (g/kg)	Starter			Pullet grower	
	UKROFS	EU incl. amino acids	EU excl. amino acids	UKROFS	EU excl. amino acids
Grain	400	373	228	282	299
Wheat middlings	100	100	100	300	300
Brewer's/distiller's grains	–	4.0	–	126	5.0
Peas/beans	150	150	106	18.0	101
Soybeans	178	167	317	–	–
Oilseeds	–	50	124	100	100
Dried grass/lucerne	50	50	50	100	100
Fishmeal	15	–	–	–	–
Vegetable oil	3.0	1.0	–	30.0	28.0
Yeast	39.0	36.0	18.0	3.0	15.0
Ca/P sources	30.0	33.0	27.0	16.0	19.0
Salt (NaCl)	29.0	30.0	28.0	23.0	31.0
Mineral/vitamin premix	3.0	3.0	3.0	2.0	2.0
Lysine/methionine	2.0	3.0	–	–	–
Calculated analysis (g/kg unless stated)					
CP	211	201	250	176	150
ME (MJ/kg)	11.5	11.5	11.5	11.0	11.0
Lysine	13	13	16	7.0	8.0
Methionine	6.0	6.0	5.0	3.0	3.0
Linoleic acid	17.0	18.0	22.0	29.0	29.0
Calcium	12.0	12.0	12.0	8.0	8.0
Non-phytate P	5.0	5.0	5.0	5.0	5.0

Table 5.3. Example diets and nutrient specifications for organic layer feeding in the UK (from Lampkin, 1997).

Ingredients (g/kg)	UKROFS	EU incl. amino acids	EU excl. amino acids
Grain	202	303	237
Wheat middlings	300	297	300
Brewer's/distiller's grains	63	6	–
Peas/beans	148	150	150
Soybeans	–	–	63
Dried grass/lucerne	50	50	50
Vegetable oil	77	34	36
Yeast	36	50	45
Ca/P sources	92	82	87
Salt (NaCl)	29	25	29
Mineral/vitamin premix	3	2	2
Lysine/methionine	1	1	–
Calculated analysis (g/kg unless stated)			
CP	160	160	170
ME (MJ/kg)	11	11	11
Lysine	8	8	10
Methionine	3	3	3
Linoleic acid	49	27	31
Calcium	35	35	35
Non-phytate P	5	5	5

Fig. 5.2. Well-fed broilers

Table 5.4. Example diets for organic layer feeding in Canada (from Bennett, 2013).

Ingredients (g/kg)	Diet with 160 g/kg CP	Diet with 140 g/kg CP	Soybean meal available	Soybeans and peas available
Wheat	474	561	744	526
Peas	333	327	–	220
Soybeans, cooked	77	–	–	147
Soybean meal	–	–	150	–
Ground limestone	92	92	84	83
Dicalcium phosphate	14.3	10.8	10.5	11.2
Salt (NaCl)	3.1	2.7	2.6	3.0
DL-methionine	1.6	1.5	–	–
Vitamin/mineral premix	5.0	5.0	10.0	10.0

local organic standards, especially those with home-grown supplies of grain and protein feedstuffs. This researcher observed that diets without supplementation with pure methionine were likely to result in smaller and fewer eggs and an increased incidence of stress and cannibalism in the layers. The diet with 160 g CP/kg was recommended for feeding from the onset of lay until egg production dropped to 85%, after which the diet with 140 g CP/kg was recommended.

Meat chickens

Lewis *et al.* (1997) compared the production of broilers fed to Label Rouge standards. The compositions of the diets are shown in Table 5.5.

The diets did not claim to be organic but could be formulated to organic standards quite readily. Surprisingly the diets contained the anticoccidial drug salinomycin and also the amino acid pure methionine, suggesting that these additives are permitted in the Label Rouge system.

Table 5.5. Composition and calculated nutrient content of broiler diets formulated to Label Rouge standards (from Lewis *et al.*, 1997).

Ingredients (g/kg)	Starter	Grower	Finisher
Maize	400	400	400
Wheat	220	250	300
Wheat middlings	70	110	120
Soybean hipro meal	280	210	150
Limestone flour	3.0	3.0	3.0
Dicalcium phosphate	14.0	14.0	14.0
Sodium bicarbonate	2.2	2.2	2.2
Salt (NaCl)	2.2	2.1	2.1
DL-methionine	1.0	1.0	–
Vitamin/drug premix	10.5	10.5	10.5
Calculated analysis (g/kg unless stated)			
ME (MJ/kg)	12.12	12.16	12.27
CP	202	177.5	155.9
Lysine	10.6	8.8	7.2
Methionine	3.3	2.9	2.6
Crude fibre	31.0	32.0	32.0
Ca	9.0	8.9	8.8
Non-phytate P	4.5	4.4	4.4
Sodium	1.6	1.5	1.5

Lampkin (1997) outlined a typical feeding programme in the UK for meat birds, based on diets with a reduced content of energy, protein and amino acids to result in slower growth than in conventional production. Examples of formulations are shown in Table 5.6.

Bennett (2013) published examples of organic diets designed for small-scale producers of meat-type chickens, with and without supplementation with pure amino acids (Table 5.7). The diets were formulated to provide a lower level of protein than in conventional production. This researcher observed that diets without supplementation with pure methionine contained an imbalance of amino acids, resulting in slower growth and poor feathering until the birds were 6–8 weeks of age. The programme involved roaster starter to be used for the first 4 weeks, followed by a 50:50 mixture of starter and finisher for the next 2 weeks. From then until market the birds were given finisher diet. An example formula for a simplified single diet to be fed from start to marketing and containing no supplemental amino acids was also provided.

A major challenge facing organic poultry producers worldwide is the formulation of diets containing adequate levels of methionine from natural sources. However, producers with access to adequate supplies of organic soybean and fishmeals do not face this problem, as was demonstrated in a study reported by Rack *et al.* (2009).

Feed formulas: turkeys

Modern hybrid strains of turkeys require diets with very high levels of protein and the amino acids necessary to support a large development of breast muscle, as outlined in Chapter 3 (NRC, 1994). However, organic producers require feeding programmes designed mainly for heritage strains of turkeys that develop less breast muscle and take a longer time to reach market weight.

Bennett (2013) published examples of organic diets designed to achieve this objective, particularly those containing no supplemental amino acids (Table 5.8). The diets were formulated to provide a lower level of protein than in conventional production. The programme involved starter to be used

Table 5.6. Example diets and nutrient specifications for free-range broiler feeding in the UK (from Lampkin, 1997).

Ingredients (g/kg)	Starter		Grower		Finisher	
	UKROFS	EU excl. amino acids	UKROFS	EU excl. amino acids	UKROFS	EU excl. amino acids
Grain	450	312	250	143	550	614
Wheat middlings	100	100	300	300	–	–
Maize gluten meal	–	–	–	–	–	85.0
Brewer's/distiller's grains	24.0	–	–	–	5.0	–
Peas/beans	100	100	100	37.0	100	–
Soybeans	107	238	137	270	153	175
Oilseeds	–	108	7.0	98.0	14.0	91.0
Dried grass/lucerne	50.0	50.0	50.0	50.0	50.0	13.0
Fishmeal	64.0	–	16.0	–	–	–
Vegetable oil	32.0	–	50.0	28.0	29.0	3.0
Yeast	35.0	33.0	33.0	19.0	37.0	50.0
Ca/P sources	13.0	25.0	23.0	23.0	29.0	29.0
Salt (NaCl)	22.0	31.0	30.0	30.0	27.0	27.0
Mineral/vitamin premix	3.0	3.0	2.0	2.0	3.0	4.0
Lysine/methionine	1.0	–	2.0	–	1.0	–
Calculated analysis (g/kg unless stated)						
CP	207	238	189	220	171	205
ME (MJ/kg)	12.0	12.0	12.0	12.0	12.0	12.0
Lysine	13.0	14.0	11.0	14.0	11.0	10.0
Methionine	5.0	4.0	5.0	4.0	4.0	3.4
Linoleic acid	29.0	19.0	41.0	37.0	29.0	18.0
Calcium	10.0	10.0	10.0	10.0	10.0	10.0
Non-phytate P	5.0	5.0	5.0	5.0	5.0	5.0

Table 5.7. Example diets for organic roaster feeding in Canada (from Bennett, 2013).

Ingredients (g/kg)	Starter 180 g/kg CP	Finisher 140 g/kg CP peas and soybeans available	Finisher 140 g/kg CP peas available	Single diet, no supplemental amino acids
Wheat	561	760	667	768
Peas	250	100	293	–
Soybeans, cooked	146	100	–	–
Soybean meal	–	–	–	192
Ground limestone	14.1	14.4	14.5	10.8
Dicalcium phosphate	18.6	15.9	16.0	17.1
Salt (NaCl)	3.0	2.9	3.1	2.0
L-lysine HCl	0.5	0.9	0.4	–
DL-methionine	1.9	0.5	1.0	–
Vitamin/mineral premix	5.0	5.0	5.0	10.0
Enzyme mixture	0.5	0.5	0.5	–

Table 5.8. Example diets for organic turkey feeding in Canada (from Bennett, 2013).

Ingredients (g/kg)	Starter	Grower	Finisher	Finisher
Wheat	490	590	752	586
Peas	–	62.0	–	205
Soybeans, cooked	62.0	–	–	–
Soybean meal	398	310	212	172
Ground limestone	14.0	10.4	9.7	9.8
Dicalcium phosphate	24.1	14.9	14.3	14.3
Salt (NaCl)	2.9	2.8	2.6	2.9
Vitamin/mineral premix	10.0	10.0	10.0	10.0

for the first 6 weeks of age followed by grower until 10 weeks, then finisher until the hens were marketed at about 9 kg live weight and the toms at 12 kg. It was recommended that either the 170 or 150 g CP/kg finisher diet could be used, though intake of the lower-protein finisher would be higher.

Feed formulas: ducks and geese

The nutritional requirements of ducks and geese have not been established exactly; therefore, some of the formulation specifications have been extrapolated from chicken data (Tables 5.9, 5.10 and 5.11). In general, the ME values of feed ingredients for ducks are similar to those for poultry except that they are about 5% higher in ducks for fibrous feedstuffs.

It is common for meat-type ducks (e.g. Pekin) to be fed diets similar to those used for growing pullets, i.e. from hatching to 2 weeks of age a diet with 180–220 g CP/kg and 2900 kcal ME/kg, followed by a diet with 160 g CP/kg and 3000 kcal ME/kg until market age. The mineral requirements are similar to, or lower than, those for pullets, and the vitamin requirements are higher than for broilers (especially niacin). Feed consumption is almost twice that of broilers so that the growth potential is high, live weight being about 50% more than that of broilers at a comparable age. Leg weakness is a major problem in waterfowl (perhaps related to the rapid growth rate) and adequate dietary choline and niacin are necessary in an attempt to reduce the problem. Carcass quality (fatness) is still a major limitation, ducks depositing considerable quantities of subcutaneous fat. Adjustment of the balance of dietary protein

Table 5.9. Examples of low-energy diets for growing ducks (from Scott and Dean, 1999).

Ingredients (g/kg)	Starter	Grower–finisher
Ground maize	354	478
Sorghum	100	–
Wheat middlings	44	156
Wheat bran	200	151
Rice bran	100	100
Maize gluten meal	16.3	–
Sunflower meal	–	50.0
Fishmeal, menhaden	85	50
Brewer's yeast	91.7	–
Ground limestone	2.1	5.7
Dicalcium phosphate	–	1.8
Salt (NaCl)	1.5	1.4
Trace mineral mix	1.0	1.0
Vitamin mix	5.0	5.0

Table 5.10. Formulas for low-energy diets for laying and breeding ducks (from Scott and Dean, 1999), based on the above nutrient specifications.

Ingredients (g/kg)	Developer	Layer–breeder[a]
Ground maize	179	539
Sorghum	150	–
Wheat middlings	–	61.5
Wheat bran	505	100
Rice bran	100	50
Sunflower meal	4.7	31.8
Fishmeal (menhaden)	47.5	90
Brewer's yeast	–	50
Ground limestone	6.2	60.1
Dicalcium phosphate	–	10
Salt (NaCl)	1.6	1.6
Trace mineral mix	1.0	1.0
Vitamin mix	5.0	5.0

[a]Modified to avoid the inclusion of meat-and-bone meal.

to energy can be used to minimize carcass fat, the duck responding similarly to broiler chickens and turkeys. Higher-protein diets generally result in leaner carcasses.

It is common for egg-laying strains (e.g. Indian Runner) to be fed diets similar to those for brown-egg strains of chickens.

For breeding ducks, a chicken layer–breeder diet with increased vitamin fortification can be used successfully, but usually with restricted feeding to control weight gains and improve egg production and male fertility. Organic producers who wish to adopt restricted feeding should check with the certifying agency that the practice is acceptable.

Ingredients used to make duck diets are similar to those used in chicken diets, except that groundnut meal should be avoided because it is often contaminated with aflatoxin (a mycotoxin). Ducks are highly sensitive to aflatoxin, particularly when diets are low in protein. In addition, canola meal may have to be used at low levels only because ducks are more sensitive to erucic acid and goitrogens than are chickens.

Table 5.11. Example feed formulas[a] for growing and laying/breeding ducks in the UK (from Bolton and Blair, 1974).

Ingredients (g/kg)	Starter–Grower	Layer–Breeder
Barley	100	–
Maize	100	40
Sorghum	302	500
Soybean meal	185	147
Herring meal	30	61
Canola meal	50	–
Maize gluten meal	9	32
Maize germ meal	100	100
Grass meal	–	25
Fat (plant source)	49	–
Brewer's yeast	50	20
Ground limestone	9	57
Dicalcium phosphate	8	10
Salt (NaCl)	3	3
Trace mineral mix	2.5	2.5
Vitamin mix	2.5	2.5

[a]Modified to avoid the inclusion of groundnut meal, meat-and-bone meal and pure methionine in the original formulas.

The diets can be offered in mash or pelleted form, and pellets are generally recommended to reduce wastage and build-up of caked feed on the ducks' bills. Alternatively, dry mash can be mixed with water and offered as a wet mash. Scott and Dean (1999) recommended a pellet of 4.0 mm diameter and 7.9 mm maximum length for newly hatched ducklings, and a pellet of 4.8 mm diameter and of 12.7 mm maximum length for Pekin-type ducks from 2 weeks of age onwards.

Geese are grazing birds but only utilize the soluble constituents in forage efficiently. Pelleted duck diets are usually recommended for goose feeding, especially on small farms. If fully fed on prepared diets, geese will reduce their forage intake; therefore, a common feeding programme is to allow goslings for the first 3 weeks to consume all the feed they wish in 15 min, 3–4 times per day. From then until market the feeding periods are reduced to twice per day. These recommendations assume that the pasture available is of good quality. Geese prefer clover, bluegrass, orchard grass, timothy and bromegrass to lucerne and narrow-leaved tough grasses.

Geese to be marketed should be fed a turkey or broiler finisher diet for 3–4 weeks before processing, if they have been fed mainly on pasture.

Small-scale producers who wish to adopt a simplified feeding programme based on information available in the 1970s and which might be more applicable to heritage strains of ducks and geese than current recommendations could use a single starter–grower diet and a single layer–breeder diet, as suggested by Bolton and Blair (1974; Table 5.11).

Pheasants and game birds

The protein requirements of pheasants and quail and guinea fowl are similar to those for turkeys; therefore, turkey diets can be used for these species.

Ratites

Aganga *et al.* (2003) described diets for growing ostriches (Table 5.12). Their recommendations did not include diets for breeding birds.

Table 5.12. Diets for growing ostriches (from Aganga *et al.*, 2003).

| Ingredients | Composition (air-dry basis) of ostrich diets at various stages (g/kg) | | | |
	1–6 weeks	3–13 weeks	13–40 weeks	Maintenance
Maize meal	600	520	550	550
Fishmeal	100	65	16.3	–
Groundnut oil cake	200	130	32.5	–
Meat meal[a]	–	30	15	10.0
Lucerne	60	260	370	410
Lysine HCl	2.5	1.63	0.41	–
DL-methionine	2.0	1.22	0.31	–
Vit/min premix	2.5	1.22	0.41	1.0
Monocalcium phosphate	13.68	1.7	13.3	17.0
Limestone	10.6	–	5.0	8.0
Salt (NaCl)	1.5	2.0	2.2	2.3

[a]For organic diets this ingredient would have to be replaced.

Feed Formulation

The following information is required for the formulation of diets when the formula is not fixed and needs to be derived based on the available ingredients.

1. Target values for energy and nutrients in the formula, based on standards for the class of poultry in question.
2. The nutrient composition of the ingredients available, in terms of nutrient bioavailability where possible.
3. Constraints on inclusion levels of certain ingredients, based on non-nutritive characteristics such as palatability and pelleting properties.
4. Costs of available ingredients (unless the cost of the diet is unimportant).

This is quite a complicated process, usually requiring advice from a nutritionist.

Mixing Complete Diets

The first step in preparing a dietary mixture is to list the available ingredients and their composition. The next step is to formulate an appropriate mixture of ingredients, to obtain the target values in the mixture for nutrients required by the class of poultry in question. The third step is to prepare the ingredients in a form and in (preferably weighed) amounts suitable for mixing together. The final step is manufacture (compounding) of the dietary mixture.

Selection of ingredients

The first step in manufacturing high-quality feed is to use high-quality ingredients. Grain should be free of caking, moulds, insects, stones, etc. and should contain a low proportion of broken kernels, which are more likely to encourage mould growth than intact kernels. Grains should contain a maximum of 120–140 g moisture/kg for safe storage. Rodent damage and mould contamination are two major problems associated with stored grain. Rodents will eat grain and contaminate it with their droppings, which will reduce the palatability of grains and feed, reduce intake of feed and possibly contaminate it with salmonella. Mould growth is enhanced when grain is stored with high moisture content. Certain moulds produce toxins called mycotoxins, which are deleterious to poultry.

A main factor influencing selection of ingredients is availability. Another factor affecting choice of ingredients is quality. This latter feature should be addressed using a quality control programme. Since a detailed chemical analysis of each batch of ingredients is not possible, assumed values for nutrient content (such as those in the tables shown in Chapter 4) are often used. Where possible, it is advisable to make allowances in formulations for variations in ingredient composition.

Surveys show that feed-quality problems are common. Often the actual CP, calcium and phosphorus levels in on-farm mixed feeds vary considerably from the intended values.

Where possible, separate vitamin and trace mineral premixes should be used. Vitamins in contact with minerals over a prolonged period of time in a hot and humid environment lose potency, possibly resulting in vitamin deficiencies and reduced performance. In countries with high ambient temperatures it may be necessary to provide cooled storage facilities. When vitamins and minerals are combined in a single premix, it should be used within 30 days of purchase. Vitamin and trace mineral premixes should be stored in the dark in dry sealed containers. Stabilizing agents are helpful in maintaining premix quality, but must meet organic standards.

Suggested premixes for use in organic diets are shown in Tables 5.13 to 5.18, based on the 1994 NRC requirements. Separate premixes for breeding and growing birds are appropriate, since their requirements are different. The premixes supply all of the required vitamins and trace minerals and assume no contribution from the main dietary ingredients since these vary and may be low in potency or bioavailability. Producers who wish to include a lower level of premix on the basis that some vitamins and minerals are provided by the main ingredients should do so with care, observing the birds closely for signs of deficiencies and ensuring that their welfare is not compromised. Vitamin D can be omitted from summer formulations, provided that the birds have access to sunshine. Vitamin D is recommended in formulations for use in regions north of latitude 42.2° N at all other times of the year. The carrier used in the premixes can be a material such as ground oat hulls. Premixes

Table 5.13. NRC (1994) estimated vitamin and trace mineral requirements of growing and laying chickens, amount/kg diet (90% moisture basis).

Ingredients	0–6 weeks	6–12 weeks	12–18 weeks	Laying (assuming an intake of 120 g/day)
Trace minerals (mg/kg)				
Copper	5.0	4.0	4.0	ND
Iodine	0.35	0.35	0.35	0.03
Iron	80	60	60	38
Manganese	60	30	30	17
Selenium	0.15	0.1	0.1	0.05
Zinc	40.0	35.0	35.0	29
Vitamins (IU/kg)				
Vitamin A	1500	1500	1500	2500
Vitamin D$_3$	200	200	200	250
Vitamin E	10.0	5.0	5.0	4.0
Vitamins (mg/kg)				
Biotin	0.15	0.1	0.1	0.08
Choline	1300	900	500	875
Folacin	0.55	0.25	0.25	0.21
Niacin	27.0	11.0	11.0	8.3
Pantothenic acid	10.0	10.0	10.0	1.7
Pyridoxine	3.0	3.0	3.0	2.1
Riboflavin	3.6	1.8	1.8	2.1
Thiamin	1.0	1.0	0.8	0.6
Vitamins (µg/kg)				
Cobalamin (vitamin B$_{12}$)	9.0	3.0	3.0	4.0

ND = not determined

Table 5.14. Composition of a trace mineral premix to provide the NRC (1994) estimated trace mineral requirements in chicken grower and layer diets when included at 5 kg/t.

Ingredients	Amount/kg premix	Supplies/kg diet
Trace minerals		
Copper	1 g	5.0 mg
Iron	16 g	80.0 mg
Manganese	12 g	60.0 mg
Selenium	30 mg	0.15 mg
Zinc	8 g	40.0 mg
Carrier	To 1 kg	

Note: Iodine is usually provided in the form of iodized salt.

Table 5.15. Composition of a vitamin premix to provide the NRC (1994) estimated vitamin requirements in chicken grower and layer diets when included at 5 kg/t.

Ingredients	Amount/kg premix	Supplies/kg diet
Vitamins (IU)		
Vitamin A	500,000	2,500
Vitamin D$_3$	50,000	250
Vitamin E	2,000	10.0
Vitamins		
Biotin	30 mg	0.15 mg
Choline	260 g	1,300 mg
Folacin	110 mg	0.55 mg
Niacin	6 g	30 mg
Pantothenic acid	2.0 g	10 mg
Riboflavin	750 mg	3.75 mg
Vitamins		
Cobalamin (vitamin B$_{12}$)	2 mg	10.0 µg
Carrier	To 1 kg	

Note: Vitamin D$_3$ can be omitted from the summer formula for birds with access to direct sunshine.

meeting these specifications can be obtained from feed manufacturers.

Some producers prefer to use 'macro-premixes' instead of premixes of the type suggested. These contain calcium, phosphorus and salt, in addition to trace minerals, and are supplied with mixing formulas. Their inclusion rate in diets is determined largely by the varying need for supplementary calcium and phosphorus. Organic producers have much more control of the dietary formulation when the premix and major minerals are included in the diet separately. However, producers wishing the convenience of a combined mineral and trace mineral premix should adopt the option of using macro-premixes and follow the mixing directions provided by the supplier to mix appropriate diets.

The premixes outlined above can be used for other stock, i.e. the chicken premixes can be used for waterfowl diets and the turkey premixes can be used for quail and ratite diets. Alternatively, specific premixes can be formulated for these species, based on the estimated requirements set out in Chapter 3.

Formulation

This involves deciding on the appropriate mixing formula (recipe). As outlined above, some producers use fixed formulas such as those suggested in this volume. Others who have access to a range of feed ingredients have to change the mixing formula as the ingredients change, e.g. barley being available

Table 5.16. NRC (1994) estimated vitamin and trace mineral requirements of growing and breeding turkeys, amount/kg diet (90% moisture basis).

Ingredients	0–4 weeks	4–8 weeks	8–14 weeks	14 weeks–	Breeding
Trace minerals (mg/kg)					
Copper	8.0	8.0	6.0	6.0	8.0
Iodine	0.4	0.4	0.4	0.4	0.4
Iron	80.0	60.0	60.0	50.0	60.0
Manganese	60.0	60.0	60.0	60.0	60.0
Selenium	0.2	0.2	0.2	0.2	0.2
Zinc	70.0	65.0	50.0	40.0	65.0
Vitamins (IU/kg)					
Vitamin A	5000	5000	5000	5000	5000
Vitamin D_3	1100	1100	1100	1100	1100
Vitamin E	12.0	12.0	10.0	10.0	25.0
Vitamins (mg/kg)					
Biotin	0.25	0.2	0.125	0.125	0.2
Choline	1600	1400	1100	1100	1000
Folacin	1.0	1.0	0.8	0.8	1.0
Niacin	60.0	60.0	50.0	50.0	40.0
Pantothenic acid	10.0	9.0	9.0	9.0	16.0
Pyridoxine	4.5	4.5	3.5	3.5	4.0
Riboflavin	4.0	3.6	3.0	3.0	4.0
Thiamin	2.0	2.0	2.0	2.0	2.0
Vitamin K	1.75	1.5	1.0	0.75	1.0
Vitamins (µg/kg)					
Cobalamin (vitamin B_{12})	3.0	3.0	3.0	3.0	3.0

Table 5.17. Composition of a trace mineral premix to provide the NRC (1994) estimated trace mineral requirements in turkey grower and breeder diets when included at 5 kg/t.

Ingredients	Amount/kg premix	Supplies/kg diet
Trace minerals		
Copper	1.6 g	8.0 mg
Iron	16 g	80.0 mg
Manganese	12 g	60.0 mg
Selenium	40 mg	0.2 mg
Zinc	14 g	70.0 mg
Carrier	To 1 kg	

Note: Iodine is usually provided in the form of iodized salt.

in place of maize. Producers who use commercial supplements and premixes should have mixing instructions available on the feed tag. Mixing instructions should be followed precisely.

Standards for dietary formulation

Standards for organic poultry production can be derived from the NRC (1994) estimates of nutritional requirements outlined in Chapter 3. However, the requirement values estimated by NRC are designed for modern, highly productive birds with an increased potential for meat development and a high rate of egg production. The birds used in organic production are preferably heritage types, which develop more slowly and lay fewer eggs than modern hybrids. Therefore, it is logical to modify the NRC values by lowering the dietary ME value to a level that is more suited to the energy needs of this type of stock, at least for chickens and turkeys. Reducing the energy content of the diet also minimizes the need for restricted feeding of breeder stock. In reducing the ME value of the diet it is logical to reduce the protein and amino acid contents by a similar proportion so that the same ratios of energy to protein and amino acids are maintained. Bellof, G. and Schmidt (2014) adopted this approach in relation to feeding organic turkeys and found it successful.

However, it is recommended that the NRC (1994) derived values for mineral and

Table 5.18. Composition of a vitamin premix to provide the NRC (1994) estimated vitamin requirements in turkey grower and breeder diets when included at 5 kg/t.

Ingredients	Amount/kg premix	Supplies/kg diet
Vitamins (IU)		
Vitamin A	1,000,000	5,000
Vitamin D$_3$	220,000	1,100
Vitamin E	5,000	25.0
Vitamins		
Biotin	50 mg	0.25 mg
Choline	320 g	1,600 mg
Folacin	0.2 g	1.0 mg
Niacin	12 g	60 mg
Pantothenic acid	3.2 g	16 mg
Riboflavin	0.8 g	4.0 mg
Vitamins		
Cobalamin (vitamin B$_{12}$)	0.6 mg	3.0 µg
Carrier	To 1 kg	

Note: Vitamin D$_3$ can be omitted from the summer formula for birds with access to direct sunshine.

vitamin requirements be adopted without modification, to help ensure correct skeletal growth and avoidance of foot and leg problems. Conventional diets are usually formulated with higher levels of minerals and vitamins but this approach is not suggested for organic diets, to try to minimize nutrient levels above those required for normal growth and reproduction.

Standards based on this approach, and derived feed formulas, are shown in Tables 5.19 to 5.26 below. Producers using modern hybrids instead of traditional genotypes are advised instead to use the NRC (1994) values or the values suggested by the breeding company as the basis for dietary standards.

CHICKENS. Suggested dietary specifications for organic diets based on the above considerations are shown in Tables 5.19 and 5.20. Examples of dietary formulas that meet these specifications are also shown (Table 5.27).

TURKEYS. Suggested dietary specifications for organic turkey diets and based on the above considerations are shown in Table 5.21.

DUCKS AND GEESE. The nutrient requirements of ducks have not been well defined. Scott and Dean (1999) published information on the nutrition and feeding of ducks which can be applied to organic production. Diets based on these standards can be used; alternatively, chicken diets can be used (Tables 5.22 and 5.23).

RATITES. The nutrient requirements of these species have not been reviewed by the NRC (1994).

A single diet for all ages of ostriches is often more practical than separate starter, grower and breeder diets. This is because the nutrient requirements of these species are not well defined and the numbers farmed in any age class are generally too low to justify several diets. Suggested specifications for such a diet are shown in Table 5.24 (Ullrey and Allen, 1996). The diet can be pelleted and fed *ad libitum* with water on pasture or in dry lot. According to the authors the concentrations of ME, CP and essential amino acids in this diet may limit body weight gain somewhat below maximum, but are sufficient to support proportional development of the skeleton without excessive soft tissue weight in growing birds. It is therefore appropriate for organic production. A feeder containing calcareous grit or oyster shell can be used to ensure a sufficient intake of calcium by laying females.

In addition, Scheideler and Sell (1997) published guidelines on feeding of ostriches and emus which can be used as a standard (Tables 5.25 and 5.26).

Table 5.19. Suggested dietary specifications for organic pullet and layer feeding (abbreviated), g/kg unless stated.

Dietary content (900 g DM/kg basis)	Starter	Pullet grower	Layer
ME (kcal/kg)	2750	2600	2650
ME (MJ/kg)	11.5	10.9	11.1
CP	175	130	150
Lysine	8.0	5.3	6.9
Methionine + cystine	6.0	4.3	5.8
Linoleic acid	10.0	10.0	10.0
Calcium	9.0	8.0	25*
Non-phytate P	4.0	3.5	3.5
Sodium	1.5	1.5	1.5

*Supply additional calcium in the form of calcareous grit.

Table 5.20. Suggested dietary specifications for organic broiler (roaster) feeding, g/kg unless stated.

Dietary content (900 g DM/kg basis)	Starter	Grower	Finisher
ME (kcal/kg)	2850	2850	2850
ME (MJ/kg)	11.9	11.9	11.9
CP	205	178	160
Lysine	10.6	8.8	7.2
Methionine + cystine	8.0	6.4	5.4
Linoleic acid	10.0	10.0	10.0
Calcium	10.0	9.0	8.0
Non-phytate P	4.5	3.5	3.0
Sodium	2.0	1.5	1.5

Table 5.21. Suggested dietary specifications for organic turkey feeding, g/kg unless stated.

Dietary content (900 g DM/kg basis)	Starter 0–6 weeks	Grower 6–10 weeks	Finisher 10 weeks–market	Breeding hens
ME (kcal/kg)	2600	2700	2900	2900
ME (MJ/kg)	10.9	11.3	12.1	12.1
CP	260	198	150	140
Lysine	15	12	7.3	6.0
Methionine + cystine	10.0	7.2	5.0	4.0
Linoleic acid	10.0	10.0	8.0	8.0
Calcium	11.0	8.0	6.0	20*
Non-phytate P	5.0	4.0	3.0	3.5
Sodium	2.0	2.0	2.0	2.0

*Provide additional calcium in the form of calcareous grit.

Formulation

The intent of formulation is to match the nutrient values in the mixed feed to the needs of the class of poultry in question.

FORMULATION BY HAND. In hand calculation of dietary formulas, the usual procedure is to list the main energy and protein sources to be used. The quantity of each ingredient needed to make a mixture with the desired ME and CP content is calculated. The content of other major nutrients (such as lysine) in the diet mix is then calculated and the quantities of major ingredients are adjusted as necessary.

Table 5.22. Suggested nutrient specifications for growing duck diets (from Scott and Dean, 1999).

Nutrient g/kg diet unless stated	Starter low energy	Grower–finisher low energy
ME (kcal/kg)	2646	2646
CP	191	140
Amino acids		
Arginine	10.4	8.8
Histidine	3.8	3.1
Isoleucine	7.7	6.1
Leucine	12.2	11.4
Lysine	10.4	7.0
Methionine	4.1	3.1
Methionine + cystine	7.0	5.2
Phenylalanine	7.0	6.1
Phenylalanine + tyrosine	13.1	11.4
Threonine	7.0	5.2
Tryptophan	2.0	1.7
Valine	7.7	7.0
Major minerals		
Calcium	5.7	5.7
Magnesium	0.44	0.44
Non-phytate phosphorus	3.5	3.1
Potassium	5.2	5.2
Sodium	1.3	1.2
Microminerals (mg/kg)		
Copper	7	5
Iodine	0.35	0.35
Iron	70	70
Manganese	44	35
Selenium	0.13	0.13
Zinc	52	52
Vitamins (mg/kg)		
A (IU/kg)	4350	3480
D_3 (IU/kg)	522	435
E (IU/kg)	22	17
K	2	1
Biotin	0.13	0.13
Choline	1130	870
Folic acid	0.25	0.25
Niacin	44	35
Pantothenic acid	10	9
Pyridoxine	3	3
Riboflavin	4	3
Vitamin B_{12}	0.01	0.004

FORMULATION USING A COMPUTER Computers are used increasingly for formulation of diets, with commercial software based on linear programming and involving a complex set of mathematical equations (Tables 5.27 to 5.29). To attempt to do the same on a desk calculator is very tedious and time-consuming. The computer is programmed using NRC or other target nutrient values, tables of feed composition and current prices of ingredients. The mathematical solution obtained lists the amounts of selected ingredients that best meet the specifications set. In commercial production these specifications usually include constraints on upper and lower limits of nutrients allowed in the mixture, and on inclusion levels of certain ingredients. The obtained formula is usually, but not always, of least cost. A simple, free online feed formulation

Table 5.23. Suggested nutrient specifications for laying and breeding ducks (from Scott and Dean, 1999).

Nutrient g/kg diet unless stated	Breeder–developer low energy	Layer–breeder low energy
ME (kcal/kg)	2205	2646
CP	128	162
Amino acids		
Arginine	8.0	8.0
Histidine	2.8	3.4
Isoleucine	5.6	6.7
Leucine	10.4	11.3
Lysine	6.4	7.4
Methionine	2.8	3.7
Methionine + cystine	4.8	6.0
Phenylalanine	5.6	6.5
Phenylalanine + tyrosine	10.4	10.2
Threonine	4.8	5.4
Tryptophan	1.6	1.8
Valine	6.4	6.9
Major minerals		
Calcium	5.5	27.7
Magnesium	0.4	0.47
Non-phytate phosphorus	2.9	3.7
Potassium	4.8	5.6
Sodium	1.2	1.4
Microminerals (mg/kg)		
Copper	5	6
Iodine	0.32	0.37
Iron	64	65
Manganese	32	37
Selenium	0.12	0.14
Zinc	48	56
Vitamins (mg/kg)		
A (IU/kg)	3200	5580
D_3 (IU/kg)	400	558
E (IU/kg)	16	28
K	2	2
Biotin	0.13	0.13
Choline	800	930
Folic acid	0.25	0.25
Niacin	32	46
Pantothenic acid	8	11
Pyridoxine	2	3
Riboflavin	3	4
Vitamin B_{12}	0.01	0.01

programme developed at the University of British Columbia is the Abacus system which can be accessed via the CABI Open Resources page (www.cabi.org/openresources/92985). It is based on a Microsoft Excel-type matrix, using the feedstuff compositional data outlined in Chapter 4 and the suggested standards for energy and nutrients to be provided in organic poultry diets that are outlined in this chapter.

Preparation, weighing, batching and blending

Having obtained a suitable formula for the feed in question, the next step is preparing to mix the feed. The procedures involved in the production of quality feeds are important and should be carried out by trained personnel. Mixing errors are quite common.

Table 5.24. Suggested nutrient specifications for a single, life-cycle diet suitable for ostriches, rheas and emus[a] (from Ullrey and Allen, 1996).

Nutrient specifications	Amount g/kg unless stated (90% DM basis)
ME	2500 kcal/kg
CP	220
Lysine	12
Arginine	13
Methionine	3.5
Methionine + cystine	7.0
Tryptophan	3.0
Crude fibre	100
Linoleic acid	10.0
Calcium	16.0[b]
Non-phytate P	8.0
Iron	150 mg/kg
Copper	20 mg/kg
Zinc	120 mg/kg
Manganese	70 mg/kg
Iodine	1 mg/kg
Selenium	0.30 mg/kg
Vitamin A	8000 IU/kg
Vitamin D$_3$	1600 IU/kg
Vitamin E	250 IU/kg
Vitamin K	4 mg/kg
Thiamin	7 mg/kg
Riboflavin	9 mg/kg
Niacin	70 mg/kg
Pantothenic acid	30 mg/kg
Pyridoxine	5 mg/kg
Biotin	0.300 mg/kg
Folacin	1 mg/kg
Cobalamin	0.03 mg/kg
Choline	1600 mg/kg

[a]To be provided in pelleted form *ad libitum*.
[b]Additional calcareous grit or oyster shell may be required for breeders.

Cereal grains and possibly protein sources need to be processed before being added into the mix. To achieve optimal poultry performance using mixed diets, it is necessary to process cereal grains through a hammer mill or roller mill to reduce particle size. Reduction of particle size breaks the hard kernel and increases the surface area of the grain in the digestive tract. This improves the efficiency of digestion and utilization of the nutrients. In addition, particle size reduction allows a more uniform mixing of grain with protein, vitamin and mineral supplements and helps to prevent sorting of ingredients by the birds when the mixture is fed to poultry. Feed to be pelleted should be ground finely.

Research has been conducted on the ideal particle size. Parsons *et al.* (2006) investigated the effects of grinding maize to 781, 950, 1042, 1109 and 2242 μm to include in a soybean-based premix to create five different mash diets for feeding to broilers over a period of 3–6 weeks. Water and a commercial pellet binder were added separately to maize–soybean-based diets before steam-pelleting to create two pelleted diets differing in texture (soft and hard, respectively). Surprisingly, both soft and hard pellets had similar pellet durability (90.4% and 86.2%, respectively) and fines (44.5% and 40.3%, respectively). Increasing the particle size of maize improved nutrient retention but broiler growth performance and energy utilization decreased when maize particle size exceeded 1042 μm. Broilers fed hard pellets (1856 g of pellet breaking force) had improved nutrient retention, energy utilization and subsequent growth performance compared with broilers fed soft pellets (1662 g of pellet breaking force). These results suggest that when diets are to be pelleted the grains should be ground no finer than 1042 μm and that pellets should preferably be hard. Another reason for avoiding too fine a grind is that, although it may improve pellet quality, it will markedly increase energy consumption during milling. A very small particle size also increases problems with feed bridging in bins and feeders.

A review of findings on particle size was conducted by Amerah *et al.* (2007) which confirms the above conclusions, indicating that the optimum particle size of broiler diets based on maize or sorghum is between 600 and 900 μm. The data showed that grain particle size was more critical in mash diets than in pelleted or crumbled diets. Although finer grinding has been shown to increase the surface area of the grain for enzymatic digestion in the gut, there was evidence that coarser grinding to a more uniform particle size improved the performance of birds maintained on mash diets. This was possibly due to a more uniform intake of ingredients from mash diets when the feed particles were of similar size, also from a positive effect of increased feed particle size on gizzard development.

Table 5.25. Suggested nutritional guidelines for ostrich diets (from Scheideler and Sell, 1997).

Nutrient g/kg diet unless stated (air-dry basis)	Starter 0–9 weeks	Grower 9–42 weeks	Finisher 42 weeks–market	Holding 42 weeks–sexual maturity	Laying from 4–5 weeks prior
AME (kcal)	2,465	2,450	2,300	1,980–2,090	2,300
AME (MJ)	11.1	11.08	9.62	8.28–8.74	9.62
Crude protein	220	190	160	160	200–210
Amino acids					
Crude fibre	60–80	90–110	120–140	150–170	120–140
Lysine	9.0	8.5	7.5	7.5	10.0
Methionine	3.7	3.7	3.5	3.5	3.8
Methionine + cystine	7.0	6.8	6.0	6.0	7.0
Neutral-detergent fibre	140–160	170–200	190–220	240–270	220–240
Minerals					
Calcium	15.0	12.0	12.0	12.0	24.0–35.0
Chloride	2.2	2.2	1.89	1.89	1.89
Phosphorus (non-phytate)	7.5	6.0	6.0	6.0	7.0
Sodium	2.0	2.0	2.0	2.0	2.0
Trace minerals (mg/kg)					
Copper	33	33	33	33	44
Iodine	1.1	1.1	0.9	0.9	1.1
Manganese	154	154	154	154	154
Zinc	121	121	88	88	88
Vitamins (IU)					
Vitamin A	11,000	8,800	8,800	8,800	11,000
Vitamin D$_3$	2,640	2,200	2,200	2,200	2,200
Vitamin E	121	55	55	55	110
Vitamins (mg)					
Choline	2,200	2,200	1,892	1,892	1,892
Vitamins (µg)					
Cobalamin (vitamin B$_{12}$)	40	20	20	20	40

Table 5.26. Suggested nutritional guidelines for emu diets (from Scheideler and Sell, 1997).

Nutrient g/kg diet unless stated (air-dry basis)	Starter 0–6 weeks	Grower 6–36 weeks	Finisher 36–48 weeks	Breeder holding 48 weeks– sexual maturity	Breeder from 3–4 weeks prior
AME (kcal)	2,685	2,640	2,860	2,530	2,400
AME (MJ)	11.23	11.05	11.97	10.59	10.04
Crude protein (g)	220	200	170	160	200–220
Amino acids (g)					
Crude fibre	60–80	60–80	60–70	60–70	70–80
Lysine	11.0	9.4	7.8	7.5	10.0
Methionine	4.8	4.4	3.8	3.6	4.0
Methionine + cystine	8.6	7.8	6.5	6.0	7.5
Neutral-detergent fibre	140–160	140–170	100–130	140–160	160–180
Minerals (g)					
Calcium	15.0	13.0	12.0	12.0	24.0–35.0
Phosphorus (non-phytate)	7.5	6.5	6.0	6.0	6.0
Sodium	2.0	2.0	2.0	2.0	2.0
Trace minerals (mg/kg)					
Copper	33	33	33	33	33
Iodine	1.1	1.1	1.1	1.1	1.1
Manganese	154	154	154	154	154
Zinc	110	110	110	110	110
Vitamins (IU)					
Vitamin A	15,400	8,800	8,800	8,800	8,800
Vitamin D_3	4,400	3,300	3,300	3,300	3,300
Vitamin E	99	44	44	44	99
Vitamins (mg)					
Choline	2,200	2,200	2,200	2,200	1,980
Vitamins (μg)					
Cobalamin (vitamin B_{12})	44	22	22	22	44

Table 5.27. Example formulas for organic diets that meet the suggested specifications for layer stock (900 g/kg DM basis), with (AA) and without (no AA) supplemental amino acids (generated using the UBC Abacus programme).

| Ingredients (kg/t) | Starter | | Grower | Layer/breeder |
	AA	No AA	No AA	No AA
Barley	–	–	279	100
Maize	100	125	–	–
Oats	–	–	25	25
Wheat	395	345	330	420
Wheat middlings	180.5	180	230	50
Soybean meal	124	150	–	–
Soybeans	–	–	–	50
Canola meal	–	–	50	100
Faba beans	–	–	50	100
Peas	160	160	–	75
DL-methionine	0.5	–	–	–
Ground limestone	14	14	14	57*
Dicalcium phosphate	12	12	8	9
Salt (NaCl)	4.0	4.0	4.0	4.0
Trace min premix	5.0	5.0	5.0	5.0
Vitamin premix	5.0	5.0	5.0	5.0
Calculated analysis (abbreviated; kg/t unless stated)				
ME (kcal/kg)	2750	2750	2617	2665
ME (MJ/kg)	11.5	11.5	11.1	11.3
CP	179	185	146	174
Lysine	8.5	9.1	5.5	8.0
Methionine + cystine	6.4	6.0	5.4	5.8
Calcium	9.1	9.1	8.2	25.1
Non-phytate P	4.2	4.1	3.7	3.6

*Supply additional calcium in the form of calcareous grit.

Note: The calculated analyses of the mixtures above cannot be guaranteed, due to ingredient availability. Producers using such formulas should have the diets analysed to ensure acceptability.

A more developed gizzard is associated with increased grinding activity in this organ, resulting in increased gut motility and improved digestion of nutrients.

Sieving tests to determine particle size range should be carried out regularly in order to check that the feed is being ground to the required degree of fineness and that no material is unground. Segregation problems can be avoided by processing grain so that it is comparable in particle size to other ingredients in the batch.

The particle size of grain processed in a hammer mill is affected mainly by the size of the screen used, though the speed of the hammer mill and the flow rate also affect it.

Equipment suitable for accurately weighing and dispensing small quantities of ingredients should be available. Care is necessary in handling premixes. Before being added into the mix these should be admixed with ground cereal to make up at least 5% of the total mix. This is to ensure that the premix disperses uniformly through the mix.

Weighed amounts of ingredients should be used where possible. This helps to ensure that the ingredients are added to the mix in the correct proportions. Some types of on-farm feed-mixing equipment are based on volume rather than weight. In these cases it is essential that the equipment be calibrated frequently to account for changes in bulk density of ingredients.

Table 5.28. Example formulas for organic diets that meet the suggested specifications for growing broilers (roasters) (900 g/kg DM basis), with (AA) and without (no AA) supplemental amino acids (generated using the UBC Abacus programme).

Ingredients (kg/t)	Starter		Grower		Finisher	
	AA	No AA	AA	No AA	AA	No AA
Maize	280	280	96	123	50	25
Wheat	280	201	520	465	600	517
Wheat middlings	–	–	25	74	100	105
Canola meal	–	55	50	100	100	115
Faba beans	–	–	25	–	–	–
Soybean meal	269	320	84	120	–	–
Peas	75	50	110	80	115	150
Sunflower meal dehulled	50	50	50	–	–	–
Soybeans	–	–	–	–	–	55
L-lysine HCl	–	–	1	–	1	–
DL-methionine	1.0	–	–	–	–	–
Ground limestone	13	12	15	15	14	14
Dicalcium phosphate	17	17	9.0	9.0	6.0	5.0
Salt (NaCl)	5.0	5.0	5.0	4.0	4.0	4.0
Trace min premix	5.0	5.0	5.0	5.0	5.0	5.0
Vitamin premix	5.0	5.0	5.0	5.0	5.0	5.0
Calculated analysis (abbreviated; kg/t unless stated)						
ME (kcal/kg)	2893	2857	2860	2850	2850	2853
ME (MJ/kg)	12.1	11.9	11.9	11.9	11.9	11.9
CP	213	236	188	189	163	183
Lysine	11.1	13.0	9.1	8.9	7.3	8.2
Methionine + cystine	8.0	7.8	6.4	6.5	5.8	6.3
Calcium	10.1	10.1	9.0	9.3	8.0	8.0
Non-phytate P	4.6	4.7	3.5	3.6	3.2	3.2

Note: The calculated analyses of the mixtures above cannot be guaranteed, due to ingredient availability. Producers using such formulas should have the diets analysed to ensure acceptability.

Mixing and further processing

The feed ingredients have to be mixed thoroughly before feeding, to blend the ingredients uniformly and ensure an adequate intake of nutrients by the animals. In using ingredients it is important to ensure that stocks are rotated correctly, those held longest being used first.

The manufacturers of the mixing equipment should be able to advise on optimum mixing time but tests should be carried out periodically as a check.

A general recommendation with vertical mixers is that about 15 min mixing time is adequate after the last ingredient has been added to the mixer. With horizontal and drum mixers the mixing time is shorter, about 5–10 min after addition of the last ingredient.

After mixing, the feed should be stored in a clean, dry container to maintain quality and to protect it from dampness and insect and rodent infestation. Bulk storage bins frequently need to be emptied completely at regular intervals so that stale feed does not remain in the bin.

Feeder management also plays an important role in delivering quality feed to poultry. Feed becomes stale and of reduced palatability if allowed to remain in feeders for long periods. Feeders should be filled often enough to provide fresh feed at all times, unless restricted feeding is being used.

Pelleting

The feed may be pelleted or cubed after mixing. The usual pelleting process involves

Table 5.29. Example formulas for organic diets that meet the suggested specifications for market turkeys (900 g/kg DM basis), with (AA) and without (no AA) supplemental amino acids (generated using the UBC Abacus programme).

Ingredients (kg/t)	Starter		Grower		Finisher		Breeding hens
	AA	No AA	AA	No AA	AA	No AA	No AA
Maize	–	–	–	437	202	202	153
Wheat	215	150	489	–	525	477	500
Wheat middlings	252	158	125	76	49	49	–
Soybean meal	–	120	95	200	–	50	–
Canola meal	50	–	150	150	40	40	–
Soybeans	150	100		–	50	50	100
Peas	125	100	50	50	49	49	125
Faba beans	–	100	–	–	–	–	50
Sunflower meal dehulled	100	150	50	50	50	50	–
Whitefish meal	78	100	–	–	–	–	–
L-lysine HCl	2	–	3.0	–	2.0	–	–
DL-methionine	1	–	–	–	–	–	–
Ground limestone	9.0	6.0	11	10	8.0	8.0	45*
Dicalcium phosphate	4.0	2.0	12	12	10	10	12
Salt (NaCl)	4.0	4.0	5.0	5.0	5.0	5.0	5.0
Trace min premix	5.0	5.0	5.0	5.0	5.0	5.0	5.0
Vitamin premix	5.0	5.0	5.0	5.0	5.0	5.0	5.0
Calculated analysis (abbreviated; kg/t unless stated)							
ME (kcal/kg)	2730	2690	2720	2740	2988	2985	2940
ME (MJ/kg)	11.4	11.2	11.4	11.5	12.5	12.6	12.3
CP	261	305	213	240	160	174	156
Lysine	15	16.8	11.9	12.0	7.6	7.3	7.0
Methionine + cystine	10	10	7.5	8.2	5.7	6.1	5.0
Calcium	11.3	11.3	8.9	8.7	7.7	7.57	20.5
Non-phytate P	6.0	6.0	4.5	4.5	4.0	4.0	3.7

*Supply additional calcium in the form of calcareous grit
Note: The calculated analyses of the mixtures above cannot be guaranteed, due to ingredient availability. Producers using such formulas should have the diets analysed to ensure acceptability.

treating ground feed with steam and then passing the hot, moist mash through a die under pressure. The pellets are then cooled quickly and dried by means of forced air. Sufficient water should be applied so that all feed is moistened. Optimum moisture content of a feed required for good pelleting will vary with the composition of the feed; however, a range of 150–180 g moisture/kg is usually desirable. Feeds containing a high concentration of fibre require a higher level of moisture, while feeds low in fibre require less.

Poultry will readily consume a diet in either mash or pelleted form; therefore, pelleting is generally not economic but does offer some advantages since the increased bulk density of the pelleted diets helps in feed handling and storage. Also wastage in feeding is reduced, which is of relevance when outdoor feeding is employed. Other benefits include the prevention of segregation, i.e. the separation of ingredients from the rest of the mix. This problem may occur during transport of the feed from the mixer to the feeder. The major factors involved in segregation are particle size, shape and density. Segregation can result in uneven intake of nutrients by the birds and uneven growth and egg production. A further benefit of pelleting is the prevention of sorting (i.e. when some feed particles in the feeder are selected by the birds and others refused). One way of minimizing this problem is to allow the feeders to be emptied completely before they are refilled.

An advantage of pelleting is that it may improve both intake and the efficiency of feed utilization, particularly with fibrous ingredients such as wheat milling by-products. A further advantage of pelleting is that it can help to reduce microbial counts in the feed.

Bioavailability of nutrients may be influenced beneficially or adversely by pelleting. For example, the availability of phytate phosphorus in grains was found to be increased by steam-pelleting (Bayley et al., 1975). On the other hand, there may be destruction of heat-labile nutrients and components, such as phytase enzyme in wheat or vitamin A.

Normally the feed is preconditioned with steam before entering the pellet mill. Steam releases natural adhesive properties in feeds to facilitate pelleting. It softens ingredient particles, so that they can more readily bind with each other under pressure. Heat and moisture cause starch gelatinization, helping to bind the particles together, as well as improving starch and fibre digestion.

Some ingredients possess properties that aid the pelleting process. Wheat and wheat by-products contain endosperm proteins that allow the feed particles to bind well during pelleting. Wheat gluten when moist has a gum-like consistency. The endosperm proteins of triticale, rye and barley also react with water to increase viscosity, but those of maize, sorghum, millet, rice and oats do not. The glucans and pentosans in barley, rye and oats have viscous properties when wet, improving pellet quality. Fat added at levels above 50 g/kg tends to cause pellet crumbling.

Pellets must then be cooled and dried down to a safe moisture content as rapidly as possible after processing, using air cooling.

Extrusion of mixed feed has been found to be beneficial in improving digestibility. However, the process is generally limited to pet and fish feeds because of cost. Pelleting or extrusion of mixed feed has additional benefits relating to environmental and consumer safety aspects. Coliform bacteria, *Escherichia coli*, salmonellae and moulds are greatly reduced by pelleting and eliminated by the extrusion process.

Pellet binders

Diets based on ingredients such as maize and soybean meal are difficult to pellet, especially when fat has been added, and a pellet binder may have to be included in the formula to achieve acceptable pellet quality. Pellet binders are permitted in organic diets and producers should check on those that are acceptable. Producers mixing feed soon learn which formulations are difficult to pellet. Inclusion levels of wheat or wheat by-products at more than 100 g/kg diet should generally result in good pellets and avoid the need for a pellet binder. The type of binder approved for organic use is colloidal clay, with inclusion levels of around 5–12 kg/t. Some forms of clay have been shown to be effective in binding aflatoxin. Molasses can also be used as a pellet binder at inclusion levels of 20–50 g/kg. Unlike other binders it also contributes nutrients to the diet.

While all types of heat processing have beneficial effects on the feed, it needs to be recognized that some destruction of nutrients may occur, particularly some vitamins and amino acids (Coehlo, 1994). For most vitamins, pelleting can be expected to result in an 8–10% loss of potency. Extrusion, however, which usually employs much higher temperatures, can lead to a 10–15% loss of most vitamins.

Quality Control

Quality control is a way of ensuring that ingredients are of an acceptable quality, so that mixed feeds are of the required nutritive value and contain minimal or zero levels of contaminants. It is also a way of ensuring that ingredients that are not to be used immediately can be stored without risk of deterioration. Quality control is particularly important in the production of organic feed, since the ingredients used are often more variable in composition than regular feed ingredients.

The ingredient buyer must ensure that acceptable quality is a prerequisite when ingredients are to be purchased. A quality control protocol must be drawn up and followed as each batch of feed ingredients is received at the feed mill. Ingredients grown on the farm should be analysed periodically and purchased ingredients should (where possible) be accompanied by a guaranteed analysis.

Ingredients and any accompanying documentation need to be examined when they are received at the feed mill to ensure that the products meet organic standards. Each ingredient needs to be inspected visually for evidence of dampness, and for the presence of deleterious substances, such as stones and soil, and of storage pests. The moisture content of cereals should be determined by a moisture meter. Any batch containing more moisture than 120–140 g/kg should be rejected unless it can be dried to this level of moisture or less. This is to prevent spoilage during storage.

Every quality control programme should include periodic laboratory analysis of ingredients and mixed feed. Care should be taken to ensure that samples are representative, by taking samples from various points in the batch, mixing them and then drawing a subsample of 500–1000 g for analysis. For bagged material a common approach is to collect a sample with a probe from 10–15% of the bags in the batch. It is advisable to use only one-half of the sample for laboratory analysis and to retain one-half as a back-up in case of subsequent problems, discrepancies or disputes.

Tests

A fairly standard analysis to be conducted on feed ingredients is to determine the moisture, CP, oil, crude fibre (more recently acid-detergent fibre and neutral-detergent fibre) and ash (mineral matter) using approved chemical or physical methods. The results (so-called proximate composition) indicate whether the feedstuff meets acceptable guidelines. It will not be possible to analyse all batches of ingredients. Only periodic tests will be possible.

Additional tests may be carried out, for instance to determine the ratio of ash soluble and insoluble in acid, which indicates the amount of sand or soil contaminating the feed. This is a useful test for root crops such as cassava.

Tests for mycotoxins may be required if harvesting conditions or clinical signs are indicative of possible mycotoxin contamination (see Chapter 7). Regular tannin analyses are advised in countries that use high-tannin sorghums.

Mixed feed should be sampled and analysed at least every 2–3 months for dry matter, CP, calcium, phosphorus and diet particle size (mash diets). In general, tests should be carried out frequently if problems persist, and less frequently when no problems occur.

Producers using purchased complete feed do not have to pay as much attention to quality control. Feed manufacturers are required by feeds legislation in most countries to supply feed of a high standard, which requires them to test ingredients and finished products routinely. In addition, regulatory officials check commercial feeds at random for label compliance. However, it is still advisable for the organic producer to carry out routine checks on any purchased feeds and to communicate regularly with suppliers to ensure that the purchased products meet the standards expected.

Sampling

Samples for laboratory analysis need to be representative. The easiest way to take a representative sample of grain grown on the farm is to take it during harvest. Several 500 g samples should be taken from each truckload and bulked. The bulk should then be mixed thoroughly and a 500 g sample drawn. Mixed feed should be sampled as it is discharged from the mixer, bulked and sampled as with grain.

Small production units without access to laboratory facilities should seek the help of feed suppliers in carrying out periodic analysis of ingredients. Other agencies such as government, university or commercial laboratories may have to be used.

Addendum

The calculated analyses of the mixtures above cannot be guaranteed, due to ingredient variability. Producers using such formulas should have the diets analysed periodically to ensure acceptability. In order to ensure maximal utilization of the nutrients contained in grains and protein feedstuffs it is helpful that an appropriate supplement of enzymes be included in diets, according to the directions of the supplier and the prevailing organic regulations.

References

Aganga, A.A., Aganga, A.O. and Omphile, U.J. (2003) Ostrich feeding and nutrition. *Pakistan Journal of Nutrition* 2(2), 60–67.

Amerah, A.M., Ravindran, V., Lentle, R.G. and Thomas, D.G. (2007) Feed particle size: implications on the digestion and performance of poultry. *World's Poultry Science Journal* 63, 439–455.

Bayley, H.S., Pos, J. and Thomson, R.G. (1975) Influence of steam pelleting and dietary calcium level on the utilization of phosphorus by the pig. *Journal of Animal Science* 46, 857–863.

Bellof, G. and Schmidt (2014) Effect of low or medium energy contents in organic feed mixtures on fattening and slaughter performance of slow or fast growing genotypes in organic turkey production. *European Poultry Science* 78, ISSN 1612–9199.

Bennett, C. (2013) Organic Diets for Small Flocks. Manitoba Agricuture publication online. Available at: https://www.gov.mb.ca/agriculture/livestock/production/poultry/organic-diets-for-small-poultry-flocks.html (accessed 17 July 2018).

Blair, R., Dewar, W.A. and Downie, J.N. (1973) Egg production responses of hens given a complete mash or unground grain together with concentrate pellets. *British Poultry Science* 14, 373–377.

Bolton, W. and Blair, R. (1974) *Poultry Nutrition*. Bulletin 174, Ministry of Agriculture, Fisheries and Food. HMSO, London.

Coehlo, M.B. (1994) Vitamin stability in premixes and feeds: a practical approach. *BASF Technical Symposium*, Indianapolis, Indiana, 25 May, pp. 99–126.

Lampkin, N (1997) *Organic Poultry Production*. Final report to MAFF. Welsh Institute of Rural Studies, University of Wales, Aberystwyth, UK.

Lewis, P.D., Perry, G.C., Farmer, L.J. and Patterson, R.L.S. (1997) Responses of two genotypes of chicken to the diets and stocking densities typical of UK and 'Label Rouge' production systems: I. Performance, behaviour and carcass composition. *Meat Science* 45, 501–516.

NRC (1994) *Nutrient Requirements of Poultry*, 9th revised edn. National Research Council, National Academy of Sciences, Washington, DC.

Parsons, S., Buchanan, N.P., Blemings, K.P., Wilson, M.E. and Moritz, J.S. (2006) Effect of corn particle size and pellet texture on broiler performance in the growing phase. *Journal of Applied Poultry Research* 15, 245–255.

Rack, A.L., Lilly, K.G.S., Beaman, K.R., Gehring, C.K. and Moritz, J.S. (2009) The effect of genotype, choice feeding, and season on organically reared broilers fed diets devoid of synthetic methionine. *Journal of Applied Poultry Research* 18, 54–65.

Scheideler, S.E. and Sell, J.L. (1997) *Nutrition Guidelines for Ostriches and Emus*. Publication PM-1696. Extension Division, Iowa State University, Ames, Iowa.

Scott, M.L. and Dean, W.F. (1999) *Nutrition and Management of Ducks*. M.L. Scott of Ithaca, Ithaca, New York.

Ullrey, D.E. and Allen, M.E. (1996) Nutrition and feeding of ostriches. *Animal Feed Science and Technology* 59, 27–36.

6

Choosing the Right Breed and Strain

A wide range of breeds and strains of poultry is available for organic production internationally, displaying greatly different growth, meat and egg production characteristics and responding differently to diet composition and feeding system. In contrast with other farm stock, most of the modern strains available are being bred by a few international companies, the birds being developed for large-scale production in specialized units. These poultry-breeding companies have so far paid little attention to the special need of birds for organic production, mainly because of the size of the industry. Some of the modern strains are suitable for organic production. In other cases traditional, relatively unimproved strains are more suitable. Therefore, the dietary regime and feeding programme need to be tailored to the particular genotype of bird selected.

The following considerations are relevant to the choice of genotype.

Consumer Attitudes

Organic production is largely consumer-driven; therefore, it is important to take into account consumer attitudes in selecting the appropriate breeds and strains for organic poultry production. Some consumers purchase whole chickens for meat, others prefer chicken parts. A yellow-skinned or coloured-skin bird is preferred by some consumers, while others prefer white-skinned birds. Some consumers prefer white-shelled eggs, while others prefer eggs with tinted or coloured shells. Some consumers prefer eggs with highly pigmented yolks.

There are four major sectors of poultry produce: chicken meat; turkey meat; eggs; and niche products such as game birds, waterfowl (ducks and geese), ratites (ostriches and emus), squabs (pigeons), silkie chickens, quail and quail eggs and game birds (pheasants, partridges, tinamou). All are produced organically in various parts of the world, the sectors differing in economical value in different regions.

One of the most striking consumer trends in recent years has been the increasing demand for natural and healthy foods where ethical issues (such as animal welfare and health) are also taken into consideration (Andersen *et al.*, 2005). Safety has also become a very important issue of concern in modern food production, prompted mainly by several health crises (hormones, bovine spongiform encephalopathy (BSE), antibiotics, dioxin contamination of feed, etc.).

The purchase of organic poultry produce appears to be governed by two main factors: (i) the perceived quality based on appearance, price, presentation and labelling; perceived freedom from chemical residues, etc.; and the actual quality experienced after cooking and eating; and (ii) ethical and philosophical considerations such as the housing and welfare of the birds.

The relative importance of these factors appears to vary according to region or country, as exemplified in Scandinavian findings. A study conducted in 1996 showed that in Denmark, 22%, 11%, 33%, 24% and 19%, respectively, of consumers purchased organically produced bread, meat, eggs, vegetables and dairy products (Borch, 1999). The corresponding figures were 13%, 12%, 19%, 19% and 13% of consumers in Sweden and 11%, 9%, 17%, 16% and 11% in Norway. The proportions of consumers in the three countries who never purchased organic food were 33%, 35% and 49%, respectively. There were considerable differences among the three countries in the stated reasons for organic purchases, with Danish and Norwegian consumers stating that the main reason for buying organic food was their belief that it was healthier and of better quality than intensively produced food, whereas the main motive for Swedish consumers was their concern for the environment and animal welfare. Swedish and Norwegian consumers trusted the accuracy of the organic trademark, though the Danes were more sceptical. Consumers in all three countries were willing to pay a higher price for organically produced food.

Consumers in the USA also are increasingly interested in both free-range and organic chicken (Alvarado et al., 2005). In a study conducted by these authors, consumers were asked to compare the eating quality and shelf life of meat from organic free-range broilers to meat from commercially raised broilers. Free-range chicken breasts were significantly larger (153 g) than commercial chicken breasts (121 g), attributed to increased exercise and increased age of bird. No significant differences in breast fillet tenderness or composition were noted between the two types of meat. Breast fillets from the free-range birds had higher pH values (5.96 versus 5.72, respectively) and were darker (49.14 versus 53.46 units, respectively) than fillets from the commercially raised birds. Free-range fillets had significantly higher aerobic plate count (APC) and coliform count and exhibited signs of spoilage earlier than commercial fillets – findings of potential consumer importance. Consumers found no difference in fillet juiciness, tenderness or flavour. Commercial breast fillets, however, were preferred to free-range fillets. Trained panellists found no difference in tenderness or flavour of drumsticks, but found meat from free-range birds to be juicier and with a stronger attachment of the meat to the bone. Commercial and free-range chicken had many similarities in meat quality and sensory attributes, but meat from free-range poultry had a shorter shelf life than meat from commercially raised chickens, confirming other findings related to the higher content of polyunsaturated fatty acids (PUFA) in organic poultry meat.

Castellini et al. (2002a) noted that organic chickens had higher yields of breast and drumsticks and lower levels of abdominal fat. The muscles had lower ultimate pH and water-holding capacity, which resulted in a higher cooking loss. They also found that lightness values, shear values and contents of iron and PUFA were higher. The sensory quality of breast muscle was also found to be higher than in conventionally raised broilers.

A UK study showed that consumers bought organic eggs because they were perceived as being healthier, free of chemicals and genetically modified materials, and because they tasted better (Stopes et al., 2001). In addition, consumers expected that the laying flocks were maintained under more humane and improved welfare conditions. This research confirms that consumer acceptance of organic poultry products depends to some extent on the nature of the production system and that consumers expect a land-based production system, particularly one based on a small flock size.

Several studies have been conducted in Europe to investigate consumer perceptions of the welfare of laying hens. A French study was a qualitative study in which 38 consumers participated in group meetings (Mirabito and Magdelaine, 2001). The second was a public opinion poll of 982 consumers. More than 95% of those interviewed stated that freshness and safety were the main criteria for egg purchase, but the first study also showed the importance of packaging, quality and brand. Generally, the ideal production system was considered to be one based on relatively few hens, natural feed and freedom of movement. Eighty-five per cent of respondents were of the opinion that free-range systems resulted in fresh (or safe) eggs compared with 27% for battery systems. Ninety-five percent of respondents interviewed thought that keeping laying hens outside was the best system to improve bird welfare. Respondents were divided on concerns over bird welfare and their willingness to pay extra for free-range eggs. Eighteen per cent were unconcerned about bird welfare and were not willing to pay extra for free-range eggs; 39% were concerned and prepared to pay 0–50% more; 27% were very concerned, were already buying organic products and were willing to pay 50% more.

Public awareness of current food issues and economic purchasing power are other factors related to the purchase of organic meat and eggs. O'Donovan and McCarthy (2002) examined the Irish consumer preference for organic meat and identified three groups of consumers. Respondents who purchased or intended to purchase organic meat placed higher levels of importance on food safety when purchasing meat, compared with those with no intention of purchasing organic meat. Furthermore, purchasers of organic meat were more concerned about their health than non-purchasers. Purchasers of organic meat also believed that organic meat was superior to conventional meat in terms of quality, safety, labelling, production methods and value. Availability and the price of organic meat were the key deterrents to the purchase of organic meat. Higher socio-economic groups were more willing to purchase organic meat. It was concluded that increasing awareness of food safety and pollution issues were important determinants in the purchase of organic meat but that securing a consistent supply of organic meat was paramount to ensuring growth in this sector.

The above studies suggest several important conclusions. Firstly, organic poultry meat and eggs should be produced in such a way that they meet the expectations of consumers both before and after purchase. Secondly, the willingness of consumers to pay a premium for organic produce is not unlimited. These conclusions indicate that organic poultry producers need to strive to produce a high-quality product as economically as possible.

Types of Poultry

Chickens are the most abundant birds in the world and provide the most popular choice of meat worldwide.

The many breeds of modern chicken found worldwide today are believed to be descendants of the red junglefowl (*Gallus gallus*). The red junglefowl can still be seen in the wild in the forests of South-east Asia, Pakistan and India, providing an excellent tool for studying the genetic changes that have occurred with domestication and genetic selection.

There is a growing movement among alternative and organic poultry producers to use heritage breeds. A list of heritage breeds suited to specific areas can typically be obtained from regional associations involved in the preservation of endangered breeds.

Over the decades, there has been intensive selection for traits preferred by humans. Today, commercial chickens can be divided into two categories: those kept for table egg production and those raised for meat.

The choice of chicken breed for use in an organic production system should take into consideration the ability of the breed or strain to adapt to the local conditions. Most organic regulations require, or encourage, providing poultry with access to the outdoors.

When compared with other domesticated species (both animal and plant), poultry has had the most intense selection because of the increased generation turnover. Careful selection can reverse this process as animals adapt to new environments such as those found in organic farming (Boelling *et al.*, 2003).

Egg production

Genotypes suitable for organic production

Egg production in many countries is now highly commercialized, using breeding stock

developed by a few multinational companies. For white egg production, especially in
North America, the available genetic strains
have been developed from the Single Comb
White Leghorn. For brown egg production,
the genetic strains have been developed
mainly from a cross of Rhode Island Red
and Barred Plymouth Rock.

Most of the currently available commercial egg-laying breeds have been selected for
high egg production in a cage-housing system and may not be suitable for the management practices used in organic production.
The body weight of the commercial laying
hen is twice that of the wild junglefowl and
it has an annual egg production more than
ten times that of the junglefowl. Jensen (2006)
compared the behaviour of junglefowl and
commercial strains of Leghorn hens and
observed important differences, reporting
that the Leghorn hens were less active as
noted by reduced foraging and exploratory
behaviour. These hens also showed a lower
frequency of social interactions and a less
intense reaction to predators. While the
Leghorn is the chosen breed for commercial
white egg production in many areas of the
world, the reduced foraging behaviour suggests that it is not the ideal choice of breed
for an organic operation or any operation
that involves outdoor access.

Commercial strains of egg layers have
adapted to produce many eggs while housed
in cages. This environment results in the
suppression of many normal behavioural
traits (Boelling et al., 2003), some of which
reappear when the hens are housed in a
free-roaming or free-ranging management system. One of the most detrimental behaviours
is feather pecking with subsequent cannibalism and mortality. Additionally, hens selected
for production of a high number of eggs in a
cage appear to lose their need to go to a particular nest for egg-laying. As a result, cage-
adapted hens tend to lay a large number of
eggs on the floor (Sørensen, 2001).

Feather pecking is a common problem
in commercial poultry production systems.
Huber-Eicher and Audigé (1999) surveyed
Swiss pullet-growing farms having more
than 500 birds. The association between
feather pecking and several management

variables was compared. On the basis of their
findings these authors recommended low
bird densities during rearing of less than
10 birds/m2 and the provision of elevated
perches (≥ 35 cm).

Researchers have demonstrated that some
undesirable behavioural traits related to an animal's ability to adapt to its environment have
a genetic component. Thus, an assessment of
fear, dominance ability and social behavioural
traits should be introduced into a breeding
programme in order to make organic production more economically sound and more
animal welfare friendly (Jones and Hocking,
1999; Boelling et al., 2003).

Feather pecking is a problem in all management systems but is harder to control in a
free-range system. It has been suggested that
feather pecking in a chicken flock is the result
of redirected ground pecking related to foraging. The research of Rodenburg et al. (2004),
however, contradicts this theory. They studied feather pecking in young chickens and
found that feather pecking could be observed
as early as 1 day after hatching, when ground
pecking is not fully developed and dust-bathing
is rarely observed. They concluded that
feather pecking was an exploratory behaviour
that was important in the development of the
social life of chicks. Rodenburg et al. (2004)
also studied the feather pecking behaviour
of older chickens. They found that feather
pecking increased with the likelihood of
encountering unfamiliar chickens. The pecks
were preferentially directed at unfamiliar
chickens. The frequency of feather pecking
decreased with time after encountering unfamiliar chickens. Consequently Rodenburg
et al. (2004) concluded that since feather
pecking is a normal behaviour that plays a
role in social exploration it is less likely to be
a redirected behaviour.

Su et al. (2005) demonstrated that it is
possible to select for, or against, feather-
pecking behaviour in chickens. After only
one generation, lines were developed with
large differences in production and incidence of feather pecking.

Rodenburg et al. (2004) also compared
behavioural, physiological and neurobiological
characteristics of high feather-pecking (HFP)
and low feather-pecking (LFP) populations.

The two lines of chickens were not selected for feather pecking but instead originated from different selection criteria and the difference in the level of feather pecking was coincidental as a result of the selection programme. They observed that the HFP line had a proactive coping behaviour while the coping behaviour for the LFP line was reactive. In addition, the LFP chickens showed higher levels of feed and foraging behaviour, suggesting that their behaviour is more externally motivated.

Most research related to feather pecking has focused on the activity of the aggressive chickens that are the peckers. Interest has shifted to the victims and whether or not such a chicken might be predisposed to be the recipient of feather pecking. Kjaer and Sorenson (1997) reported that the likelihood of being a victim of feather pecking was heritable in young chickens but not in adults. Similarly, Buitenhuis et al. (2003b) demonstrated that the genes that regulate gentle feather pecking at 6 weeks of age are different from those involved in feather pecking at 30 weeks of age. They suggested that the feather pecking problem could be resolved using molecular genetics (Buitenhuis et al., 2003a).

Jensen (2006) was able to identify the genetic mutation responsible for the white phenotype of chickens and one of these genes was also found to have an influence on feather pecking. Specifically, Jensen (2006) found that the plumage colour was related to the risk of being a victim of feather pecking. In his research, chickens that were homozygous for the wild genotype were significantly more likely to become victims of feather pecking when compared with heterozygous and homozygous mutants (both of which were white in colour). Jensen (2006) concluded that lack of feather pigmentation reduced the risk of being a victim of feather pecking and speculated that this is the reason for the development of domesticated white phenotypes for commercial production. Organic poultry systems typically do not use white-coloured breeds, which may explain the increase in feather pecking often observed with these systems.

Su et al. (2006) showed that a breeding programme focusing on selection for reduced feather pecking resulted in a change in egg production, egg quality and feed efficiency. The LFP line showed an increase in egg numbers and feed conversion efficiency. It was suggested that the improved feed conversion efficiency was the result of the lower requirement for maintenance energy for the chickens with better plumage cover. The researchers did note, however, that the HFP line had greater egg weight, albumen height, shell thickness and yolk percentage.

Cannibalism is another problem reported in organic laying hens. It is known that several factors are involved in triggering cannibalism in a flock. These include breed, feed composition, rearing environment, external parasites and other management factors (Berg, 2001).

Some organic egg producers are also interested in producing value-added eggs such as those with increased omega-3 content. Scheideler et al. (1998) reported genetic differences in feed consumption and utilization and storage of dietary fats, and also strain–diet interactions affecting egg yolk composition. These interactions may be relevant in a breeding programme when the production on an organic, low-lipid egg is the objective.

Specific breeds and strains

The use of local breeds in a breeding programme should be encouraged. These breeds and strains are better adapted to the local environment and have been shown to have better disease resistance and a better ability to escape predators (Sørensen, 2001). One country that may be ahead of other countries in developing layers for organic production is Denmark. Until 1980 Denmark did not allow for the housing of hens in cages. Since then selection of breeding stock has been based on the performance of hens kept in floor systems. One of the results of this selection process is the Danish Skalborg breed.

Sørensen (2001) compared performance of the Danish Skalborg and international hybrids (Shaver and Lehman) in both cage and floor systems. The hybrids housed in cages produced 8% more eggs than those housed in the floor system. The Skalborg hens produced at the same rate in both systems. The

Skalborg, however, had five times the rate of mortality when housed in cages than in the floor system. The hybrids housed on the floor had 1.5 times the mortality of those housed in cages. Sørensen and Kjaer (1999) also demonstrated that cannibalism was less in non-commercial alternative breeds raised under organic conditions. From this research it is clear that the Skalborg breed is genetically adapted to a floor system and more suitable for organic production systems than conventional hybrids.

Swedish researchers selected strains for egg production, using a diet mixed with home-grown cereals or diets with a low protein level (130 g/kg). The hens were a Rhode Island Red × White Leghorn, which resulted in the development of the SLU-1329 hen (Abrahamsson and Tauson, 1998) (Table 6.1). This Swedish hen was tested in aviary and free-range conditions and compared with conventional hybrids (Lohmann LSL, Hisex white and Hisex brown) using a low-protein diet and housed in a floor system. The Swedish hen had the same or higher egg production than the hybrids, but had better overall feed conversion efficiency (Sørensen, 2001).

North American organic egg producers use a variety of different breeds and/or strains. In an on-farm research project, Peterson (2006) compared alternative breeds using Leghorns as the standard for comparison. The breeds tested included Speckled Sussex, Silver Gray Dorking and Buff Plymouth Rock. The Leghorn hens had the highest egg production, followed by the Speckled Sussex and the Buff Plymouth Rock. The Dorking had the lowest level of egg production. The three alternative species were all larger than Leghorns and, as a result, production costs for these breeds were approximately 1.5–2 times as much as those for Leghorns.

In general, traditional breeds of laying hens have a higher body weight than modern hybrids and a lower rate of lay (Rizzi and Chiericato, 2010). Egg quality is also likely to differ, as exemplified in Table 6.2 (Rizzi and Marangon, 2012). In the trial (Table 6.2) in question the eggs from modern hybrids were heavier than from traditional breeds, but the percentage of yolk was higher, the percentage of albumen was lower and the percentage of shell lower in the eggs from traditional breeds. Yolk cholesterol content was higher in the eggs from traditional breeds than in the eggs from hybrids (258 versus 219 mg/yolk).

Several standard-sized alternative egg-laying breeds are available worldwide. In the USA this includes Rhode Island Red, New Hampshire, Barred and White Plymouth Rocks, and Buff Orpington. There appears to be a lack of published peer-reviewed research comparing the productivity of these breeds in an organic production system. Many organic egg producers in the USA keep a

Table 6.1. Comparison of the Danish Skalborg hen with international hybrids in 1978 in a floor system, and in 1982 in a cage system. All breeds of White Leghorn type. (From Abrahamsson and Tauson, 1998.)

	Test in floor system 1978		Test in cage system 1982	
Hybrids	Eggs in 365 days/hen placed	Eggs in 365 days/hen placed	Eggs in 365 days/hen placed	Eggs in 365 days/hen placed
Shaver	265	274	278	298
Babcock	259	264	–	–
Hisex	264	267	–	–
Lohmann	259	268	276	285
Dekalb	–	–	264	292
Average of above hybrids	262	268	273	292
Skalborg	262	267	240	266

Table 6.2. Effects of hen genotype on egg weight, egg components and egg quality (from Rizzi and Marangon, 2012).

	Hy-Line Brown	Hy-Line White	Ermellinata Di Rovigo	Robusta Maculata
Egg weight, g	62.9[A]	60.4[B]	54.4[D]	56.5[C]
% albumen	64.7[A]	63.5[B]	60.2[C]	60.5[C]
% yolk	24.6[D]	26.2[C]	29.9[A]	28.4[A]
% shell	10.8[B]	10.4[C]	9.98[D]	11.1[A]
Albumen height, mm	8.06[A]	7.97[A]	6.77[A]	7.00[A]
Albumen pH	8.11[C]	8.11[D]	8.23[D]	8.16[B]
Haugh units	88.8[A]	88.7[A]	83.1[A]	84.0[A]

A–D: Means within same rows followed by a different superscript are significantly different ($P < 0.01$).

flock for multiple years and run multiple ages together. The choice of a different coloured breed each year makes it possible to keep track of the age of each of the hens in these flocks.

Chicken meat production

Genotypes suitable for organic production

Traditionally chickens have been selected for growth rate and feed conversion efficiency. The result has been an abundant supply of chicken meat at a price affordable for most of the general population. Unfortunately, such selection has had negative consequences (Emmerson, 1997). There has been an increased incidence of ascites (also known as water belly) and sudden death syndrome (SDS), a reduction in reproductive performance and immune competence, and an increase in skeletal abnormalities. Kestin *et al.* (1999) compared the incidence of leg weakness in four commercial broiler crosses. Large differences in walking ability and other measures of leg weakness were found. Incidence of ascites has been reported to be heritable and correlated positively with body weight (Moghadam *et al.*, 2001). As a result, selection programmes based on body weight are likely to increase, or at best maintain, current levels of this metabolic disorder.

Most commercial chicken meat production has become industrialized and in many cases involves enclosed housing without access to outdoors. In such conditions the environment is strictly controlled and genotype–environment interactions can be disregarded. These interactions become more important when the birds are raised organically.

Incidence of ascites and SDS has been shown to increase in suboptimal conditions such as heat or cold stress. In addition, growth rate is depressed during heat stress. This suggests that strains with higher potential growth rate under normal conditions are more likely to suffer from ascites under cold stress. In addition, there is a negative correlation between growth under heat stress and incidence of ascites, indicating that strains whose growth is depressed under heat stress are more likely to suffer from ascites under cold stress (Deeb *et al.*, 2002).

Castellini *et al.* (2002b) compared the carcass and meat quality of the same breed and strain of broiler chickens (Ross male) raised conventionally (an indoor pen allowing 0.12 m²/bird) or organically (an indoor pen allowing 0.12 m²/bird with access to a grass paddock allowing 4 m²/bird). At 56 and 81 days of age, 20 chickens per group were slaughtered to evaluate carcass traits and the characteristics of breast and drumstick muscles (m. pectoralis major and m. peroneus longus). Results showed that carcass yield was significantly different between the two management systems in that organically produced carcasses had a higher proportion of breast and drumstick meat and a lower level of abdominal fat (Table 6.3). In terms of meat quality, the organic chickens had lower water-holding capacity, increased cooking loss and increased muscle shear value (indicating increased

Table 6.3. Comparison of the carcass and meat quality of the same breed and strain of broiler chickens (Ross male) raised conventionally and organically (from Castellini et al., 2002b).

	Conventional		Organic	
	56 days	81 days	56 days	81 days
Live weight (g)	3219	4368	2861	3614
kg feed/kg gain	2.31	2.89	2.75	3.29
Eviscerated weight (g)	2595	3529	2314	2928
Abdominal fat (g/kg)	19.0	29.0	9.0	10.0
Breast (g/kg)	220	235	232	252
Drumstick (g/kg)	148	150	149	155
Breast measurements				
Moisture (g/kg)	755.4	748.5	762.8	757.8
Lipids (g/kg)	14.6	23.7	7.2	7.4
pH	5.96	5.98	5.75	5.80
Cooking loss (%)	31.1	30.3	34.0	33.5
Shear value (kg/cm^2)	1.98	2.10	2.25	2.71

toughness). A sensory panel ranked organic chicken higher in terms of juiciness and overall acceptability. The organic chickens had a higher proportion of saturated and reduced levels of monounsaturated fatty acids (MUFA) in the meat. Most importantly, the meat of organic chickens had higher levels of PUFA, specifically levels of eicosapentaenoic (EPA), docosapenaenoic (DPA) and total n-3 fatty acids. These omega-3 fatty acids have been shown to be beneficial to human health and development. However, such a fatty acid profile is associated with a shorter shelf life, due to oxidative rancidity. Castellini et al. (2002b) speculated that the higher level of omega-3 fatty acids was the result of grass consumption.

Specific breeds and strains

European organic regulations set specific rules with regard to breed choice. According to EEC-Regulation 1804/1999:

> When selecting the breed or strains, the livestock's ability to adapt to the conditions of their surroundings, their vitality and their resistance to diseases shall be taken into account. Breeds and strains used in intensive livestock production that are prone to diseases and other health problems must be avoided. Preference must be given to indigenous breeds and strains of breeds adapted to local conditions.

In addition, poultry must be adapted to the outdoor environment and have a long rearing period with a minimum slaughter age of 81 days (European Commission, 2007).

In many countries it is difficult to obtain slow-growing broiler strains. As a result, most producers are using those breeds that have been selected for rapid growth rate and high feed conversion efficiency. Fortunately, many poultry-breeding companies have an increased interest in the development of slow-growing strains suitable for organic production systems (Katz, 1995; Saveur, 1997).

The choice of slower-growing meat strains may ultimately result in a return to the type of bird used in the past. Havenstein et al. (2003) compared the production characteristics of 1957 and 2001 strains of broilers fed diets representative of those fed in 1957 and 2001. The results showed that genetic selection had had a much greater influence on broiler growth than diet, about 85–90% of the change in growth characteristics being attributed to selected genotype and only 10–15% to improved nutrition.

The choice of breed also depends on the final market for the birds. The consumer preference for chicken meat varies dramatically throughout the world. For example, in North America consumers prefer yellow skin colour while in Europe they prefer white skin. Similarly, in East Asia and Europe consumers prefer a more tasty chicken meat produced in less confined conditions (Yang and Jiang, 2005). In North America the breed of chicken most readily available to producers is a white-feathered Cornish cross. In many areas, however, breeds have been developed that are slower-growing and have coloured feathers.

Slow-growing strains raised with outdoor access and harvested at a greater age have been shown to have firmer meat and be more flavourful than those in conventional production. In a European consumer study they were preferred to conventional poultry meat (Touraille *et al.*, 1981).

There are two distinct markets for meat-type chickens: live birds and processed carcasses. For the live-bird market the most important characteristics for the consumers include feather colour, skin and shank colour, the redness and size of the comb, and body conformation. Although the Australorp was originally developed as an egg layer in Australia, its black feathers and shanks make it popular for ethnic groups shopping at live-bird markets. More recently, coloured-feather Cornish crosses have become available and are being raised for the live-bird markets. Chickens with black skin and shanks are believed by some consumers to have medicinal properties. The silkie chicken has been raised to meet this consumer preference. They are reported to have a concentration of phosphoserine (with aphrodisiac effect) 11 times that of conventional broilers (Lee *et al.*, 1993).

In China, the Three Yellow (3Y) breed is popular for chicken meat production (Yang and Jiang, 2005). The 3Y designation refers to the yellow feathers, yellow skin and yellow shanks which are popular in Southern China. Yellow is a traditional symbol of fortune and luck in most parts of China. Conversely, the colour white is considered unlucky. The 3Y is considered a slow-growing breed in that it takes 100 days to reach a market weight of 1.2–1.5 kg. The meat is harvested close to sexual maturation and is considered to be more flavourful than meat from conventional broilers, as the meat is firm but not tough.

Label Rouge production in France also involves slow-growing breeds. This production system has the following components, allowing the produce to be given the Label Rouge mark under a 1960 French government law (King, 1984): (i) a slow-growing chicken; (ii) feed low in fat and high in cereals; (iii) low stocking rates; (iv) a minimum rearing period of 81 days; and (v) strict processing conditions and quality grading.

Under this system the chickens reach a market weight of 2.25 kg in 12 weeks. The resulting carcass is longer than that of conventional broilers and has a smaller breast and larger legs (Yang and Jiang, 2005). Label Rouge chicken production is a pasture-based system. The mild climate of France allows for year-round production, but this would not be possible in many parts of the world because of the colder winters (Fanatico and Born, 2002).

Lewis *et al.* (1997) compared the production of Label Rouge (ISA 657) and conventional broilers (Ross 1) in the UK. The slower-growing ISA 657 birds weighed 1534 g on average at 48 days, whereas Ross 1 birds weighed 2662 g. At 83 days of age the respective weights were 2785 g and 4571 g. Based on market ages for the two groups (83 and 48 days, respectively) the production values were: live weight 2785 g and 2662 g; feed intake 8257 g and 5046 g; and feed required per kilogram gain 3.01 kg and 1.96 kg. Mortality was notably different at 0% and 11.3%, respectively. These data provide important economic information for producers planning to adopt the Label Rouge system.

A few entrepreneurs have worked towards introducing a Label Rouge type of poultry production system in the USA, using the Redbro Cou Nu breed, which is a red-feathered naked-neck chicken with a distinctive taste, a thin, translucent skin, an elongated breast, a high keel bone and long legs. Another region-specific chicken meat product is Poulet de Bresse, 'Gauloise' blue-legged chickens raised in the Bresse region of France. Collaboration between a Canadian poultry breeder and an American poultry producer has resulted in an American facsimile – referred to as the Blue Foot chicken. Like the French counterpart, the Blue Foot chicken has a red comb, white feathers and steel-blue feet. The blue-coloured feet are typically left on when the cooked chicken is placed on the table.

In an Italian study, the meat quality of three breeds available for chicken meat production was compared under organic production systems – the fast-growing Ross, medium-growing Kabir and slow-growing Robusta maculate (Castellini *et al.*, 2002a). The Ross and Kabir chickens were slaughtered at 81 days of age but the Robusta maculate required 120 days to reach market weight (> 2 kg). Chickens of the slower-growing breed showed a better adaption to the extensive rearing conditions. The fast-growing chickens, however, were reported to show an 'unbalanced muscle development and reduced oxidative stability'.

Where no slow-growing broiler strains have been developed for use in specific areas, some local pure breeds should be considered. These breeds may be more suitable for extensive production systems and their use may also help to prevent them from becoming extinct. Lists of these local heritage breeds can be obtained from area preservation societies.

Dual-purpose breeds

For many producers the ideal chicken for organic production is one of the dual-purpose breeds, i.e. one of those developed for both egg and meat production (Fig. 6.1). These breeds fit into organic poultry systems more appropriately than breeds developed specifically for egg or meat production. Many are heritage breeds and have a much lower incidence of health problems such as ascites and SDS than found with commercial meat breeds. The meat from dual-purpose breeds is not available until the end of the laying period when the hens are replaced. Surplus cockerels, however, can be marketed earlier.

In general, they are more docile and hardier than Leghorn-type birds but require more feed. Most produce brown-shelled or coloured eggs. The meat from these breeds at the end of lay is considered to have more flavour than the meat from young birds. Among dual-purpose breeds are the following:

Fig. 6.1. Example of a dual-purpose breed of chickens.

- Rhode Island Red. This popular breed has been used to produce many cross-bred varieties available today. It is a good producer of large brown eggs and is quiet and easy to handle. Both males and females have dark red plumage. At the end of lay the hens weigh approximately 2.5 kg.
- Barred Plymouth Rock. This is another heritage breed and is still used in some countries because of its good meat qualities, combined with good production of brown eggs. Both hens and cockerels are grey-barred, the hens weighing 2.5–2.75 kg.
- New Hampshire × Barred Rock. This cross is from two of the oldest heritage breeds and produces a very hardy chick. The birds are reported to be very quiet, with attractive plumage. The hens have a red comb and a jet-black body laced with brown on the neck and breast. They lay brown-shelled eggs and reach about 2.75 kg at the end of lay. Males have dark-coloured bars.
- Rhode Island Red × Columbian Rock. This bird is reputed to be a very hardy

dual-purpose breed and has had excellent performance in small flocks over the past 30 years. The pullets are reddish-brown in colour, very quiet and easy to handle, and have a body weight of about 2.75 kg at the end of lay. The eggs are rich brown in colour, with good shell texture and interior quality. The cockerels are white with black markings.

- Other dual-purpose breeds developed by large breeding companies include the Shaver Red Sex-Link and the Harco Black Sex-Link, which is reported to be one of the best producers of large brown eggs.
- Chantecler. This breed was developed in the Canadian Province of Quebec and has an interesting history as a dual-purpose breed (Cole, 1922). A monk, Brother Wilfred Chatelain of the Cistercian Abbey in Oka, Quebec, set out to create a breed of chicken that could stand the harsh climate of that part of Canada, and that could be used for both egg and meat production. From the French *chanter*, 'to sing', and *clair*, 'bright', the Chantecler was the first Canadian breed of chicken. Although work began on this breed in 1908, it was not introduced to the public until 1918, and admitted to the American Poultry Association Standard of Perfection in 1921. The Chantecler was created by first crossing a Dark Cornish male with a White Leghorn female, and a Rhode Island Red male with a White Wyandotte female. The following season pullets from the first cross were mated to a cockerel from the second cross. Then selected pullets from this last mating were mated to a White Plymouth Rock male, producing the final cross. A small comb and wattles allow this breed to withstand the cold eastern Canadian winters without the problem of frostbite. In addition to being very hardy, the breed is noted for being an excellent layer of brown eggs and has a fleshy breast. It is also noted for being calm and easy to manage. Like several other heritage-type poultry, the stocks are critically low.

- Favorelle. This is a white-skinned chicken that was initially developed in France as a dual-purpose breed. It has excellent laying qualities and its performance does not change much with the different seasons. Favorelles are hardy and active and adapt easily to free-range systems.

Turkeys

Genotypes suitable for organic production

Despite the name, turkeys have no relation to the country of Turkey. Turkeys are native to North America and were first domesticated by the Aztecs. The many varieties of the domesticated turkey that exist today are descendants of the wild turkey (*Meleagris gallopavo*). The other species of turkey is *Meleagris ocellata*, the ocellated turkey (the tail feathers of both sexes are bluish-grey with an eye-shaped, blue-bronze spot near the end with a bright gold tip), which is found in southern Mexico. Wild turkeys were hunted by the early American colonists.

When the Spanish explorers reached South America they found that turkey was a major source of protein (meat and eggs) for the Aztecs. They also used the feathers for decorative purposes. The Spanish took the turkey back to Europe, where genetic selection resulted in the development of different varieties (e.g. Spanish Black and Royal Palm).

Genetic selection for commercial turkey production has resulted in broad-breasted varieties that have rapid growth and high feed conversion efficiency. The modern varieties have lost their natural ability to fly or forage for feed. Artificial insemination is routinely used with turkey breeders to avoid injury of the hens by the much larger toms and because the broad breasts have resulted in most males having a conformation that renders them incapable of natural mating.

Wild turkeys still exist in many parts of the USA. While they are the precursors of the modern turkey, they taste very different. In wild turkeys almost all the meat is considered dark, including the breast.

Several factors are involved in creating the flavour of poultry meat – the natural flavour of the meat, the age of the bird and how it was raised. Meat from older turkeys has more flavour than from younger birds. Heritage turkeys grow more slowly than commercial birds and, as a result, tend to have a more intrinsic flavour. They are typically slaughtered at 7–8 months of age, whereas commercial turkeys are ready for market in 3–4 months. With increased physical activity, as would occur in an organic production system, there is an increase in the flavour of the meat. In addition, the meat of turkeys consuming green grass, plants and insects that are available when the birds have access to pasture is reported to have a stronger taste than from those raised exclusively on a grain-based diet.

To be classified as a heritage turkey (Fig. 6.2), the bird must meet the following criteria:

- Reproduce and be genetically maintained through natural mating. Expected fertility rates should be between 70% and 80%.
- A long productive lifespan. Breeding hens are typically productive from 5 to 7 years, and breeding toms from 3 to 5 years.
- A slow to moderate rate of growth, reaching marketable weight in 26–28 weeks. This gives the birds time to develop strong skeletal systems and healthy organs before building muscle mass.

Specific breeds and strains

Commercial turkeys, or Large Whites, have a rapid growth rate and require less feed to

Fig. 6.2. A heritage type of turkey.

reach market weight than heritage types. In North America most consumers prefer the breast, or white meat, of a turkey. As a result, many organic turkey producers continue to use the commercial strains of turkeys. There is a growing interest in heritage turkeys. These turkeys do not grow as quickly as commercial breeds, giving their meat a stronger turkey flavour. Since they are single-breasted, rather than double-breasted like the commercial varieties, they have less white meat. A typical Large White, for instance, has nearly 70% white meat; the heritage breeds have about 50:50 white to dark meat.

Heritage turkeys adapt well to organic production systems since they are more disease-resistant and are good foragers. In addition, they are strong flyers and foragers, and can mate naturally and raise their young successfully. Several heritage turkey strains are available and the preference varies from area to area. The strains are usually distinguished according to their colour and region of origin. They include Standard Bronze, Narragansett, Bourbon Red, Jersey Buff, Slate, White Holland, Beltsville Small White and Royal Palm. The Narragansett is one of the oldest varieties available and once served as the foundation of the New England turkey industry. Royal Palm (also known as Crollweitzer or Pied) was also popular prior to the era of commercial turkey production. Both varieties were traditionally grown on family farms.

The Royal Palm is one of the smallest turkey varieties available. The breed was originally developed as an ornamental bird. Royal Palm turkeys are active and thrifty, excellent foragers and good flyers. They are most suitable for small production units, however, because they have the reputation of being high-strung. Royal Palm turkeys have been used in some areas as a biological means of insect control.

Waterfowl

The choice of waterfowl species depends on their intended use (i.e. meat, eggs, weeding, herding or guard animal).

Several duck breeds have been developed for meat (Muscovy, Pekin, Rouen) or egg production (Khaki Campbell, Indian Runner). Meat ducks are also raised for the production of paté de foie gras (fatty liver), a system not approved in organic production. Indian Runners have been used to train herding dogs since they move rapidly and are flock animals. The amount and quality of the space available will influence the choice of waterfowl species. In general, ducks are smaller and require less space than geese. The domesticated duck requires a grain supplement year-round while geese can do well with limited grain supplementation provided that they have a sufficient grazing area.

The different breeds of ducks are believed to have originated from the wild Mallard (*Anas platyrhynchos*). The Muscovy (*Cairina moschata*) is often referred to as a duck, but it is distinctly different. The Muscovy is believed to have originated in South America, though there are records of a similar domesticated species in ancient Egypt. Since the Muscovy originates from the southern hemisphere, its meat is leaner than that of the very fat common duck. Some breeds are capable of laying more than 230 eggs a year.

In many areas of the Orient, duck and rice production are linked. Over several generations, native ducks have been selected for their ability to obtain most of their feed requirements from the levees, swamps and waterways associated with wetland rice production. The ducks may also feed on the broken rice left in harvest fields. In some of the rural areas a flock of ducks may be a major source of income. In Indonesia the Alabio and Bali breeds are common while in China the native Maya is used.

There are believed to be two distinct sources of domesticated geese. Domestic breeds of western origin are believed to have developed from the Greylag goose (*Anser anser*) while those of eastern origin are believed to have developed from the Swan goose (*Anser cygnoides*). Geese are used primarily for meat production, with the main breeds being Emden, African and Pilgrim. They are also used for the production of paté de foie

gras. Geese have not been selected for egg production. Goose eggs are extremely high in cholesterol (> 1200 mg/egg) and fat, and so are not considered a healthy choice for the human consumer. Chinese geese are commonly raised as weeder geese and fit well into some organic production systems as biological control of weeds. Geese have a preference for grass and broadleaf weeds. They also have a loud, harsh call when startled, so they are sometimes used as 'watch' or guard geese. Geese can be quite aggressive and are seldom bothered by predators.

The goose is one of the most ancient of domesticated birds now bred commercially. While the greatest concentration of the world's geese is to be found in Asia, there is considerable breed diversity in Europe. Geese enterprises are most successful, with highly productive breeds well adapted to local conditions. An important feature of geese is their ability to consume green forages and crop residues. However, it is not clear as to how well they utilize these feedstuffs.

Specific breeds and strains

The breeds of duck typically raised for meat production are the Muscovy and Pekin. Pekin ducks grow rapidly and reach 3.2 kg in 7 weeks. Different strains of Pekin ducks have been developed and are used in commercial production systems. In many of the organic duck farms in the UK the Aylesbury breed is commonly used.

The Muscovy is also raised commercially in several areas of the world. Muscovies and common ducks mate naturally, but the fertility rate is usually very low. These hybrids are typically sterile and referred to as mule (Muscovy male × common female) or hinny (common male × Muscovy female) ducks. These hybrids are often raised commercially for meat production. In many areas of the world the Kaiya duck is popular, being produced from the cross of Pekin and Tsaiya. It was developed in Taiwan, where it is a traditional native breed (Lee, 2006). The Pekin is a meat duck while the Tsaiya, a native duck in Taiwan, is used for egg production.

Chartrin *et al.* (2006) used Muscovy, Pekin and their cross-bred hinny and mule ducks, overfeeding them from 14 days to 12 weeks of age to test the effects of these factors on the quantity and quality of lipid deposition in adipose and muscle (*pectoralis major* and *iliotibialis superficialis*) tissue. Pekin ducks were found to have higher amounts of abdominal fat and higher lipid levels in muscles (+105% and +120% in *p. major* and *i. superficialis*, respectively) than Muscovy ducks. Muscovy ducks showed the lowest triglyceride and phospholipid levels in muscles and Pekin ducks the highest levels. In addition, the Muscovy ducks had the lowest cholesterol levels in *i. superficialis* muscles. Muscovy ducks had the highest levels of saturated fatty acids (SFA) and PUFA in muscle and adipose tissues and the lowest levels of MUFA. Pekin ducks showed the opposite effect. For all parameters, the crossbred ducks had intermediate values.

Overfeeding induced an accumulation of lipids in adipose and muscle tissues (1.2–1.7-fold, depending on muscle type and genotype). This increase was higher in *p. major* than in *i. superficialis* muscles. The increase in the amount of abdominal fat was 1.7–3.1-fold, depending on genotype. This increase in lipid levels in peripheral tissues was mainly the result of triglyceride deposition. It was accompanied by a considerable increase in the proportion of MUFA (particularly oleic acid) at the expense of PUFA (particularly arachidonic acid) and SFA. The researchers concluded that genotype has a major influence on both the quantity and quality of fat deposition in peripheral tissues in the duck depending on the inherent ability of the liver to synthesize and export lipids.

Duck eggs are larger (about 65 g) than chicken eggs and have a stronger flavour. They also have a higher fat content and more cholesterol than chicken eggs. Common egg-laying breeds include the Khaki Campbell and Indian Runner used in North America and Europe. The Tsaiya breed is common in Asia. Some flocks have been able to produce 300 eggs per duck per year.

There are a number of duck breeds considered multi-purpose in that they produce a large number of eggs but also have a meatier carcass than most egg-laying breeds. These include the Aylesbury, Cayuga and the Maya (China).

It is important to choose a breed of duck that best suits particular needs and resources. For example, the Nagoya and Mikaw are local breeds available in Japan. In China the Shao (Shaoxing brown duck), Gaoyou, Jinding, Baisha and Yellow Colophony are often raised.

Geese are divided into three categories: light, medium and heavy. The most common geese raised for meat are in the heavy category and include the Toulouse, Embden, African and Pilgrim. The Toulouse and Embden breeds are most popular in the USA.

There is a growing niche market for goose eggs and they can be found with increasing frequency in farmers' markets. They have a high percentage of egg white and are high in protein. Goose eggs, however, have a higher content of cholesterol than chicken or duck eggs.

Egg production in geese is seasonal and is determined by the number of hours of daylight per day. The laying period can be altered by the use of supplementary lighting in an open-house management system, but does not appear to affect total egg number.

In the past few decades there have been attempts at developing a goose breed for increased egg production. Shalev *et al.* (1991) reported the results of an 8-year goose-breeding programme in Israel. Based on local Egyptian and Israeli breeds, two lines of geese were developed. Grey-feathered lines were created with local and Toulouse breeds and white-feathered lines were created with local and Embden breeds. Imported geese breeds were used to increase genetic variability – the Landaise from France and the Rhenish for the Grey and White lines, respectively. The White line was superior in egg production, producing 11.1–13.6 more eggs per year.

Quail

Two types of quail have been domesticated and raised as food animals: Japanese and Bobwhite quails.

Japanese quails (*Coturnix japonica*) are native to Asia. They are also known as Coturnix quail and Manchurian quail. Intensive quail production began in Japan in the 1920s and the stock was successfully introduced into North America, Europe and Asia between the 1930s and 1950s (Minvielle, 2004). Breeding programmes have developed lines of Japanese quail specific for egg and meat production. The first egg lines were developed through selection programmes. Quail eggs are much smaller than chicken eggs, though the flavour is similar. About five quail eggs equal the volume of one chicken egg. Japanese quail production for meat occurs primarily in Europe, while production of quail eggs occurs primarily in Asia and South America. Processed quail cuts and meat are appearing more frequently on shelves in Europe (Minvielle, 2004).

Bobwhite quails (*Colinus virginianus*) are native to the USA and are primarily raised for slaughter and sale as quail meat or for release in hunting preserves. There are a number of varieties that differ in body size. The smaller varieties tend to lay more eggs than the larger varieties (Giuliano et al. (2016)).

Ostriches and Emus

The emu (*Dromaius novaehollandiae*) and ostrich (*Struthio camelus*) are ratites, i.e. flightless birds with broad, rounded breast plates missing the keel to which the breast or flight muscles attach. Both are now being farmed in several countries.

Ostriches are native to South Africa where they have been raised commercially for more than 100 years. In the late 19th century, South African farmers raised almost a million ostriches to meet the needs of the fashion industry at that time. In the 1980s ostrich farming again became popular, with a growing demand for ostrich products, including meat and leather. Ostriches breed between 3 and 4 years of age. Chicks reach maturity within 6 months, adults weighing 95–175 kg and measuring 2–3 m in height. Thus, they require careful handling. Ostriches are farmed mainly for meat, hide (leather) and feathers. In addition, the eggshells can be carved into ornaments or containers. One ostrich egg is equal in volume to 20–24 large chicken eggs. The meat is favoured by health-conscious consumers seeking leaner and healthier food. It has a texture and colour similar to that of beef and is low in fat, calories and sodium. It has fewer calories, less fat and less cholesterol than beef, emu, chicken or turkey. It is also a good source of iron and protein.

Emus are native to Australia. The original inhabitants consumed emu meat and used the oil for medicinal purposes. Until the early 1990s, the Australian government prohibited commercial emu farming, but has now licensed emu farms. The USA first imported emus between 1930 and 1950 but commercial emu farming in North America did not begin until the late 1980s.

The female emu begins to breed between 18 months and 3 years of age, and may continue to produce eggs for more than 15 years. The emu grows to full size within 2 years, when it is 1.5–1.8 m and weighs up to 65–70 kg. Emu products include leather, meat and decorative eggshells. Emu oil is sold for cosmetic and pharmaceutical purposes. Emu meat, like ostrich meat, is similar in texture and colour to beef.

Wang *et al.* (2000) studied the lipid characteristics of emu meat and tissues, using samples collected from farms and supermarkets. Their results confirmed the appeal of emu meat for the health-conscious consumer. The content of total lipid in emu leg meat was low at about 3%. Phospholipid constituted the major lipid class in emu and chicken meat at 64%, higher than beef (47%). The emu drumsticks contained higher levels of linoleic, arachidonic, linolenic and docosahexaenoic acids than chicken drumsticks or beef steak. The ratio of PUFA to SFA in emu meat was 0.72, higher than in chicken meat (0.57) and beef (0.3). The ratio of n-6 to n-3 fatty acids did not differ among the three sources of meat. The abdominal fat and backfat samples contained over 99% triacylglycerols. MUFA constituted about 56% in these fat samples, with SFA about 31% and PUFA about 13%. Oleic acid was the predominant MUFA at 48%.

Sources of Information on Endangered Poultry Breeds

American Livestock Breeds Conservancy: http://albc-usa.org/

Association for Promotion of Belgian Poultry Breeds: http://users.pandora.be/jaak.rousseau/index.htm

Rare Breeds Trust of Australia: http://www.rbta.org/

References

Abrahamsson, P. and Tauson, R. (1998) Performance and egg quality of laying hens in an aviary system. *Journal of Applied Poultry Science* 7, 225–232.

Alvarado, C.Z., Wenger, E. and O'Keefe, S.F. (2005) Consumer perceptions of meat quality and shelf-life in commercially raised broilers compared to organic free range broilers. *Proceedings of the XVII European Symposium on the Quality of Poultry Meat and XI European Symposium on the Quality of Eggs and Egg Products*. Doorwerth, The Netherlands, 23–26 May 2005, pp. 257–261.

Andersen, H.J., Oksbjerg, N. and Therkildsen, M. (2005) Potential quality control tools in the production of fresh pork, beef and lamb demanded by the European society. *Livestock Production Science* 94, 105–124.

Berg, C. (2001) Health and welfare in organic poultry production. *Acta Veterinaria Scandinavica* Supplement 95, 37–45.

Boelling, D., Groen, A.F., Sørensen, P., Madsen, P. and Jensen, J. (2003) Genetic improvement of livestock for organic farming systems. *Livestock Production Science* 80, 79–88.

Borch, L.W. (1999) Consumer groups of organic products in Scandinavia. *Maelkeritidende* 112, 276–279.

Buitenhuis, A.J., Rodenburg, T.B., van Hierden, M., Siwek, M., Cornelissen, S.J.B. *et al.*(2003a) Mapping quantitative trait loci affecting feather pecking behavior and stress response in laying hens. *Poultry Science* 82, 1215–1222.

Buitenhuis, A.J., Rodenburg, T.B., Siwek, M., Cornelissen, S.J.B., Nieuwland, M.G.B. *et al.* (2003b) Identification of quantitative trait loci for receiving pecks in young and adult laying hens. *Poultry Science* 82, 1661–1667.

Castellini, C., Mugnai, C. and Dal Bosco, A. (2002a) Meat quality of three chicken genotypes reared according to the organic system. *Italian Journal of Food Science* 14, 321–328.

Castellini, C., Mugnai, C. and Dal Bosco, A. (2002b) Effect of organic production system on broiler carcass and meat quality. *Meat Science* 60, 219–225.

Chartrin, P., Bernadet, M.D., Guy, G., Mourot, J., Duclos, M.J. and Baéza, E. (2006) The effects of genotype and overfeeding on fat level and composition of adipose and muscle tissues in ducks. *Animal Research* 55, 231–244.

Cole, L.J. (1922) Chantecler poultry. A new breed of poultry – developed to meet the winter conditions of the north. *Journal of Heredity* 13, 147–152.

Deeb, N., Shlosberg, A. and Cahaner, A. (2002) Genotype-by-environment interaction with broiler genotypes differing in growth rate. 4. Association between responses to heat stress and to cold-induced ascites. *Poultry Science* 81, 1454–1462.

Emmerson, D. (1997) Commercial approaches to genetic selection for growth and feed conversion in domestic poultry. *Poultry Science* 76, 1121–1125.

European Commission (2007) Council Regulation EC No 834/2007 on organic production and labelling of organic and repealing regulation (EEC) No 2092/91. *Official Journal of the European Communities* L 189205, 1–23.

Fanatico, A. and Born, H. (2002) Label Rouge: pasture-based poultry production in France. An ATTRA Livestock Technical Note. Available at: http://attra.ncat.org/attra-pub/PDF/labelrouge.pdf (accessed 15 August 2007).

Giuliano, W. M., Selph, J. F., Hoffman, R. and Schad, B.J. (2016) Supplemental Feeding and Food Plots for Bobwhite Quail http://edis.ifas.ufl.edu/uw264 accessed July 17, 2018.

Havenstein, G.B., Ferket, P.R. and Quereshi, M.A. (2003) Growth, livability and feed conversion of 1957 versus 2001 broilers when fed representative 1957 and 2001 broiler diets. *Poultry Science* 82, 1500–1508.

Huber-Eicher, B. and Audigé, A. (1999) Analysis of risk factors for the occurrence of feather pecking in laying hen growers. *British Poultry Science* 40, 599–604.

Jensen, P. (2006) Domestication – from behaviour to genes and back again. *Applied Animal Behaviour Science* 97, 3–15.

Jones, R.B. and Hocking, P.M. (1999) Genetic selection for poultry behaviour: big bad wolf or friend in need? *Animal Welfare* 8, 343–359.

Katz, Z. (1995) Breeders have to take nature into account. *World's Poultry Science Journal* 11, 124–133.

Kestin, S.C., Su, G. and Sørensen, P. (1999) Different commercial broiler crosses have different susceptibilities to leg weakness. *Poultry Science* 78, 1085–1090.

King, R.B.N. (1984) *The Breeding, Nutrition, Husbandry and Marketing of 'Label Rouge' Poultry*. A Report for the ADAS Agriculture Service Overseas Study Tour Programme for 1984/85. Ministry of Agriculture, Fisheries and Food Agricultural Development and Advisory Service, London.

Kjaer, J. and Sørensen, P. (1997) Feather pecking behaviour in White Leghorns, a genetic study. *British Poultry Science* 38, 335–343.

Lee, H.F., Lin, L.C. and Lu, J.R. (1993) Studies on the differences of palatable taste compounds in Taiwan Native chicken and broiler. *Journal of the Chinese Agricultural Chemistry Society* 31, 605–613.

Lee, Y.P. (2006) Taiwan country chicken: a slow growth breed for eating quality. *2006 Symposium COA/INRA Scientific Cooperation in Agriculture*, Tainan, Taiwan, 7–10 November.

Lewis, P.D., Perry, G.C., Farmer, L.J. and Patterson, R.L.S. (1997) Responses of two genotypes of chicken to the diets and stocking densities typical of UK and 'Label Rouge' production systems: I. Performance, behaviour and carcass composition. *Meat Science* 45, 501–516.

Minvielle, F. (2004) The future of Japanese quail for research and production. *World's Poultry Science Journal* 60, 500–507.

Mirabito, L. and Magdelaine, P. (2001) Effect of perceptions of egg production systems on consumer demands and their willingness to pay. *Sciences et Techniques Avicoles* 34, 5–16.

Moghadam, H.K., Macmillan, I., Chambers, J.R. and Julian, R.J. (2001) Estimation of genetic parameters for ascites syndrome in broiler chickens. *Poultry Science* 80, 844–848.

O'Donovan, P. and McCarthy, M. (2002) Irish consumer preference for organic meat. *British Food Journal* 104, 353–370.

Peterson, S. (2006) Comparing alternative laying hen breeds. In: *The Greenbook 2006*. The Minnesota Department of Agriculture's Agricultural Resources Management and Development Division (ARMD). Available at: http://www.mda.state.mn.us/news/publications/protecting/sustainable/greenbook2006.pdf (accessed 17 July 2018).

Rizzi, C. and Chiericato, G.M. (2010). Chemical composition of meat and egg yolk of hybrid and Italian breed hens reared using an organic production system. *Poultry Science* 89, 1239–1251.

Rizzi, C. and Marangon, A. (2012) Quality of organic eggs of hybrid and Italian breed hens. *Poultry Science* 91, 2330–2340.

Rodenburg, T.B., van Hierden, Y.W., Buitenhuis, A.J., Riedstra, B., Koene, P. *et al.* (2004) Feather pecking in laying hens: new insights and directions for research? *Applied Animal Behaviour Science* 86, 291–298.

Saveur, B. (1997) Les critères et facteurs de la qualité des poulets Label Rouge. *Production Animal* 10, 219–226.

Scheideler, S.E., Jaroni, D. and Froning, G. (1998) Strain and age effects on egg composition from hens fed diets rich in n-3 fatty acids. *Poultry Science* 77, 192–196.

Shalev, B.A., Dvorin, A., Herman, R., Katz, Z. and Bornstein, S. (1991) Long-term goose breeding for egg production and crammed liver weight. *British Poultry Science* 32, 703–709.

Sørensen, P. (2001) Breeding strategies in poultry for genetic adaption to the organic environment. In: *Proceedings of the 4th NAHWOA Workshop*, Wageningen, The Netherlands, 24–27 March.

Sørensen, P. and Kjaer, J.B. (1999) Comparison of high yielding and medium yielding hens in an organic system. *Proceedings of the Poultry Genetics Symposium*, 6–8 October, Mariensee, Germany, p. 145.

Stopes, C., Duxbury, R. and Graham, R. (2001) Organic egg production: consumer perceptions. In: Younie, D. and Wilkinson, J.M. (eds) *Proceedings of a Conference on Organic Livestock Farming*. Heriot-Watt University, Edinburgh and University of Reading, UK, 9 and 10 February 2001, pp. 177–179.

Su, G., Kjaer, J.B. and Sørensen, P. (2005) Variance components and selection response for feather-pecking behavior in laying hens. *Poultry Science* 84, 14–21.

Su, G., Kjaer, J.B. and Sørensen, P. (2006) Divergent selection on feather pecking behavior in laying hens caused differences between lines in egg production, egg quality and feed efficiency. *Poultry Science* 85, 191–197.

Touraille, C.J., Kopp, J., Valin, C. and Ricard, F.H. (1981) Chicken meat quality. 1. Influence of age and growth rate on physico-chemical and sensory characteristics of the meat. *Archiv für Geflügelkunde* 45, 69–76.

Wang, Y.W., Sunwoo, H., Sim, J.S. and Cherian, G. (2000) Lipid characteristics of emu meat and tissues. *Journal of Food Lipids* 7, 71–82.

Yang, N. and Jiang, R.S. (2005) Recent advances in breeding for quality chickens. *World's Poultry Science Journal* 61, 373–381.

7

Integrating Feeding Programmes into Organic Production Systems

One of the aims of organic production is to manage poultry in such a way as to mimic as closely as possible the natural state. Thus, the production system is quite different from that used in conventional production, and the practical implication of these differences on organic production techniques needs to be recognized and quantified. The main differences between organic and conventional poultry production relate to housing system, access to outdoor areas, genotype, range of feedstuffs available for dietary use and disease prevention measures. Most of the research relating to this issue has been conducted with chickens (layers and meat birds) and the findings have to be extrapolated to other species when these are lacking.

Denmark is a leader in organic production systems and in organic food sales to consumers, therefore it is useful to review findings from that country. The EU regulations mandate a maximum flock size for layers of 3000 and for growing chickens of 4800. These flock sizes are below those often found in conventional free-range poultry production in some regions but are still much higher than what can be considered as natural flock sizes. The birds have to be kept under free-range conditions, i.e. having access to a hen-yard providing at least 4 m² per laying hen. Also, medication to control coccidiosis cannot be included in the feed, no beak trimming is allowed and

age at slaughter for meat chickens has to be at least 81 days.

In spite of these restrictions, organic egg production in Denmark has been shown to be quite efficient and amounts currently to about 18% of total production, although feed consumption is usually considerably higher than in conventional production (Kristensen, 1998). Survey results from farms in Denmark (Hermansen *et al.*, 2004) indicate a lower rate of lay in organic farms in comparison with conventional layer farms when calculated on the basis of initial number of hens in the flock (Table 7.1). This has been attributed mainly to a higher mortality from cannibalism in organic systems due to the type of genotype used, also intestinal problems such as gizzard compaction due to forage ingestion. Although egg production is lower, profitability is increased on organic farms due to a higher selling price for the eggs.

These findings illustrate some of the features of organic production likely to be experienced by organic egg producers (Fig. 7.1).

Housing System

One consequence of allowing poultry access to outdoors is that they can no longer be housed in temperature-controlled buildings and instead are subject to ambient temperatures.

Table 7.1. Average productivity and prices in the period 1995–2002 per hen housed as reported by the Danish Poultry Council, 2003 (from Hermansen *et al.*, 2004).

	White layers in cages (21–76 weeks)	Organic brown layers (21–68 weeks)
Feed/day (g)	112	131
Rate of lay (%)	86.8	73.5
Mortality (%)	4.9	14.8
Feed (kg)/eggs (kg)	2.07	2.81
Egg price (DKK/kg)	5.89	14.21
Price relation egg/feed	4.17	6.39

Fig. 7.1. An organic flock of chickens on pasture.

Energy requirements of outdoor poultry are generally higher because of increased exercise and exposure to outside temperatures. As a result feed intake is increased. For instance, van Wagenberg *et al.* (2017) conducted a review of the sustainability of organic systems on behalf of the World Association of Animal Production and showed that the feed/gain ratio of organic laying hens was about 1.06–1.28 times the value for conventional laying hens. A more

detailed outline of their findings (based on quite limited data) is shown in Table 7.2.

Conversely, at high ambient temperatures voluntary feed intake may be reduced to a level that is inadequate to meet the energy requirements for high production. Therefore, in situations of low and high ambient temperatures, changes may have to be made to the feed, otherwise both animal welfare and productivity may suffer. Reliable quantitative data on this subject for organic poultry are lacking.

The body temperature of an adult chicken is 40.6–41.7°C and the thermoneutral zone is accepted as 18–24°C, the ambient temperature range which allows chickens to maintain their body temperature without a change in metabolism. In practical terms this means that birds housed at temperatures lower than 18°C may have to consume more feed to maintain body temperature. Conversely, when the ambient temperature increases above the comfort level of the bird (24°C) voluntary feed intake is likely to be reduced, resulting in reduced weight gain and lowered production of eggs. The explanation for the effect of high temperature on feed intake is that chickens have no sweat glands to promote cooling. Since eating increases body temperature, chickens reduce their feed intake during hot weather. Chickens begin panting at 29.4°C to help dissipate heat and drink more to avoid dehydration. Heavy birds are more susceptible to heat stress than light birds, because they have a relatively lower surface area for heat dissipation per unit of body weight. The effects in broilers of increasing ambient temperature over

Table 7.2. Environmental impact of organic production on the productivity and health of broilers and laying hens (from van Wagenberg et al., 2017).

	Broilers	Laying hens
	Organic relative to conventional (conventional = 100)	
Weight gain	76–84	
Feed/gain ratio	140–153	106–128
Egg production		87–99
Global warming potential	72–150	56–130
Land acidification potential	150–196	110–154
Land eutrophication potential	200–240	130–185
Land use	189–315	166–220
Energy use	86–159	87–140

the range of 7.2–37.8°C were quantified by Howlider and Rose (1987). Within the range of 7.2–21°C, both growth rate and feed intake declined by 0.12% for each degree rise in environmental temperature. Another important finding was that total body fat and abdominal fat increased by 0.8% and 1.6%, respectively, for each degree rise in temperature, related perhaps to effects on mobility.

Producers need to be aware of these effects, possibly adjusting the level of metabolizable energy (ME) and nutrients in the diet when feed intake is lower than expected, and ensuring an adequate supply of drinking water (preferably cooled) in close proximity to the birds during periods of high ambient temperature.

The observations above do not apply to recently hatched chicks. They have not yet developed the ability to regulate body temperature. As a result, they are sensitive to heat stress and are especially prone to becoming chilled, usually requiring an external heat source.

Most organic producers, at least in the temperate regions, do not make changes in the composition of the diet to account for winter temperatures. During cold weather birds fed *ad libitum* simply eat more to maintain body processes and, if given restricted amounts of feed, have to receive a higher feed allowance. Producers accept higher feed intakes and lower efficiencies of feed utilization during cold periods. It is logical also to increase the allowance of forage at this time. The increased intake of fibrous

materials results in an increased heat of fermentation in the gut, which helps to keep the birds warm. However, as overall efficiency of organic production becomes more important, producers may have to alter the composition of dietary mixtures for birds with access to outdoors during periods of cold weather. It would be logical to reduce the concentration of protein, amino acids and micronutrients to take into account the increased intake of feed and maintain the target daily intake of nutrients. For instance, if intake is increased by 10%, then the concentration of protein, amino acids and micronutrients could be reduced by about 10%. Another approach could be to raise the energy level of the diet in relation to protein and other nutrients, perhaps by the use of fat, in such a way as to result in the correct target intake of amino acids but an increased intake of energy. Such changes should be made following advice from a nutritionist. Few organic producers are in a position to make these changes, though those using purchased feed could do so in collaboration with a feed manufacturer.

Periods of high heat/high humidity can be expected to result in the opposite effect: a decrease in voluntary feed intake, leading to slower growth, reduced egg production and a lower efficiency of feed conversion. In these situations it is advisable to consider altering the feed formulation to a higher nutrient density designed to ensure the correct intake of nutrients in a lower total intake. In addition, the diet should be formulated, if possible, during periods of hot weather to minimize the inclusion of less-well-digested feedstuffs including forage, to avoid raising the body temperature by the increased heat of fermentation of fibre. Strategies such as the provision of cooled drinking water and feeding during the cooler parts of the day should be adopted to try to encourage intake of feed during periods of hot weather.

These changes to the diet are best made following advice from a nutritionist. If such advice is unavailable, a rule of thumb is to increase the content of ME and other nutrients in the diet by 10% when feed intake is 90% of the ideal. This is to ensure that the reduced intake of feed provides a similar intake of ME and nutrients as that with 100% of intake of the regular diet.

Genotype

The genotypes available for organic production systems have been described in Chapter 6. It seems clear from that outline that a great deal of research will have to be done to identify breeds and strains best suited for organic production. At present, many producers have to use modern hybrids, which may not be the most suitable for this purpose since they are likely to have been bred for different housing and management systems. It is not sufficient to evaluate breeds and strains for organic production based solely on production characteristics. Breeds and strains for egg production may have been selected for cage-housing systems and meat birds may have been produced as hybrids with a propensity for rapid growth on diets of high nutrient density.

One important factor is the suitability of the strain for the local environment and the influence of outdoor conditions on bird health. Another main factor is the behavioural characteristics of the strain in question when managed in flocks, its propensity for pecking and cannibalism and its ability to thrive in the outdoor environment in question.

Fanatico et al. (2008) measured the growth performance, liveability and carcass yield of slow- and fast-growing meat-type genotypes fed low-nutrient or standard-type diets (formulated to NRC specifications) and raised indoors or with outdoor access. Results are shown in Tables 7.3. and 7.4. The slow-growing birds took 84–91 days to reach market weight, whereas the fast-growing birds took 56–63 days and had a lower overall feed intake. Access to outdoor conditions had a marked effect by increasing the feed requirement and reducing overall mortality, particularly when the birds were fed the low-nutrient diet. Bone mineral density was similar in both genotypes, though it was improved slightly in the birds fed the NRC-type diets.

Response to stressors in the environment is another factor related to the choice of suitable genotypes. Distinct strain and breed differences in the response of chickens to heat stress have been established, as reviewed by Leeson (1986). For example, the Bedouin fowl raised by nomads in the Negev desert of Israel are renowned for their ability to regulate not only body temperature, but also metabolic rate and acid–base balance when exposed to extreme heat stress of 37–40°C (Leeson, 1986). A cross of White Leghorn (WL) and Bedouin fowl showed

Table 7.3. Influence of diet type (low-nutrient or to NRC specifications) on the productivity and health of slow-growing and fast-growing male meat chickens (from Fanatico et al., 2008).

Genotype	Diet type	Weight gain (g)	Feed intake (g)	Feed/gain (g/g)	Bone mineral density (g/cm^2)	Mortality (%)
Slow-growing	Low-nutrient	2593[a]	7994[a]	2.96[a]	0.192[b]	1.0[b]
Slow-growing	NRC	2888[b]	7959[a]	2.76[a]	0.204[a]	3.0[b]
Fast-growing	Low-nutrient	2888[b]	6404[b]	2.22[b]	0.183[b]	9.0[a]
Fast-growing	NRC	2808[b]	5546[c]	1.97[c]	0.203[a]	19.0[a]

[a-c]: values with different superscripts differ significantly at $P < 0.05$.

Table 7.4. Influence of genotype and outdoor access on the productivity and health of male meat chickens (from Fanatico et al., 2008).

Genotype	Housing	Weight gain (g)	Feed intake (g)	Feed/gain (g/g)	Bone mineral density (g/cm²)	Mortality (%)
Slow-growing	Outdoor access	2254[b]	8459[a]	3.75[a]	0.193	3.0[b]
Slow-growing	Indoor	2105[b]	6752[c]	3.21[b]	0.189	0[b]
Fast-growing	Outdoor access	3370[a]	8087[a]	2.40[c]	0.183	11.0[a]
Fast-growing	Indoor	3389[a]	7402[b]	2.19[cd]	0.185	9.0[a]

[a-d]: values with different superscripts differ significantly at P < 0.05.

an improved heat tolerance in offspring relative to Leghorns, suggesting a genetic basis for heat resistance (Arad et al., 1975). In related studies Arad et al. (1981) found that Leghorns performed quite well at 41°C when acclimatized to this temperature, and that Sinai fowl were less affected at this temperature. Although performance of the Leghorns at 41°C was reduced by 30%, these birds still outperformed the indigenous Sinai breed in egg production.

Confounding factors involved in studies of a breed effect in relation to heat stress are body weight and general activity (Leeson, 1986). Washburn et al. (1980) showed that resistance to heat stress in strains selected for fast growth was significantly less than in slower-growing control birds. In addition, they found that restricting the feed intake of fast-growing birds resulted in a dramatic improvement in resistance to heat stress. Van Kampen (1977) reported a similar finding, namely that activity may play a role in the birds' response to heat since the lower critical temperature could be reduced by as much as 5°C when birds were made active. Wilson et al. (1975) indicated that selection for high oxygen consumption and high heat tolerance resulted in birds with similar characteristics. This situation agrees with the finding that Leghorns, with their higher metabolic rate, are more tolerant to heat stress than are heavier breeds.

The broiler chicken and turkey appear to respond differently in relation to diet selection at elevated temperatures. Cowan and Michie (1978a) showed that the growth depression of turkeys caused by high environmental temperatures could be corrected if the turkeys were allowed to adjust their protein intake. When choice-fed cereal grain and protein concentrates were maintained at elevated temperatures, they adjusted their intake of cereals in line with a reduced maintenance energy requirement and increased their intake of protein feeds. Under comparable conditions, broilers were unable to regulate their nutrient intake, and growth depression was still evident (Cowan and Michie, 1978b).

Productivity of several genotypes of laying hens under organic conditions was reported by Sørensen and Kjaer (2000) and their behavioural aspects and liveability when housed in large flocks were reported by Kjaer and Sørensen (2002).

Table 7.5 shows a considerable difference in laying capacity among the four lines, with ISA brown showing the highest egg production. During the rearing period no difference was seen in mortality, but during the laying period mortality was significantly higher in the ISA strain, due to a higher rate of cannibalism (Table 7.6). Mortality due to gizzard impaction was higher in New Hampshires and total mortality was related in part to an outbreak of coccidiosis which mainly affected the New Hampshires. Overall mortality was lower in the cross strain. Level of methionine + cystine in the diet did not influence mortality.

Studies of this type are particularly valuable in assessing the suitability of various genotypes for organic production since they are carried out under organic conditions.

Table 7.5. Productivity under organic conditions of four genotypes of laying hens from 18 to 43 weeks of age (from Sørensen and Kjaer, 2000).

	ISA brown	New Hampshire	White Leghorn	New Hampshire × White Leghorn
Rate of lay (%)	84.6	63.2	72.4	69.2
Eggs/hen placed	127.2	88.8	103.4	105.5
Age at first egg (weeks)	19.8	22.2	22.9	21.4
Egg weight (g)	59.3	54.7	58.3	57.0

Table 7.6. Mortality in four genotypes of laying hens from 16 to 43 weeks of age (from Kjaer and Sørensen, 2002).

	ISA brown	New Hampshire	White Leghorn	New Hampshire × White Leghorn
Cannibalism (%)	17.5	2.38	0	1.11
Impaction (%)	0.42	7.14	0	0.56
Others (%)	3.33	4.76	6.67	2.22
Total mortality (%)	21.25	14.3	6.67	3.89

However, the high-yielding hen, through many generations, has been selected for high performance on the basis of production capacity measured in individual cages. Thus, little attention has been paid to a genetically based ability to produce well in a large flock of hens.

As reported by Hermansen *et al.* (2004) too many cases have been observed in free-range systems with large flocks, in which hens have started to engage in feather pecking that ended with an unacceptably high rate of cannibalism. As indicated above and confirmed in other investigations, total mortality is often reported to be at least 20% annually in organic layer flocks (Kristensen, 1998). This figure covers not only cannibalism, but also deaths caused by predators and by inappropriate behaviour of the birds, which sometimes suffocate because they tend to bunch together. This high mortality rate is a major problem. There is a need to develop improved lines that are still high-yielding, but with less of a propensity to engage in unacceptable feather pecking. Small selection experiments have shown that these behavioural traits have a genetic basis (Boelling *et al.*, 2003) and need to be incorporated into a breeding programme

for lines used in organic farming, in order to make production in the farming system economically sound and acceptable from a welfare point of view.

Choosing the right breed and strain of bird for any particular situation is therefore not easy and requires that the producer receive as much advice as possible, especially from local producers, before a final selection is made.

Feeding Programmes

Suggested diets for organic poultry production have been outlined in Chapter 5. The diets are based on feed mixtures prepared on-farm to meet the suggested nutritional targets for the breed and class of bird in question, or purchased from a feed supplier. Preparing the feed mixtures on-farm has the advantage that the producer has more control over the mixing formula and can utilize home-grown feedstuffs. Consequently this system may be more cost-effective than purchasing complete feed. However, the disadvantages are that preparing the feed on-farm requires the availability of feed ingredients, storage and mixing equipment and an adequate knowledge of feed formulation.

A logical system for organic producers to adopt is choice-feeding, involving the use of whole grain that may be available on-farm (Fig. 7.2). This system approaches the natural feeding system much more closely than other feeding systems and is therefore highly appropriate for organic production. Free-choice feeding of poultry was used commonly in the past in many countries before commercial poultry production became intensified, the chickens being allowed to range free in the fields and fed mainly on scratch grains. However, for intensive production involving automated feeding systems and high-producing stock this earlier feeding system was largely abandoned in favour of the all-mash (complete) or pelleted diet (Blair *et al.*, 1973; Henuk and Dingle, 2002). One of the main reasons for selecting choice-feeding was pointed out by Blair *et al.* (1973), namely that since the bird has a digestive system capable of processing whole grain it seems illogical and unnecessary to feed it a pre-ground diet. The system also presents energy savings in feed preparation. The process of grinding requires about 20 kW/h per tonne of grain, pelleting requires a large input of electrical energy amounting to about approximately 10% of the total feed costs and additional energy is required to generate steam for the steam-pelleting process (Henuk and Dingle, 2002).

Henuk and Dingle (2002) described the practical and economic advantages of

Fig. 7.2. Choice-fed chickens.

choice-feeding systems for laying poultry (Table 7.7). The systems can be applied in other classes and species of poultry. As outlined by these authors, poultry can be fed in several ways: (i) a complete dry feed offered as a mash *ad libitum*; (ii) the same feed offered as pellets or crumbles *ad libitum*; (iii) a complete feed with added whole grain; (iv) a complete wet feed given once or twice a day; (v) a complete feed offered on a restricted basis; and (vi) choice-feeding. Of all these, a practical alternative to offering complete diets is choice-feeding, which can be applied on both a small or large scale.

One of the disadvantages of the complete diet system is that birds can only adjust intake according to their appetite for energy. When the environmental temperature falls or rises, the birds either over- or underconsume protein and minerals such as calcium.

In choice-feeding or 'free-choice feeding' the birds are usually offered a choice among three types of feedstuffs: (i) an energy source (e.g. maize, rice bran, sorghum or wheat); (ii) a protein source (e.g. soybean meal, canola meal or fishmeal) supplemented with vitamins and minerals; and (iii) in the case of laying hens, calcium in granular form (calcareous grit such as oyster-shell grit). Regular grit of a suitable granular size (to aid grinding in the gizzard) should also be made available.

The basic principle behind choice-feeding is that birds possess some degree of 'nutritional wisdom' (see Chapter 3) which allows them to select from the various feed ingredients on offer and construct their own diet according to their actual needs and production capacity. The wild ancestors of modern chickens possessed an ability to select nutrients appropriate to their requirements in a variety of environments, both tropical and temperate (Shariatmadari and Forbes, 1993). Reviews by Hughes (1984) and Rose and Kyriazakis (1991) cited strong evidence to indicate that when domestic birds are offered a range of different feedstuffs they have the ability to choose a diet that provides them with all the nutrients necessary for growth, maintenance and production (Dove, 1935) (Table 7.8).

Table 7.7. Estimated feed savings by adopting choice-feeding with laying hens (from Henuk and Dingle, 2002).

Reference	Conventional diet (g/hen/day)	Choice-feeding (g/hen/day)	Saving in feed intake (%)
Henuk *et al.*, 2000b	123.8	120.3	2.9
Blair *et al.*, 1973	116.2	108.9	6.7
Leeson and Summers, 1978	114.4	107.2	6.7
Leeson and Summers, 1979	118.4	110.7	7.0
Henuk *et al.*, 2000a	126.5	114.6	10.4
Karunajeewa, 1978	132.5	118.5	11.8

Table 7.8. Dietary self-selection by chicks and effect on nutrient intake (from Dove, 1935).

Feed ingredient offered	Intake % of total	Nutrient content of selected dietary mixture	NRC-estimated requirements (1994)
Yellow maize	52.8	Crude protein, 179 g/kg	180 g/kg
Oat meal	8.9	ME, 2729 kcal/kg	2880 kcal/kg
Wheat bran	21.3	Calcium, 13 g/kg	9.0 g/kg
Fishmeal	11.4	Phosphorus, 11 g/kg	4.0 g/kg non-phytate P
Bone meal	2.9		
Dried skim milk	2.1		
Oyster shell	0.6		

In selection of feed by birds, visual stimulation evidently plays a major role. Taste also is involved, consequently a separate vitamin/trace mineral mix should not be offered since it is likely to be avoided by birds on account of its taste.

Choice-feeding should work well with birds provided access to forage, since it would allow the birds to regulate the intake of energy and nutrients according to those supplied in the forage. However, research data on this issue are lacking.

A choice-feeding system is of particular importance to small poultry producers in developing countries, because it can substantially reduce the cost of feed. The system is flexible and can be constructed in such a way that the various needs of a flock of different breeds, including village chickens, under different climates can be met. The system also offers an effective way of using home-produced grain, such as maize, and by-products such as rice bran. A further advantage is that minimal or no mixing equipment is required since grinding of most grains is unnecessary, thus reducing feed-processing costs. When introducing birds to whole grain it is recommended

that it be done gradually over 2–3 weeks to allow development of the gizzard. Regular grit should also be provided.

Most basic choice-feeding experiments have been carried out with chickens. Such work requires that the quantity of feedstuffs offered and rejected is measured accurately, and as a result the research has mostly involved feeding by hand in either troughs of short length with internal dividers, or in two or more small separate containers. It is possible to offer all the choice feedstuffs in a single trough or in a single feeder. Results from many studies on the effect of choice-feeding on performance of laying hens suggest that, when birds are offered a choice of feed ingredients *ad libitum*, they consume less feed in total than control birds receiving a conventional complete diet (see Table 7.7).

Bennett (2006) made the following recommendations for small-scale egg producers:

1. Do not give the hens too many choices. Hens can handle up to three choices quite well (grain, 'supplement' and limestone or oyster shell). When using more than one grain, such as wheat and barley, mix them together in the same feeder.

2. Give the hens choices that are nutritionally distinct. For example, grain is high in starch and energy, supplement is high in protein and vitamins and limestone is high in calcium. When provided with such clear choices, the hens learn which feeders to go to and how much to eat in order to meet their basic nutritional needs. Some choices may not be clear enough for the hens. For example, wheat and peas both are high in starch and have moderate levels of protein. Having separate feeders containing wheat and peas may not provide a distinct enough nutritional difference for the birds to detect.
3. Introduce the whole grain and choice-feeding a month before the start of egg production (about 15 weeks of age). This adjustment period will allow the birds time to learn how to choice-feed themselves before they are exposed to the nutritional demands of egg production. It will also allow the pullets the opportunity to increase their calcium intake and build up the calcium reserves in their bones before they start to lay eggs. Finally, it takes the gizzard 3 weeks to build muscle mass and the intent is for the hens to be able to grind the grain efficiently in this organ once egg production begins.
4. Do not feed vitamins or trace minerals in a separate feeder. Use the supplement as a source of these nutrients. If vitamins or trace minerals are placed in a separate feeder, some birds may not eat them because they do not like the taste while other birds may overconsume them and suffer toxic side effects.
5. Give the birds adequate feeder space. With a large flock, several feeders are required for each ingredient. For a 100-hen barn, two hanging feeders each of grain, supplement and limestone are suggested.
6. Purchase a supplement designed to be mixed with grain or grain and limestone (or oyster shell) to provide a complete laying hen diet. A supplement formulated in this way will contain 250–400 g crude protein (CP) per kg protein. A grower supplement may be used prior to the start of egg production but a layer supplement should be used once the birds start to lay.

The birds will readily consume whole wheat, whole oats or whole barley, but they have difficulty with whole maize, which needs to be kibbled (reduced in particle size). Hens can successfully consume 70% of their diet as whole grain when it is choice-fed. It is important to note that when whole grain, supplement and limestone are mixed together into a traditional laying hen diet and offered in one feeder, the whole grain should not comprise more than 50% of the diet. The rest of the grain in the feed should be ground. At higher levels of whole grain, the hens sometimes have trouble finding the supplement in the feed mixture. When the grain, supplement and limestone are in different feeders, these separation problems are avoided.

Jones and Taylor (2001) showed that whole triticale could be utilized successfully in pelleted broiler diets with broilers, giving production results similar to (or better than) those with pelleted diets containing ground triticale. In addition, proventricular dilatation and mortality due to ascites were reduced by feeding pelleted diets containing whole grain. Also there was evidence to suggest that the addition of exogenous enzymes to broiler diets could be reduced or eliminated by incorporating whole grain into pelleted diets for broiler chickens.

One health benefit reported by Toghyani et al. (2014) in choice-fed broilers was a reduction in footpad dermatitis and in tonic immobility, attributed to a reduction in protein intake and a subsequent reduction in the ammonia content of the litter.

Research on choice-feeding with other classes and breeds was reviewed by Pousga et al. (2005). Olver and Malan (2000) reported that growing pullets offered choice-fed diets during the rearing period (7–16 weeks) gained more weight than those fed a commercial grower diet. These pullets consumed approximately 7 g more maize per day than the control pullets but had a lower total feed intake, suggesting a more efficient utilization of dietary energy. Olver and Malan (2000) also found that the pullets fed the free-choice diets consumed more than twice the amount of limestone than those offered the complete diet. Research on choice-feeding of broiler chickens suggests that these birds effectively select a combination that maximizes their biological performance. Under conditions of heat stress, with daytime temperatures rising to 33°C, Cumming (1992a) demonstrated that broiler chickens reduced their grain (energy) intake by 34% but their protein intake by only 7%, compared with similar birds in a cooler (20°C) environment. Remarkably, choice-fed birds were found to have a 'protein memory' and consumed early on the next day, before the temperature rose, the protein they did not eat during the previous hot day. Thus, the performance of choice-fed broilers was significantly better than that of the same birds fed the most sophisticated complete diets (Mastika and Cumming, 1985) in hot environments. This work demonstrated the importance of experience and group learning in free-choice broiler chickens. They took about 10 days to learn to balance their protein concentrate and whole-grain intakes accurately. They needed to be in groups of at least eight birds and be offered the protein concentrate (in mash or crumble form) and whole grain in identical, adjacent troughs or in the same feed trough.

Choice-feeding based on ground maize as the grain source does not appear to be as successful with broilers as systems based on whole grain (Rack et al., 2009). These researchers reported that a fast-growing genotype gave better growth performance and carcass characteristics than a slow-growing genotype, also that access to pasture tended to have no effect on slow-growing broilers and decreased the performance of fast-growing broilers. Growth performance was decreased in late autumn, with the onset of lower ambient temperatures.

A recent study by Fanatico et al. (2016) showed that choice-feeding could be used successfully with growing hybrid broilers (Table 7.9). Two diet regimes were used during the growing period (28–64 days): (i) a complete diet containing cracked maize 507.5 kg/t, roasted ground full-fat soybeans 312.5 kg/t, whole wheat 50 kg/t, lucerne meal 50 kg/t, fishmeal 37.5 kg/t, oystershell 12.5 kg/t and vitamin/mineral premix 30 kg/t and (ii) cracked maize and a protein concentrate containing roasted ground full-fat

Table 7.9. Effect of choice-feeding on the productivity and carcass quality of growing broilers (from Fanatico et al., 2016).

	Complete diet	Choice diet
Weight gain 28–64 days (g)	1910.9[a]	1802[b]
Feed intake/bird (g)	6997.7[a]	5827.4[b]
Feed conversion ratio	3.65[a]	3.20[b]
Weight of carcass (g)	1746.8[a]	1655.3[b]
Carcass yield (%)	67.08	66.85
Breast weight (g)	380.6[a]	349.7[b]
Breast yield (%)	21.76	21.23
Leg yield (%)	33.54	34.06

soybeans 634.5 kg/t, whole wheat 101.5 kg/t, lucerne meal 101.5 kg/t, fishmeal 76.1 kg/t, oystershell 15.4 kg/t and vitamin/mineral premix 60.9 kg/t in separate feeders to allow self-selection. Both diets were fed ad libitum. The choice-fed birds ate less and had lower weight gains but had an improved feed conversion efficiency, in comparison with birds fed the complete diet. Dressed carcass weights and breast weights were also lower in the choice-fed birds, but the proportions of breast, leg and wing were unchanged. The cost of feed was lower with choice-feeding ($0.58/kg versus $0.66/kg.) These results indicate that overall productivity and chicken quality may be improved with use of complete feeding but that it is more expensive and less convenient for the producer wishing to utilize home-grown feedstuffs.

Cumming (1992a) demonstrated genetic differences in the ability of different strains to adapt to free-choice feeding; egg-type stock adapted more quickly than broiler stock. The same author found considerable differences between adults of layer strains in adapting to free-choice feeding. Brown-egg layers seemed to adapt more readily than white- or tinted-egg layers. However, all strains of commercial layers and broilers being used in Australia were found to learn within 10–14 days to balance their energy and protein intakes very accurately and maximize production and optimize economic returns.

Emmerson et al. (1990) studied diet selection by turkey hens. Controls were given a complete feed containing 185 g CP/kg and 11.3 MJ ME/kg, or were provided with a choice between a high-protein–low-energy diet (348 g CP/kg and 7.74 MJ ME/kg) and a high-energy–low-protein diet (81 g CP/kg and 13.39 MJ ME/kg). The choice-fed turkeys consumed 10% less feed, 44% less protein and the same amount of energy, yet laid a similar number of eggs as those fed conventionally. In a further experiment of similar design Emmerson *et al.* (1991) reported no difference in egg production over a 20-week period due to feeding regime. Broodiness tended to be reduced by choice-feeding but fertility and hatchability were lower than with the control diet.

An experiment was carried out in Vietnam by Bui *et al.* (2001) in which commercial diets with differing levels of CP were fed to growing broiler ducklings *ad libitum*. It was found that the ducklings preferred high-protein to low-protein feeds, resulting in excess protein intake and higher protein conversion ratios. From this study it was concluded that choice-feeding of that type was not an economically viable system for growing meat ducks.

Forage

Two aspects of forage are of interest: how much do free-range poultry consume and how well is it utilized? The topic of forage intake and its significance in relation to nutrient needs is of practical importance. Another consideration is that one of the most important egg-quality parameters for the consumer is the yolk colour, which can be affected by forage intake and quality (Fig. 7.3).

Fuller (1962) reported that access to pasture resulted in a 6% saving of total feed consumed when pullets were fed a conventional mash–grain diet, 13% when pullets were permitted to select the grain, mineral and protein–vitamin components and 20% when pullets were forced to forage by providing them with grain and minerals only. Although laying hens are able to consume considerable amounts of roughages (Steenfeldt *et al.*, 2001), information on herbage intake from range areas by high-performance layers is scarce. Some results suggest that layers on range consume 30–35 g dry

Fig. 7.3. Good pasture is a valuable resource for organic poultry.

matter (DM) per day of herbage in addition to *ad libitum* feeding of concentrates (Hughes and Dun, 1983). However, different forage crops may vary in nutritional value and attractiveness to laying hens. Moreover, restriction in nutrient supply has been shown to increase forage intake in pullets (Fuller, 1962), which could result in a drastic reduction in intake of protein and some amino acids and a negative effect on plumage condition due to feather pecking (Ambrosen and Petersen, 1997; Elwinger *et al.*, 2002). As a result, it is necessary to ensure an adequate supply of good foraging material when access to regular feed is restricted.

Work has been conducted on this topic by researchers in Denmark (Horsted *et al.*, 2007). A main purpose of the study was to determine the feed intake of organic layers when given a normal feed mixture or a diet consisting of whole wheat plus oyster shell, both groups being allowed access to different types of forage. In experiment 1, the forage consisted of a grass/clover pasture (*Lolium perenne* and *Trifolium repens*) or a mixture of forbs (*Fagopyrum esculentum*, *Phacelia tanacetifolia* and *Linum usitatissimum crepitans*, which were expected to attract insects). In experiment 2, the forage consisted of a grass/clover pasture or chicory (*Cichorium intybus* cv. 'Grassland Puna').

The method used to evaluate intake in this study was to examine the crop contents. Limited digestion of feed occurs in this organ, making it easy to identify the feed items. Other researchers have used this approach;

for example, Jensen and Korschgen (1947) found that analysis of crop contents of quail was more accurate than analysis of droppings and gizzard contents.

Antell and Ciszuk (2006) found a linear relationship between intake of grass and the grass content in the crop at the end of the day for confined hens, indicating that daily forage intake could be estimated from the content of plant material in the crop of hens slaughtered in the evening.

The Horsted *et al.* (2007) study started when the hens (Lohmann Silver) were about 25 weeks of age. During the experiment the hens had access to forage from sunrise to sunset. The control group received a pelleted feed mixture containing 184 g CP/kg, 8.7 g lysine/kg, 4.6 g methionine/kg and 41 g calcium/kg, whereas the whole wheat contained 120 g CP/kg, 3.4 g lysine/kg, 1.9 g methionine/kg and less than 10 g calcium/kg (all values on a DM basis). Feed, water, oyster shell and insoluble grit were provided *ad libitum* outdoors.

Results showed that hens fed the pelleted feed consumed an average of 129 g feed per hen per day in experiment 1 and 155 g in experiment 2. Hens fed the wheat diet consumed 92 g and 89 g feed per hen per day, respectively, in the two experiments. Wheat-fed hens had a significantly higher intake of oyster shell than the pellet-fed hens. The number of eggs per hen per day was significantly lower for wheat-fed hens in experiment 1 (0.91 versus 0.75) but no differences were observed in experiment 2 (0.83 for both feeding regimes).

The results showed a huge difference in crop content between morning and evening, indicating that laying hens have a higher intake of most feed items at the end of the day. The main results on relative intake of forage are shown in Table 7.10.

These results showed that type of diet influenced the amount of plant material found in the crop, a higher level of plant material occurring in the crops of wheat-fed hens. They suggested that hens restricted in nutrients by being fed a grain diet increased their intake of plant material even though the wheat was of a larger particle size than the pelleted diet. In both experiments the type of feed significantly affected the amount of seeds in the crops, hens receiving the pelleted diet having a higher intake of seeds than the wheat-fed hens. It was theorized that in hens fed on wheat, the weed seeds presumably were of little additional nutritional value, thus priority was given to other feed items. The higher intake in hens fed the complete diet was attributed to a behavioural need for foraging. Hens with access to chicory had more seeds in the crops than hens with access to grass/clover. This was presumably due to more weeds being present in the chicory plots. The type of plant material ingested also differed. The plant material in the crops of wheat-fed hens consisted of a mixture of leaves, stems and roots, while that in the crops from the hens fed pelleted complete diet consisted mainly of leaves. It was speculated that this result might have been due to the nutrient-restricted hens searching and scratching the ground

Table 7.10. Amount (g) of feed items in the crop during the evening of hens fed a diet based on concentrate plus wheat and allowed access to a forage based on grass/clover or chicory (from Horsted *et al.*, 2007).

Feed item	Grass/clover		Chicory	
	Concentrate	Wheat	Concentrate	Wheat
Supplementary	22.1	28.7	20.4	27.4
Plant material	5.8	7.1	5.8	10.3
Seeds	0.105	0.047	0.282	0.121
Insects	0.048	0.036	0.029	0.048
Earthworms, larvae, pupae	0.687	0.254	0.698	0.559
Oyster shell	1.381	2.645	1.623	2.423
Grit	0.235	1.056	0.692	1.297
Soil	0	15.0	0	8.2

to find earthworms and insects as a source of nutrients. This, together with the intake of earthworms, might explain the large amount of soil found in the crops from the wheat-fed hens. The large amounts of soil found in the evening crops apparently disappeared from the crops during the night, since only relatively small amounts of soil were found in the morning crops. Surprisingly, no differences were found in the amount of earthworms and larvae in relation to type of feed. Due to the protein deficit in the wheat-fed hens, it was expected that more insects and earthworms would be found in the crops from these hens. However, since the first day of slaughter took place 9–10 days after introduction, it is possible that the foraging areas had been cleared of these feed items by the hens. The greater amount of earthworms found in the crops on the second day of slaughter was attributed to wet weather producing a larger number of earthworms in the surface of the ground.

Grit was significantly more abundant in the crops in the evening. Also, more grit was found in the crops of wheat-fed hens. This was attributed to a large amount of coarse feed in the crops of these hens, since more grit in the gizzard would be required to grind the feed into smaller particles.

A related finding was that the amount of whole wheat in the crop was much greater on the second day of slaughter, whereas the amount of concentrate remained constant. This suggested that the crop increases its capacity to retain larger amounts of coarse feed such as whole wheat when continuous access to this feed is given. Increased capacity of the crop has been reported in chickens trained to eat rapidly, since these chickens had larger quantities of feed in the crop per unit of time than untrained chickens. Moreover, crops were heavier in trained chickens, presumably reflecting their greater capacity for intake (Lepkovsky et al., 1960). This increase in the capacity of the crop presumably corresponds to the fact that whole wheat remains for a longer time in the crop than pelleted concentrate; Heuser (1945) found that wheat and whole maize were retained for a longer time in the crop than maize meal or mash.

The conclusion of the above studies is that a huge difference in crop content occurs between morning and evening crops, indicating that laying hens have a higher intake of most feed items at the end of the day, regardless of feeding strategy and type of forage vegetation offered. The results further indicate that the type of forage vegetation used in the above studies influenced the balance between feed items in the crops to a minor degree only. In contrast, the type of supplementary feed affected the intake of several feed items, suggesting that a reduced nutrient content in the supplementary feed can be used as a method of increasing foraging in the outdoor area. Thus, hens fed whole wheat and oyster shells as the only supplementary feed had more plant material, oyster shells, insoluble grit and soil in the crops than hens fed a complete feed mixture.

The study was extended by Horsted et al. (2006) to include other aspects of intake and how the feeding system affected productivity. Results on productivity and egg quality suggested that laying hens consume large amounts of foraging material when accessible. In nutrient-restricted hens (wheat-fed) the forage may yield a substantial contribution to the requirements of amino acids and ME although productivity parameters and measurements on DM in albumen showed that wheat-fed hens, on a short-term basis, were not able to fully compensate for the lack of protein and amino acids by increased foraging. Of the forage crops investigated, chicory especially appeared to contribute to the nutrient intake of the hens. Measurements on eggshell parameters showed that oyster shells together with foraging material were sufficient to meet the calcium requirements. Yolk colour clearly showed that laying hens consume large quantities of green fodder irrespective of the type of supplementary feed. Yolk colour from hens with access to chicory tended to be darker, of a redder and less yellow hue compared with grass/clover-fed hens.

Estimates for removal of herbage (except for the mixed forbs) are given in Table 7.11. The removal of herbage by the hens was particularly pronounced in the chicory plots. Visual assessment of the chicory plots revealed no signs of chicory leaves left in the plots. The remaining herbage was exclusively weeds.

Feed consumption of the two types of supplementary feed differed significantly in both experiments (Table 7.12). Hens consumed approximately 90 g wheat daily in both experiments, whereas they consumed considerably more concentrate (129 g and 155 g in experiment 1 and 2, respectively). Forage crop had no effect on supplementary feed intake. The hens on the wheat diet had a significantly higher intake of oyster shell compared with those fed concentrate in both experiments. Forage crop did not significantly affect intake of oyster shell. No differences in the intake of grit were seen.

In experiment 2, body weight was reduced when wheat-fed hens had access to grass/clover, whereas it remained constant when they had access to chicory. Type of forage crop had no overall effect on egg production or egg weight (Table 7.12). In experiment 1, hens fed wheat had a significantly lower egg production per day compared with hens fed concentrate, but no difference was observed in experiment 2. A significantly lower egg weight was seen in both experiments when hens were fed wheat. Yolk colour became significantly lighter and albumen wetter when hens were fed the wheat diet in experiment 1. Eggshell strength was not affected by dietary treatment, indicating that all the hens were able to meet their needs for calcium through an increased intake of oyster shell and forage.

The results suggested that hens found a considerable part of their nutrient requirements by foraging, even though some lost weight. The hens with access to chicory showed a relatively high egg production and did not lose weight to the same extent as those with access to grass/clover or mixed forbs. This is consistent with the amount of herbage removed from the plots and the fact that the chicory contained a relatively high content of lysine (12.1 g/kg DM) and methionine (4.0 g/kg DM).

Table 7.11. Estimated removal of herbage by laying hens in small experimental plots (from Horsted et al., 2006).

Dietary treatment	Herbage DM at start (/m^2)	Herbage DM at end (/m^2)	Herbage DM removed (g DM/hen/day)
Grass/clover			
Wheat	269	228	17
Concentrate	252	228	9
Chicory			
Wheat	423	231	73
Concentrate	372	236	51

Table 7.12. Consumption of supplementary feed and egg production in hens with access to forage crops (from Horsted et al., 2006).

Dietary treatment	Feed intake (g/day)	Intake of oyster shell (g/day)	Rate of lay (%)	Egg weight (g)
Experiment 1				
Wheat	92	5.7	75	55.2
Concentrate	129	1.7	91	59.6
Grass/clover	111	4.2	83	57.6
Herb mixture	110	3.2	83	57.2
Experiment 2				
Wheat	89	7.2	83	56.5
Concentrate	155	3.1	82	59.2
Grass/clover	119	4.6	80	58.0
Chicory	126	5.8	85	57.8

Other plant material can provide valuable nutrients to poultry. Steenfeldt *et al.* (2007) conducted an experiment (Table 7.13) to examine the suitability of using maize silage, barley–pea silage and carrots as foraging materials for egg-laying hens. Production level, nutrient digestibility, gastrointestinal (GI) characteristics, composition of the intestinal microflora and incidence of feather pecking were studied. CP content of the foraging material (g/kg DM) averaged 69 g in carrots, 94 g in maize silage and 125 g in barley–pea silage. Starch content was highest in the maize silage (312 g/kg DM), and the content of non-starch polysaccharides (NSP) varied from 196 g/kg to 390 g/kg, being lowest in carrots. Sugars were at trace levels in the silages, whereas carrots contained on average 496 g/kg DM. Egg production was highest in hens fed either carrots or maize silage, whereas hens fed barley–pea silage produced less (219 versus 208). Intake of forage was high at 33%, 35% and 48%, respectively, of the total feed intake. Hens fed maize silage had energy intakes similar to the control group (12.61 and 12.82, respectively), whereas access to barley–pea silage and carrots resulted in slightly lower values (12.36 and 12.42, respectively). Hens receiving silage had greater gizzard weights than the control or carrot-fed groups. The dietary supplements had only minor effects on the composition of the intestinal microflora. Important findings of this study were that mortality was reduced significantly in the groups given forage (0.5–2.5%) compared with the control group (15.2%) and showed decreased pecking damage, a reduction in severe feather-pecking behaviour and an improved quality of plumage at 54 weeks of age.

The above findings indicate that good-quality forage has the potential to supply a significant proportion of the nutrient needs of poultry. Producers often ask, therefore, whether the micronutrients in the feed can be reduced. Given the present state of knowledge this is not advised, though producers with access to good-quality forage in sufficient quantity could perhaps experiment during the summer period by reducing the vitamin and trace mineral premixes by 10–25%. Close monitoring of the flock would indicate whether the reduction could be maintained or cancelled. The integrity of the welfare of the birds has to be maintained even if, at times, the intake of micronutrients is in excess of requirements.

One question raised in relation to pasture access is the effect on meat quality. This probably depends on the quality and nature of the pasture available. Ponte *et al.* (2008) fed RedBro Cou × RedBro M broilers on a diet based on maize, wheat and soybean meal either in indoor pens or in pens with access to two types of clover pasture. Results showed that weight gain and feed intake were higher in the birds given access to pasture. In addition carcass yield was higher in the birds allowed access to pasture. The meat of the indoor-raised birds was found to be more tender, but the meat from the birds allowed access to pasture had a higher overall rating by a taste panel.

Table 7.13. Production, body weight and mortality of laying hens fed a layer diet with or without supplements of maize silage, barley–pea silage or carrots (from Steenfeldt *et al.*, 2007).

	Layer diet	Layer diet + maize silage	Layer diet + barley–pea silage	Layer diet + carrots
Egg production (%)	89.9	91.4	87.2	92.0
Egg weight (g)	61.5	61.1	61.5	61.9
Feed/hen/day				
Silage/carrots (%)	130.1	177.7	165.4	221.7
	–	33.4	35.1	48.5
Body weight (g)				
Start	1750	1742	1718	1726
End	1813	1787	1805	1917
Mortality (%)	15.3	1.5	2.5	0.5

Sales (2014) reviewed a large number of studies conducted to examine this issue and concluded that 'consumer preference for poultry meat from free-range birds is not justified by scientific evidence'. He concluded that, except for a lower relative fat concentration in commercial poultry cuts, a meta-analysis of available reports presented little evidence that a poultry production system with outdoor access has any impact on performance, carcass composition and meat quality.

The conclusions of Sales (2014) confirm some other conclusions that genotype and feeding method have greater effects in meat birds on productivity and meat quality than access to pasture per se. For instance, Cömert et al. (2016) found that the main factor affecting the carcass composition and carcass yield was genotype, but that organic feeding enhanced the quality of the meat. Other research does confirm that chicken meat quality can be improved by providing access to pastures and organic feed that provide an enhanced intake of n-3 fatty acids (Sirri et al., 2011).

One of the results of access to soil and forage is that birds ingest arthropods such as insects and earthworms. These may provide an additional source of nutrients.

Intake of insects by junglefowl chicks and wild turkey poults can exceed 50% of their diet, and adult females increase their intake of insects at the time of reproduction (Klasing, 2005). In the case of junglefowl, termites and bamboo mast are preferred feeds in the area of South-east Asia where domestication likely occurred (Klasing, 2005). However, as shown by Hossain and Blair (2007) and discussed in Chapter 4, the chitin contained in the hard outer shell of insects is difficult for domestic poultry to digest, although the high chitin content of insect meals does not appear to have detrimental effects on poultry performance (Ravindran and Blair, 1993). Because insects are part of the natural diet of poultry, it has been suggested that some birds may utilize chitin more efficiently than other animals, but evidence is lacking. Chitinase does occur in the stomach of some insect-eating birds. Austin et al. (1981) reported that the microbial digestion of chitin in poultry could be increased by feeding a source of lactose such as milk by-products.

Research findings reviewed by Ravindran and Blair (1993) suggested that the type of insect preferred by poultry is soft-bodied. Insects are high in CP, with contents ranging from 420 g/kg to 760 g/kg (Ravindran and Blair, 1993). Accurate determination of protein levels in insect meals, however, requires correction for the non-protein nitrogen contributed by the chitin.

Several studies have evaluated housefly (Musca domestica) pupae meal as a poultry feed ingredient (Ravindran and Blair, 1993). Results showed that it could successfully replace soybean meal in poultry diets and that its inclusion had no adverse effects on the taste of the meat. Inaoka et al. (1999) reported no significant differences in body weight gain, feed conversion ratio, dressing percentage or meat quality in broiler chicks fed diets containing ground dried housefly larvae at 70 g/kg or a similar level of fishmeal. In a more recent study the nutritive value of housefly larvae as a feed supplement for turkey poults was investigated by Zuidhof et al. (2003). The poults were fed either dehydrated larvae or a commercial diet. Gross energy, apparent ME and CP were 23.1, 17.9 MJ/kg and 593 g/kg, respectively, in the larvae, and 17.0, 13.2 MJ/kg and 318 g/kg, respectively, in the commercial diet. Digestibility of the larvae was found to be high, suggesting that the product compared favourably with soybean meal as a protein supplement for poultry.

Silkworm pupae meal has been used to replace fishmeal completely in layer diets and to replace up to 50% of the fishmeal in chick diets (Ravindran and Blair, 1993). The pupae contain about 480 g CP/kg and 270 g crude fat/kg, and require de-oiling to improve their keeping quality. De-oiling is also necessary to remove the highly unsaturated fats that affect the flavour of poultry meat (Gohl, 1981). De-oiled silkworm pupae meal may contain as much as 800 g CP/kg. These findings will be of particular interest to producers in Japan, since silkworm pupae are on the list of approved feed ingredients for organic production in that country.

Meals prepared from grasshoppers contain as much as 760 g CP/kg, but the amino acid profile is poorer than that of fishmeal (Ravindran and Blair, 1993). Moreover, the CP fraction is only 62% digestible. Feeding trials have shown that partial replacement of fish or soybean meals with grasshopper meal was feasible and that the taste of the meat was unaffected by its inclusion.

Wang *et al.* (2005) found that a meal made from adult field crickets (*Gryllus testaceus* Walker) contained (g/kg DM basis) 580 g CP, 103 g ether extract, 87 g chitin and 29.6 g ash, respectively. The total amounts of methionine, cystine and lysine in the meal were 19.3, 10.1 and 47.9 g/kg, respectively, and their true digestibility coefficients were 0.94, 0.85 and 0.96, respectively. The TME_n of this insect meal was 2960 kcal/kg. When maize–soybean meal diets were formulated on an equal CP and true ME basis, it was found that up to 150 g cricket meal/kg could replace the control diet without any adverse affects on broiler weight gain, feed intake or gain/feed ratio from 8 to 20 days post hatching.

Makkar *et al.* (2014) reviewed findings on the nutritional composition of black soldier fly larvae, housefly maggots, mealworms (*Tenebrio molitor* larvae), locusts, grasshoppers, crickets and silkworm meal and their use as replacements for soymeal and fishmeal in the diets of poultry, pigs, fish species and ruminants. In general, the CP and oil contents were reported to be high at 420–630 g/kg and up to 360 g/kg, respectively. The concentration of unsaturated fatty acid was found to be high in housefly maggot meal, mealworm and house cricket (60–70%), and lowest in black soldier fly larvae (19–37%). The reviewed studies confirmed their palatability as alternative feeds and their ability to replace 25–100% of soybean meal or fishmeal, depending on the animal species. Except for silkworm meal, other insect meals were found to be deficient in methionine and lysine. Most insect meals were found to be deficient in Ca. Composition of these meals was found to vary, depending on the substrate used in their production. Salmon and Szabo (1981) evaluated meal made from dried, spent bees as a feedstuff for growing turkey poults. Although higher in CP and differing in amino acid composition, dried bee meal was found to be similar to soybean meal in content of total amino acids and true ME. Diets containing dried bee meal at 150 g/kg or 300 g/kg resulted in a decrease in live weight gain by poults. The researchers speculated that the adverse effects might be related to non-protein nitrogen in bee meal or to toxicity of dried bee venom.

Research on termites suggested that chickens may be able to utilize termite meal more efficiently than rats. Other research found that meal made from snails could partially replace fish or meat meals in poultry diets (Ravindran and Blair, 1993).

Earthworms are a natural feed source for poultry kept under free-range systems and, live or dried, are highly palatable to poultry. They may represent an opportunity as an alternative feed source for organic poultry production.

Meal made from earthworms contains about 600 g CP/kg, with an amino acid composition comparable to that of fishmeal (Ravindran and Blair, 1993). It can replace fishmeal in chick and layer diets but care must be taken to balance the dietary calcium and phosphorus contents, since these minerals are low in earthworms due to the absence of an exoskeleton. Moreover, earthworms are known to accumulate toxic residues, particularly heavy metals and agrochemicals. According to Ndelekwute *et al.* (2016) in addition to providing a good source of nutrients for poultry, earthworms secrete cellulase, lichenase, chitinase and cellulolytic microorganisms which enable them to degrade plant waste.

Reinecke *et al.* (1991) fed broilers from 10–17 days of age on diets containing 45, 90 or 135 g protein/kg from commercial fishmeal or from earthworm meal made from the worms *Eisenia fetida*, *Eudrilus eugeniae* or *Perionyx excavatus*, respectively. No differences were found in protein utilization or growth. Son (2006) reported that supplementing the diet of 55-week-old laying hens with earthworm meal at 3 g/kg improved their rate of lay and egg quality. A possible concern noted by these researchers was the presence of heavy metals in the earthworm meal (arsenic, cadmium, chromium and lead at 4.41, 1.23, 1.18 and 3.39 mg/kg, respectively), which were not detected in the control diet.

Earthworm meal has also been shown to be a useful protein source for Japanese quail. Das and Dash (1989) fed 1-week-old male and female Japanese quails on a maize-based diet containing fishmeal at 60 g/kg or earthworm meal at 60 g/kg. After 56 days, total weight gains were 96.1 g and 98.5 g, feed intakes were 533 g and 511 g and feed conversion ratios 5.54 g/g and 5.19 g/g for the control and earthworm diets, respectively. Egg quality was not influenced by the diets.

In a recent investigation Bahadori *et al.* (2017) tested the effects of earthworm (*Eisenia foetida*) meal with vermi-humus (VH, the medium remaining after the worms were grown) on growth performance, haematology, immunity, intestinal microbiota, carcass characteristics and meat quality of hybrid broiler chickens. Five diets were fed during the starting, growing and finishing stages, i.e. a maize/soybean meal control diet; or a diet modified to contain vermi-humus at 10 g/kg supplemented with earthworm meal (EW) at 0, 10, 20, or 30 g/kg, DM basis. The rationale for including vermi-humus was that it is a source of humic acid, resulting from the decomposition of organic materials in soil which has the ability to inhibit bacterial and fungal growth.

Feed intake was reduced with the two highest levels of supplementation but weight gain was unchanged. As a result feed conversion efficiency was improved with supplementation with the earthworm products. Serum total protein, albumin, Ca and P concentrations were found to be lower in chickens fed the control diet, and they increased linearly with increase in dietary EW. In addition, humoral immune response (except heterophil/lymphocyte ratio) and relative weights of immune organs (spleen, thymus and bursa of Fabricius) were found to be lower in chickens fed the control diet. Increasing supplementation of EW also resulted in increased total counts of lactic acid bacteria and a reduced population of pathogenic microbiota in the ileal contents. Tests on the quality of the meat showed that the antioxidant activity during storage was increased by supplementation with EW.

The above findings indicate that earthworms can provide a useful amount of protein for poultry. Concerns are that the earthworms may concentrate heavy metals and contaminants present in the soil and may act as intermediate hosts for cestode worms and disease vectors such as those causing blackhead in turkeys. This concern about disease spread is minimized in some tropical countries by collecting the earthworms and sun-drying them before feeding to poultry. In developed countries the concern is addressed by adequate heat-treatment of these products.

The significance of the above results for organic producers is that ranged poultry may be able to obtain a substantial proportion of their nutrient needs from insects, earthworms, etc. However, the intake is difficult to quantify; therefore, the most appropriate way to manage the situation is to choice-feed the flock on grain and supplement. By feeding in this way the birds can adjust their intakes of protein and energy according to the amounts received from insects, earthworms and other soil organisms.

Health and Welfare

Health and disease problems in organic flocks can be categorized into those affecting the birds directly and those that affect eggs and meat and may pose problems for the human consumer. These will be addressed separately, in relation to dietary treatments that have been shown to be useful in their control.

Health and welfare problems in organic flocks

Disease prevention in organic farming is based on the principle that an animal that is allowed to exhibit natural behaviour is not subject to stress, and if fed an optimal (organic) diet will have a greater ability to cope with infections and health problems than animals reared in a conventional way

(Kijlstra and Eijck, 2006). Fewer medical treatments are then necessary and if an animal becomes diseased, alternative treatments should be used instead of conventional medication. However, strict biosecurity is needed to help prevent diseases such as avian influenza. Other measures to avoid or reduce disease risks include the use of an all-in, all-out management system, based on the principle that depopulation at the end of a flock reduces the pathogen load because some pathogens die when there is no host. A related measure is the principle not to mix ages or species. Older birds can carry disease while showing no signs of infection and can spread the disease to young birds. Likewise, domestic ducks and geese can carry diseases that infect chickens.

Providing birds with access to outdoors has the advantages of providing exercise and fresh air but has the disadvantages of exposing the birds to predators and disease threats in soil and water, and from wild birds and other animals in the environment. Appropriate housing and yard design and approved veterinary treatments such as vaccination should be adopted so that these threats are minimized.

Lampkin (1997) identified coccidiosis, feather pecking, cannibalism and external parasites as potential problems in organic poultry production. A Danish survey of large organic layer flocks reported high mortality rates (15–20%) 2–3 times higher than in layers in battery cages (Kristensen, 1998). A Dutch study of organic layer flocks reported severe feather pecking in 50% of the flocks. Koene (2001) concluded that a combination of improved stockmanship, better nest design and identification of genetically suitable strains for extensive rearing without beak trimming were likely to produce a sustainable solution to feather pecking and cannibalism in organic egg production. There are also indications that the incidence of helminth infections (*Ascaridia galli*, *Heterakis gallinarum* and *Capillaria* spp.) in laying hens is considerably higher in organic production systems than in conventional systems (Hovi *et al.*, 2003) but that coccidiosis is not a major problem in organic broilers, despite a ban on the routine use of anticoccidial agents.

More recent data confirm that losses are unacceptably high in organic production, at least with laying flocks. Bestman and Maurer (2006) reported that in The Netherlands the mean mortality rate in organic laying hens was 11% (0–21%), caused by infectious diseases such as *Escherichia coli*, infectious bronchitis, coccidiosis and brachyspira. This situation was attributed possibly to a high infection pressure in The Netherlands (25 million layers kept mainly in two areas) and perhaps an inadequate disease resistance. In Switzerland, the mean mortality rate in organic layer flocks was reported as 8% (range 3–25%). The hybrids used in organic production in these studies were mostly the same as in conventional poultry husbandry and some vaccination was practised. Maurer *et al.* (2002) reported that hens on organic farms had more parasite problems than birds kept under conventional conditions.

Bestman and Maurer (2006) suggested feather pecking as a good measure of animal welfare in organic laying hens since degree of feather pecking is associated with stress and is mainly correlated with reduced use of the outdoor run. In The Netherlands feather pecking had been reported in 70% of laying flocks and in 54% of pullet flocks. In rearing farms feather pecking was correlated with high densities of chicks combined with poor environments. However, on many poultry farms the runs were not used well, indicating that the birds did not feel safe in them.

It seems clear from the above findings that the environmental conditions on many organic layer farms need to be improved.

In The Netherlands and in Switzerland organic broilers are less important than laying hens. The main health problems of these birds appear to differ from those in laying hens (Bestman and Maurer, 2006). Diseases with a long incubation period (e.g. ascarids) do not usually occur, even in the relatively long-living organic broilers. Intestinal diseases such as diarrhoea are more important than in layers and vaccination is recommended on organic farms against coccidiosis, the other main health problem reported by these authors and by Lampkin (1997). According to these authors, the use of slow-growing hybrids has reduced the incidence of skel-etal lesions in organic broilers. Predators (hawks, foxes and martens) were identified as a cause of broiler losses in free-range systems.

An additional health problem associ-ated with commercial poultry production is foot-pad dermatitis (FPD) which is an infec-tion affecting the plantar region of the feet of chickens. FPD is closely related to 'hock burns', a condition in which the skin on the hock becomes dark brown. In Sweden and Denmark, scores of foot health are being used to evaluate the health and welfare of broiler flocks. There is a relatively high her-itability for FPD and it has a low genetic cor-relation to body weight (Kjaer *et al.*, 2006). This suggests that selection for decreased incidence of FPD should be possible with-out negative effects on growth rate.

A variety of bacterial infections can be a problem in poultry production. Colibacillosis results from a coliform infection, the most common organism being *E. coli*. Problems range from severe acute infections with high mortality to chronic mild infections with low mortality. Ask *et al.* (2006) reported consid-erable genetic variation in susceptibility to colibacillosis, which suggests that selection against this disease is possible and perhaps an alternative to antibiotics, which cannot be used in organic poultry production.

Intestinal parasites can be a worldwide problem in all poultry production systems, but are particularly problematic in cage-free systems where the birds have access to their faecal material. Abdelqader *et al.* (2007) suggested that the natural resistance in native poultry breeds can provide an alternative to chemical treatment in organic production systems. They compared the resistance of different isolates of *Ascaridia galli* in a local chicken breed from Jordan and in Lohmann LSL white chickens. The *A. galli* isolates were from different geo-graphical locations – German isolates were used in the first trial and Jordan isolates in the second. Their results suggested that the difference in genetic background between the two chicken breeds is involved in resist-ance to *A. galli* infection. In addition, they reported that *A. galli* isolates from different

geographical areas differ in their ability to infect different chicken genotypes. The local Jordanian breed harboured significantly fewer worms than the Lohmann breed and the female worms from the Jordanian breed were less fecund than female worms isolated from the Lohmann chickens. The reduced fecundicity indicates that the *A. galli* isolate from naturally infected local chicken in Jordan is less infectious than *A. galli* from Germany for both strains of birds.

As income in the more developed coun-tries increases, consumers are becoming more selective over their food items. While poultry-breeding companies are focusing on growth rate and meat yield, consumers are becoming more concerned about meat qual-ity. In addition, the shift from the marketing of whole birds to further-processed prod-ucts has highlighted problems with meat quality as assessed by toughness and cohe-siveness, colour, and water-holding proper-ties (Sosnicki and Wilson, 1991).

A relationship between muscle growth and poultry meat quality has been docu-mented by a number of researchers. It has been suggested that improving breast weight through genetic selection has the potential to result in lighter-coloured breast meat with lower water-holding capacity (Berri *et al.*, 2001). Fast-growing animals have more and larger muscle fibres than slow-growing strains (Dransfield and Sosnicki, 1999). Smaller fibre diameters are believed to allow for higher pack-ing density of the fibres, and thus increased toughness of the meat. This has been shown to be the case with fish (Hurling *et al.*, 1996) but the effects on pork and beef have not been conclusive (Dransfield, 1997). With increased growth rate, muscle fibres have been shown to become more glycolytic and such fibres have a more rapid development of rigor mortis. Increased rate of rigor mor-tis development results in an increased likelihood of paler colour and reduced meat quality (Dransfield and Sosnicki, 1999).

Diet and infectious diseases in birds

The effect of diet on disease severity in poul-try has been investigated in several studies.

It has been shown that feeding diets high in n-3 fatty acids is protective against some protozoan infections. Allen *et al.* (1997) reported that feeding diets supplemented with flaxseed as a source of n-3 fatty acids was beneficial in reducing lesions caused by one type of coccidial organism (*Eimeria tenella*, which attacks the caecum) but had no effect on lesions caused by another coccidial organism (*Eimeria maxima*, which attacks the small intestine). These findings are of potential interest to producers of 'designer' eggs, produced from hens fed diets containing flaxseed.

Organic producers can take several nutrition-related steps to control problems related to health of their flocks, with the banning of routine medication, including the use of antibiotics, in feed. These can be summarized as below, and readers are directed to veterinary publications for a more detailed outline of the appropriate procedures.

As stated above, a main problem in poultry is GI disease. Relevant approaches to this problem include the improvement of immunity, using whole grains in the diet to encourage gizzard development and adding fibrous ingredients to the diet to encourage fermentation in the large intestine. Another approach is the supplanting of disease organisms in the gut with beneficial organisms (competitive exclusion) to allow the intestinal epithelium and host microflora to act as natural barriers to damage from pathogenic bacteria, antigens and toxic substances inside the gut. The usual approach in organic poultry production is to manage the outdoor area so that the number of coccidia is reduced, keeping infection at a minimum challenge until natural immunity is established in the birds. Alternatively, vaccination is recommended where permitted.

Mycotoxin problems

Mycotoxins are defined as secondary metabolites of mould growth which are generally believed to be produced in response to stress factors acting on the fungus. Mycotoxins are estimated to affect as much as 25% of the world's crops each year (CAST, 1989). Both cereal and protein crops can be affected. For instance, groundnuts are susceptible to aflatoxin contamination. Mycotoxins cause significant economic losses in animal agriculture worldwide. These fungal toxins are particularly important because they can be transmitted from animals to humans in milk or in meat products and because some of them are potent carcinogens and teratogens (Hesseltine, 1979). Therefore, it is important to keep them out of the food chain, and procedures to minimize the risk to livestock and humans are an important part of the feed-quality programme.

They occur sporadically both seasonally and geographically (CAST, 1989). Some areas are at a higher risk for specific mycotoxins than others because of local conditions such as early frost, drought and insect damage. The mycotoxins that may be found in feeds that come from different global locations occur naturally in a wide variety of crops used as feedstuffs. However, only a few mycotoxins occur significantly in naturally contaminated feeds: ochratoxin A (OA), patulin, zearalenone, trichothecenes, citrinin and penicillic acid (Jelinek *et al.*, 1989). Blending of grains, damage in transit and improper storage conditions can lead to increased contamination. High concentrations are found in damaged, light or broken grain such as those which occur in screenings. Grain that is above optimum moisture content for storage may continue to respire and produce water. This increase in moisture in portions of a storage bin can allow mould growth and production of toxin. Alternating warm and cool weather conditions can favour water migration and condensation within a storage bin, again creating conditions favourable for local mould growth and toxin production. For these reasons it is very important that grain be stored at a moisture content not exceeding 140 g/kg and preferably in insulated storage bins. Susceptibility to mould growth increases when grains are ground, because the protective seedcoat is broken. For this reason, care should be taken that ground feeds are stored in cool dry areas.

Lawlor and Lynch (2001) showed that cereals grown in Ireland may be contaminated with vomitoxin, zearalenone, fusaric acid or ochratoxin. The presence of aflatoxins in animal feeds in Ireland is considered likely to be due to the importation of feed ingredients from warmer climates.

Individual moulds, fungi or mycotoxins rarely occur in isolation and two or more mycotoxins together may have a greater toxic effect than any one alone (Pasteiner, 1997). The presence or absence of toxin-producing fungi is not a good indicator of the presence or absence of mycotoxins (Osweiler, 1992; Pasteiner, 1997). High temperatures and high pressures during the drying and milling of cereals may reduce the fungal load but mycotoxins are resistant to the temperatures that kill moulds and they can persist in grains in the absence of evidence of fungal contamination (Osweiler, 1992). In fact, most mycotoxins are chemically stable and persist long after the fungi have died (Pasteiner, 1997). Fungi that grow in crops prior to harvest are classified as field fungi; those that develop in storage are called storage fungi. Some fungi have the ability to be both storage and field fungi, e.g. *Aspergillus flavus*. Fusarium is a field fungus. It requires high relative humidity (> 90%) and heat (> 23°C) for growth and so rarely grows after harvest as storage conditions are generally not suitable (Osweiler, 1992). In the field, the fungus causes death of ovules, shrivelling of grain and weakening or death of embryos. This process is described as 'weathering' (Osweiler, 1992). Storage fungi may be pathogenic or saprophytic, including the genera *Aspergillus* and *Penicillium* which produce many of the mycotoxins that are important in poultry production, and which can grow at moisture contents ranging from 140 to 180 g/kg and at temperatures that range from 10°C to 50°C (Osweiler, 1992).

Effects of mycotoxins in poultry

Mycotoxins generally affect the liver (hepatotoxins) or the kidney (nephrotoxins). The problem of mycotoxins in poultry was highlighted in 1960 when more than 100,000 turkey poults in the UK died from what was called 'Turkey X disease' (Blount, 1961). At post-mortem the turkeys were found to be suffering from enteritis with engorged kidneys and/or hepatitis. Bacteriological examinations were negative. The problem was eventually found to be due to aflatoxin contamination in groundnut meal imported from Brazil.

Among poultry, ducks are the most susceptible to aflatoxin, followed by turkeys, broilers, laying hens and quail. Aflatoxins are liver toxins, causing hepatocyte degeneration, necrosis, fatty changes and altered liver function. Suppression of protein synthesis in the liver results in growth suppression and reduced egg production. Aflatoxin is also known to interfere with vitamin D metabolism, contributing to reduced bone strength and leg weakness. By reducing bile salt production, aflatoxin reduces the absorption of lipids and pigment. In addition, the metabolism of other minerals including iron, phosphorus and copper is also affected by aflatoxin. Aflatoxin increases the fragility of capillaries, reducing prothrombin levels and resulting in an increased incidence of bruising in carcasses and carcass downgrading.

Aflatoxin is the most commonly known mycotoxin in poultry feeding, but a number of other mycotoxins can result in adverse effects. For example, the mycotoxin OA is approximately three times more toxic to poultry than is aflatoxin, and when present in combination, the depressing effects can be even more severe.

The severity of the effects of mycotoxins depends mainly on the specific mycotoxin present and the level of contamination. High levels can result in very high mortality within a short period but generally the adverse effects of subacute mycotoxin contamination represent the greatest problem to the poultry industry. Routine testing for mycotoxins is helpful in managing the risk of contamination.

Dealing with mycotoxin contamination

Blending contaminated feedstuffs with uncontaminated feedstuffs to dilute the concentration of mycotoxin and the use of binding agents (such as bentonite clays and yeast mannanoligosaccharide (MOS)) might be used to lower the risk of mycotoxicosis, if acceptable to the local organic certifying agency. Binding agents such as bentonite, aluminosilicates, spent canola oil, bleaching clays and lucerne fibre have been used in feeds containing mycotoxins to prevent intestinal absorption of the toxins (Smith and Seddon, 1998). Even if effective, a binding agent is likely to be effective only against a specific mycotoxin. The absorbing clays or binding agents may also bind vitamins, making them unavailable for absorption (Dale, 1998). Modified yeast cell wall MOS has been reported to effectively bind aflatoxin, and to bind ochratoxin and the fusariotoxins to a lesser degree. This product has advantages over other binding agents in that it does not bind vitamins or minerals (Devegowda et al., 1998).

Preventing mould growth and subsequent mycotoxin production during storage of feedstuffs is a more certain way to avoid problems, as recommended by Lawlor and Lynch (2001). Cleaned grain should be stored at a moisture content less than 140 g/kg in clean, preferably insulated, bins. If grain must be stored at a higher moisture content or if storage conditions are poor, a suitable mould inhibitor (e.g. propionic acid) should be used, if permitted under the local organic regulations.

Whole grain and health

Several research findings have demonstrated the beneficial effects of whole-grain feeding on the digestive microflora and the overall health of poultry (Santos et al., 2008). They show that a better-developed gizzard has an important function as a barrier organ in preventing pathogenic bacteria from entering the distal digestive tract. As a result, some researchers have suggested that whole grain could be considered as an effective alternative to antibiotic growth promoters.

Engberg et al. (2004) reported an increase in intestinal counts of some beneficial Lactobacillus species, Glünder (2002) showed a decrease in E. coli counts, and Engberg et al. (2002, 2004) reported a lower number of pathogens such as Salmonella spp. or Clostridium perfringens (responsible for necrotic enteritis) as a result of whole-grain feeding. Gracia et al. (2016) confirmed these effects by showing that supplementation of the diet of broilers with whole wheat or oat hulls significantly reduced caecal Campylobacter jejuni colonization at 42 days after a challenge with this organism. C. jejuni is the leading cause of bacterial gastroenteritis in humans worldwide, though infection with this organism is not currently considered to be pathogenic in poultry.

Taylor and Jones (2004) reported that the incidence of proventricular dilatation and mortality from ascites was reduced in broilers when whole grain was incorporated into pelleted diets at 200 g/kg. Enzyme supplementation did not affect the incidence of proventricular dilatation. Evans et al. (2005) studied the effects of feeding laying hens on diets containing wheat in whole or ground form on coccidial oocyst output after being challenged with coccidiosis. Birds fed on the diet with whole wheat had a significantly lower (2.5 times) oocyst output than birds fed on the diet with ground wheat, suggesting that an active functioning gizzard can play a role in resistance to coccidiosis. Other work has shown that chickens fed free-choice on a high (420 g CP/kg) protein concentrate and whole wheat were more resistant to coccidiosis than those on complete high-fibre diets (Cumming, 1989). In male broiler chickens, oocyst output has been shown to be negatively correlated with relative gizzard size, both with a conventional complete diet or free-choice feeding (Cumming, 1992b). In related work this author showed that the feeding of insoluble grit reduced the output of oocysts from chickens fed either complete diets or whole grains and supplement free-choice.

Bjerrum et al. (2005) reported the results of an experiment involving broilers infected with a rifampicin-resistant Salmonella typhimurium strain at 15 days of age. Lower numbers of the organism were found in the gizzard and ileum of birds receiving whole wheat compared with pellet-fed birds, confirming the beneficial effects of inclusion of whole grain in the diet. Feeding whole or coarsely ground grains has also been shown to result in decreased caecal salmonella populations in 42-day-old broilers (Santos *et al.*, 2008).

Modification of the intestinal microflora

The phenomenon by which the normal GI microflora protects the host against invading pathogens is termed competitive exclusion. It implies the prevention of entry or establishment of one bacterial population into the GI tract by a competing bacterial population already occupying potential attachment sites. To be able to succeed, the latter population must be better suited to establish or maintain itself in that environment or must produce compounds inhibitory to its competitors.

Numi and Rantala (1973) were the first to apply the concept to domestic animals, mainly poultry. These authors had observed that introduction of gut contents containing viable anaerobic bacteria and originating from adult birds could protect young birds against salmonella infection. The concept was therefore designed originally to reduce salmonella infections in growing chickens, but was later extended to other enteropathogens such as pathogenic *E. coli*, *C. perfringens*, *Listeria monocytogenes* and *Campylobacter* spp. Those products approved for use in organic production are listed in Chapter 4. They are mixed cultures derived mainly from the caecal contents and/or gut wall of domestic birds. The treatment is normally given to newly hatched chicks or turkey poults as soon as possible after hatching, either by spraying at the hatchery or at the farm or by addition to the first drinking water.

Competitive exclusion also has an important role in maintaining the health of older poultry. Microorganisms in the gut compete with the bird for digestion products. Gut health and enteric disease resistance are often dependent upon the composition of the diet and the digestibility of feed ingredients. Poorly digested feedstuffs result in an increased amount of undigested material in the hindgut, leading to a proliferation of breakdown bacteria in this part of the intestine, which can result in an increase in toxic metabolites that compromise gut health. This explains why antibiotics are most effective in birds fed diets containing high levels of indigestible protein (Smulders *et al.*, 2000). Similarly, poultry fed diets containing high levels of poorly digested NSP from wheat, barley or rye are more susceptible to enteric disease such as necrotic enteritis (Burel and Valat, 2009). Langhout (1999) observed that dietary NSP significantly increased gut populations of pathogenic bacteria at the expense of beneficial bacteria. It is thus possible to shift the microbial population from harmful to non-harmful bacteria or even beneficial bacteria by changing the diet (Burel and Valat, 2009). As a result, these authors suggested that, in the absence of antibiotics in the diet, feedstuffs containing high levels of NSPs should be used with care. Presumably supplementation with an appropriate enzyme mixture would reduce or prevent this potential problem.

As described in Chapter 4, adding whey and other milk products is being investigated as a possible way of reducing salmonella colonization in organic poultry, birds given access to outdoors being more prone to salmonella infection than conventional poultry.

The digestive microflora can also be modified by processing of the diet (Engberg *et al.*, 2002). Pelleting of the feed contributes to an increase in coliforms and enterococci in the ileum and a reduction of *C. perfringens* and lactobacilli in the hindgut. The temperature of the conditioning process and presence of steam are known to affect the microflora in the gut, suggesting that feed processing can be used to control and manage the GI microflora.

Prebiotics

Prebiotics have been defined as non-digestible or low-digestible feed ingredients that benefit the host organism by selectively stimulating the growth or activity of beneficial bacteria (bifidobacteria and some Gram-positive bacteria) in the hindgut (Burel and Valat, 2009). Belonging to this group are chicory and Jerusalem artichoke, which contain inulin-type fructans in the sap and roots.

Lactulose, galacto-oligosaccharides, fructo-oligosaccharides (FOS), malto-oligosaccharides and resistant starch have also been used as prebiotics. Part of the reasoning for the use of dietary fibrous sources that ferment in the large intestine is that they may produce butyrate, a short-chain fatty acid (SCFA). Butyrate and other SCFA are important in relation to the absorption of electrolytes by the large intestine and may play a role in preventing certain types of diarrhoea (and cancer in humans). Medium-chain fatty acids (MCFA) may be more effective than SCFA in poultry. Some herbs such as allium, thymus, anthriscus and ferule are also known to stimulate acid production by lactobacilli and might be useful prebiotics in animal and human nutrition.

The effects of dietary FOS on the GI microflora of poultry are well documented. Hidaka *et al.* (1991) found that consumption of 8 g FOS per day increased numbers of bifidobacteria, improved blood lipid profiles and suppressed putrefactive substances in the intestine. Patterson *et al.* (1997) found that caecal bifidobacteria concentrations increased 24-fold and lactobacilli populations increased sevenfold in young broilers with FOS. Bifidobacteria may inhibit other microbes because of a high production of volatile fatty acids (VFAs) or the secretion of bacteriocin-like peptides (Burel and Valat, 2009). The improvement in gut health status by dietary FOS supplementation often results in improved growth performance. Ammerman *et al.* (1988) demonstrated that the addition of dietary FOS at a level of 2.5 or 5.0 g/kg diet improved feed efficiency over the period from 1 to 46 days of age. Mortality was reduced with the higher level. However, Waldroup *et al.* (1993) found that supplementing the diet of broilers with 3.75 g FOS/kg had few consistent effects on production parameters or carcass salmonella concentrations.

Several other experiments have been conducted with dietary oligosaccharides. Li *et al.* (2007) conducted a study with laying hens fed a diet containing 20 mg zinc bacitracin/kg plus 4 mg colistinsulfate/kg or 2000, 4000 or 6000 mg FOS/kg. The results showed improvements in egg production, feed consumption and feed conversion of layers when 2000 mg FOS/kg was added to the diets, as well as an increase in eggshell thickness, yolk colour and Haugh unit and a decrease in yolk cholesterol concentration. However, the larger doses of FOS did not improve the performance of layers.

Data suggested that novel oligosaccharides with improved anti-pathogen effects can be synthesized (Burel and Valat, 2009).

One of the disadvantages of using diets with partly indigestible carbohydrates is that they can lead to increased parasite infections. For instance, Petkevicius *et al.* (2001) found that diets that led to high numbers of *Oesophagostomum dentatum* in pigs were characterized by having high levels of insoluble dietary fibre and a relatively low digestibility. In contrast, a diet composed of highly degradable carbohydrates decreased worm establishment, size and female fecundity. This result suggests that poultry producers could use highly digestible diets during outbreaks of helminth infestation and, where possible, should use liquid whey as a dietary supplement. This product is known to be useful in helping to control ascarid infestations. Grazing management should also be used. Most helminths are strictly host-specific and mixed grazing is known to be useful in helminth control.

Probiotics

Certain probiotics have been approved for use in organic diets, provided that they have not been derived using genetic modification (GM) technology. A probiotic has been defined as 'a preparation or a product containing viable, defined microorganisms in sufficient number, which alter the microflora (by implantation or colonization) in a compartment of the host, and by that, exert beneficial health effects on the host' (Roselli *et al.*, 2005). The description implies that probiotics should be able to survive exposure to the digestive juices and that an adequate dose is necessary to have beneficial effects. The most known characteristics of probiotics are the following: (i) a capacity to adhere to the intestinal mucosa and to inhibit pathogen adhesion; (ii) the ability to colonize and proliferate in the intestine; (iii) the ability to prevent some intestinal diseases such as diarrhoea; and (iv) the ability to modulate the immune system of the host (Roselli et al., 2005). The rationale for probiotic use is that probiotics are able to restore normal gut microflora.

The mechanisms by which probiotics (and prebiotics) produce beneficial effects on the gut have not yet been fully elucidated. However, at least three mechanisms of action have been proposed: (i) antibacterial agents produced by probiotic organisms may have an inhibitory effect on pathogenic microbes; (ii) immune responses may be enhanced to suppress potential pathogens; and (iii) competition in the gut epithelium may allow lactic acid bacteria and bifidobacteria to supplant pathogenic organisms.

The effect of probiotics is well documented in the literature (Burel and Valat, 2009). Hollister et al. (1999) reduced salmonella colonization in chicks by feeding a live caecal culture from salmonella-free poultry. Gram-positive bacteria, including those of *Lactobacillus*, *Enterococcus*, *Pediococcus*, *Bacillus* spp., and bifidobacteria, and fungi such as *Saccharomyces* (yeast) are often fed after antibiotic therapy as a means of reintroducing a beneficial microflora to the gut of affected animals (Burel and Valat, 2009). They appear to act by enhancement of competitive exclusion of the GI microflora against exogenous pathogenic microorganisms in the gut, allowing lactobacilli and bifidobacteria to multiply and reduce the pathogenic bacterial population by simple competition.

Probiotics have also been shown to have some effectiveness as anti-coccidial agents (Hessenberger et al., 2016).

A current problem with probiotics, at least in North America, appears to be that commercial veterinary probiotic preparations are not accurately represented by label claims (Weese, 2002). In this investigation, quantitative bacteriological culture was performed on eight veterinary probiotics and five human probiotics and isolates identified by biochemical characteristics.

It was found that the label descriptions of organisms and concentrations accurately described the actual contents of only two of 13 products. Five veterinary products did not specifically list their contents. Most products contained low concentrations of viable organisms. Five products did not contain one or more of the stated organisms, and three products contained additional species. Some products contained organisms with no reported probiotic effects, some of which could be pathogens. The authors concluded that quality control appears to be poor for commercial veterinary probiotics.

Zinc is an important element in fighting infections and is sometimes used for disease control in conventional production. Use of this trace mineral is not approved for that purpose in organic production and poultry producers are advised to use phytase in their dietary formulations to help ensure that the maximal amount of dietary zinc is available to the animal and not bound in the dietary ingredients with phytate.

Food Safety Issues for the Consumer

Infections such as *Salmonella* or *Campylobacter* may not have marked effects on bird health, but when they occur in eggs or meat they can present a risk for the human consumer.

Campylobacter jejuni is the most common enteric bacterial pathogen reported in developed countries and is considered to be of food-borne origin. Sporadic cases of *Campylobacter* infections during the summer months are mainly attributed to improper handling or consumption of undercooked poultry or the consumption of raw, unpasteurized milk or contaminated water. According to a study conducted by the USDA Food Safety and Inspection Service in 1994/95, the prevalence of *Campylobacter* on immersion-chilled poultry carcasses was 88.2% (Food Safety and Inspection Service; USDA, 1996). Sulonen et al. (2007) reported that 76–84% of organic layer farms in Finland were positive for *Campylobacter* contamination, based on examination of droppings samples. However, only one of 360 eggs sampled showed contamination of the shell and no contamination was found in the yolks. In The Netherlands, a study on 31 organic farms by Rodenburg *et al.* (2004) showed a prevalence of 13% for *Salmonella* and 35% for *Campylobacter*. The incidence of *Salmonella* was lower and that of *Campylobacter* higher in organic than in conventional broiler flocks.

Current data suggest that *Campylobacter* is primarily transferred on to poultry carcasses via fluid and excreta from the GI tract of the bird, due to the high numbers of the organism found in these fluids (Franco and Williams, 2001). The organism then attaches to the skin and perseveres to final products. Davis and Conner (2000) reported that the incidence on raw, retail poultry products decreased from 76% on whole broilers to 48% on skin-on split-breast and to only 2% on boneless, skinless breast meat. As noted above, Gracia *et al.* (2016) showed that supplementation of the diet of broilers with whole wheat or oat hulls significantly reduced caecal *C. jejuni* colonization at 42 days after a challenge with this organism.

Eggs appear to pose less of a risk from *Campylobacter* to the human consumer than poultry meat. Hauser and Fölsch (2002) found no differences in the microbial quality of eggs from four different farming systems.

The leading cause of human food-borne infections associated with consumption of poultry products worldwide is *Salmonella* (Van Immerseel *et al.*, 2002). Poultry can become infected from sources such as litter, droppings, soil, insects and rodent infestations, and the most serious serotypes are those that can pass from the intestine of birds into the tissues to contaminate the meat and eggs. Prevention of infection by appropriate management protocols, including proper hygiene, is the most important control measure. Feed-related control measures that are known to be effective in helping to control *Salmonella* contamination levels include steam-pelleting of the feed, and the inclusion of approved additives such as prebiotics, probiotics and SCFA in the feed mixture. As outlined in Chapter 4, the use of whey and other milk products is being investigated as a possible way of reducing salmonella colonization in organic poultry.

Mycotoxins may contaminate the meat and eggs from poultry but are more likely to be concentrated in organs such as the liver and kidneys. As a result they can pose a threat to human health. For instance, Denmark has legal regulations for maximum levels of ochratoxin in pig meat (no similar regulation appears to exist for poultry meat). When levels reach 10–15 µg/kg (ppb) in liver or in kidney, these organs are condemned, and when levels exceed 25 µg/kg (ppb), the entire carcass is condemned. These limits have been set because of the link between ochratoxin and human kidney disease (Devegowda *et al.*, 1998).

Other countries have similar regulations. Limits have also been set on aflatoxin contamination, because of its connection with liver cancer. The FDA guidelines in the USA allow no more than 20 µg aflatoxin/kg in animal feeds. The EU regulation (SI No: 283, 1998) sets limits between 5 and 50 µg/kg (ppb) for aflatoxin B1 in animal feedstuffs, depending on the feed ingredient and the animal for which the feedstuff is intended. The limit for a complete feedstuff for poultry and pigs was set at 20 µg/kg (ppb).

The available evidence suggests that the mycotoxin risk is lower with organic production than with conventional production, at least for Fusarium toxins.

References

Abdelqader, A., Gauly, M. and Wollny, C.B.A. (2007) Response of two breeds of chickens to *Ascaridia galli* infections from two geographic sources. *Veterinary Parasitology* 145, 176–180.

Allen, P.C., Danforth, H. and Levander, O.A. (1997) Interaction of dietary flaxseed with coccidia infections in chickens. *Poultry Science* 76, 822–827.

Ambrosen, T. and Petersen, V.E. (1997) The influence of protein level in the diet on cannibalism and quality of plumage of layers. *Poultry Science* 76, 559–563.

Ammerman, E., Quarles, C. and Twining, P. (1988) Broiler response to the addition of dietary fructo-oligosaccharides. *Poultry Science* 67 (Suppl 1), 46.

Antell, S. and Ciszuk, P. (2006) Forage consumption of laying hens – the crop as an indicator of feed intake and AME content of ingested feed. *Archiv für Geflügelkunde* 70, 154–160.

Arad, Z., Moskovits, E. and Marder, J. (1975) A preliminary study of egg production and heat tolerance in a new breed of fowl (Leghorn × Bedouin). *Poultry Science* 54, 780–783.

Arad, Z., Marder, J. and Soller, M. (1981) Effect of gradual acclimation to temperatures up to 44°C on productive performance of the desert Bedouin fowl, the commercial White Leghorn and the two reciprocal crossbreeds. *British Poultry Science* 22, 511–520.

Ask, B., van de Waaij, E.H., Stegeman, J.A. and van Arendonk, J.A.M. (2006) Genetic variation among broiler genotypes in susceptibility to colibacillosis. *Poultry Science* 85, 415–421.

Austin, P.R., Brine, C.J., Castle, J.E. and Zikakis, J.P. (1981) Chitin: new facets of research. *Science USA* 212(4496), 749–753.

Bahadori, Z., Esmaielzadeh, L., Karimi-Torshizi, M.A., Seidavi, A., Olivares, J. *et al.* (2017) The effect of earthworm (*Eisenia foetida*) meal with vermi-humus on growth performance, hematology, immunity, intestinal microbiota, carcass characteristics, and meat quality of broiler chickens. *Livestock Production Science* 202, 74–81.

Bennett, C. (2006) *Choice-feeding of Small Laying Hen Flocks.* Extension Report, Manitoba Agriculture, Food and Rural Initiatives, Winnipeg, Canada, pp. 1–2.

Berri, C., Wacrenier, N., Millet, N. and Le Bihan-Duval, E. (2001) Effect of selection for improved body composition on muscle and meat characteristics of broilers from experimental and commercial lines. *Poultry Science* 80, 833–838.

Bestman, M. and Maurer, V. (2006) Health and welfare in organic poultry in Europe: state of the art and future challenges. *Proceedings of Joint Organic Congress.* Odense, Denmark, 30–31 May, 2006.

Bjerrum, L., Pedersen, A.K. and Engberg, R.M. (2005) The influence of whole wheat feeding on Salmonella infection and gut flora composition in broilers. *Avian Diseases* 49, 9–15.

Blair, R., Dewar, W.A. and Downie, J.N. (1973) Egg production responses of hens given a complete mash or unground grain together with concentrate pellets. *British Poultry Science* 14, 373–377.

Blount, W.P. (1961) Turkey 'X' disease. *Turkeys* 9, 52–61, 77.

Boelling, D., Groen, A.F., Sørensen, P., Madsen, P. and Jensen, J. (2003) Genetic improvement of livestock for organic farming systems. *Livestock Production Science* 80, 79–88.

Bui, X.M., Ogle, B. and Lindberg, J.E. (2001) Effect of choice feeding on the nutrient intake and performance of broiler ducks. *Asian-Australian Journal of Animal Science* 14, 1728–1733.

Burel, C. and Valat, C. (2009) The effect of the feed on the host-microflora interactions in poultry: An overview. In: Aland, A. and Madec, F. (eds) *Sustainable Animal Production: The Challenges and Potential Developments for Professional Farming*, 367–385. Wageningen Academic Publishers, The Netherlands.

CAST (1989) *Mycotoxins, Economics and Health Risks.* Report No. 116. Council for Agricultural Science and Technology, Ames, Iowa.

Cömert, M., Şayan, Y., Kırkpınar, F.,. Hakan Bayraktar, Ö. H. and Mert, S. (2016) Comparison of carcass characteristics, meat quality, and blood parameters of slow and fast grown female broiler chickens raised in organic or conventional production system. *Asian-Australasian Journal of Animal Science* 29, 987–997.

Cowan, P.J. and Michie, W. (1978a) Environmental temperature and turkey performance. The use of diets containing increased levels of protein and use of a choice-feeding system. *Annals of Zootechnology* 17, 175–180.

Cowan, P.J. and Michie, W. (1978b) Environmental temperature and choice feeding of the broiler. *British Journal of Nutrition* 40, 311–314.

Cumming, R.B. (1989) Further studies on the dietary manipulation of coccidiosis. *Australian Poultry Science Symposium*, University of Sydney, Sydney, Australia, pp. 96.

Cumming, R.B. (1992a) The advantages of free-choice feeding for village chickens. In: *Proceedings of XIX World's Poultry Congress.* Amsterdam, pp. 627.

Cumming, R.B. (1992b) The biological control of coccidiosis by choice-feeding. In: *Proceedings of XIX World's Poultry Congress.* Amsterdam, pp. 525–527.

Dale, N. (1998) Mycotoxin binders. Now it is time for real science. *Feed International* June, 22–23.

Das, A.K. and Dash, M.C. (1989) Earthworm meal as a protein concentrate for Japanese quails. *Indian Journal of Poultry Science* 24, 137–138.

Davis, M.A. and Conner, D.E. (2000) Incidence of Campylobacter from raw, retail poultry products. *Poultry Science* 79 (Suppl. 1), 54.

Devegowda, G., Radu, M.V.L.N., Nazar, A. and Swamy, H.V.L.M. (1998) Mycotoxin picture worldwide: novel solutions for their counteraction. In: Lyons, T.P. and Jacques, K.A. (eds) *Proceedings of Alltech's 14th Annual Symposium on Biotechnology in the Feed Industry.* Nottingham University Press, Nottingham, UK, pp. 241–255.

Dove, F.W. (1935) A study of individuality in the nutritive instincts and of the causes and effects of variations in the selection of food. *American Naturalist* 69 (Suppl.), 469–543.

Dransfield, E. (1997) When the glue comes unstuck. In: *Proceedings of the 43rd International Congress of Meat Science and Technology.* Auckland, New Zealand, pp. 52–61.

Dransfield, E. and Sosnicki, E.A. (1999) Relationship between muscle growth and poultry meat quality. *Poultry Science* 78, 743–746.

Elwinger, K., Tauson, R., Tufvesson, M. and Hartmann, C. (2002) Feeding of layers kept in an organic feed environment. In: *11th European Poultry Conference.* Bremen, Germany.

Emmerson, D.E., Denbow, D.M. and Hulet, R.M. (1990) Protein and energy self-selection by turkey hens: reproductive performance. *British Poultry Science* 31, 283–292.

Emmerson, D.E., Denbow, D.M., Hulet, R.M., Potter, L.M., and van Krey, H.P. (1991) Self-selection of dietary protein and energy by turkey breeder hens. *British Poultry Science* 32, 555–564.

Engberg, R.M., Hedemann, M.S. and Jensen, B.B. (2002) The influence of grinding and pelleting of feed on the microbial composition and activity in the digestive tract of broiler chickens. *British Poultry Science* 43, 569–579.

Engberg, R.M., Hedemann, M.S., Steenfeldt, S. and Jensen, B.B. (2004) Influence of whole wheat and xylanase on broiler performance and microbial composition and activity in the digestive tract. *Poultry Science* 3, 925–938.

Evans, M., Singh, D.N., Trappet, P. and Nagle, T. (2005) Investigations into the effect of feeding laying hens complete diets with wheat in whole or ground form and zeolite presented in powdered or grit form, on performance and oocyst output after being challenged with coccidiosis. In: Scott, T.A. (ed.) *Proceedings of the 17th Australian Poultry Science Symposium.* Sydney, New South Wales, Australia, 7–9 February 2005, pp. 187–190.

Fanatico, A.C., Pillai, P.B., Hester, P.Y., Falcone, C., Mench, J.A. *et al.* (2008) Performance, livability, and carcass yield of slow- and fast-growing chicken genotypes fed low-nutrient or standard diets and raised indoors or with outdoor access. *Poultry Science* 87, 1012–1021.

Fanatico, A.C., Owens-Hanning, C.M., Gunsaulis, V.B. and Donoghue, A.M. (2016) Choice feeding of protein concentrate and grain to organic meat chickens. *Journal of Applied Poultry Research* 25, 156–164.

Franco, D.A. and Williams, C.E. (2001) *Campylobacter jejuni.* In: Hui, Y.H., Pierson, M.D., and Gorham, J.R. (eds) *Foodborne Disease Handbook*, 2nd edn. Marcel Dekker, New York, pp. 83–106.

Fuller, H.L. (1962) Restricted feeding of pullets. 1. The value of pasture and self-selection of dietary components. *Poultry Science* 41, 1729–1736.

Glünder, G. (2002) Influence of diet on the occurrence of some bacteria in the intestinal flora of wild and pet birds. *Deutsche Tierarztliche Wochenschrifte* 109, 266–270.

Gohl, B. (1981) Tropical feeds. Feed information summaries and nutritive values. *FAO Animal Production and Health Series.* Food and Agriculture Organization of the United Nations, Rome, pp. 529.

Gracia, M.I., Sánchez, J., Millán, C., Casabuena, Ó., Vesseur, P. *et al.* (2016) Effect of feed form and whole grain (2016) feeding on gastrointestinal weight and the prevalence of *Campylobacter jejuni* in broilers orally infected. *PLoS ONE. 2016; 11(8)*:e0160858. doi:10.1371/journal.pone.0160858 online (accessed 12 October 2017).

Hauser, R. and Fölsch, D. (2002) How does the farming system affect the hygienic quality of eggs? In: *Proceedings of the Eleventh European Symposium on Poultry Nutrition*, Bremen, Germany.

Henuk, Y.L. and Dingle, J.D. (2002) Practical and economic advantages of choice feeding systems for laying poultry. *World's Poultry Science Journal* 58, 199–208.

Henuk, Y.L., Thwaites, C.J., Hill, M.K. and Dingle, J.G. (2000a) The effect of temperature on responses of laying hens to choice feeding in a single feeder. In: Pym, R.A.E. (ed.) *Proceedings of the Australian Poultry Science Symposium*, University of Sydney, Sydney, Australia, pp. 117–120.

Henuk, Y.L., Thwaites, C.J., Hill, M.K. and Dingle, J.G. (2000b) Dietary self-selection in a single feeder by layers at normal environmental temperature. In: *Proceedings of the Nutrition Society of Australia* 24, 131.

Hermansen, J.E., Strudsholm, K. and Horsted, K. (2004) Integration of organic animal production into land use with special reference to swine and poultry. *Livestock Production Science* 90, 11–26.

Hesseltine, C.W. (1979) Introduction, definition and history of mycotoxins of importance to animal production. In: *Interactions of Mycotoxins in Animal Production*. National Academy of Sciences, Washington, DC, pp. 3–18.

Hessenberger, S., Schatzmayr, G. and Teichmann, K. (2016). *In vitro* inhibition of *Eimeria tenella* sporozoite invasion into host cells by probiotics. *Veterinary Parasitology* 229, 93–98.

Heuser, G.F. (1945) The rate of passage of feed from the crop of the hen. *Poultry Science* 24, 20–24.

Hidaka, H., Hirayama, M. and Yamada, K. (1991) Fructooligosaccharides enzymatic preparations and biofunctions. *Journal of Carbohydrate Chemistry* 10, 509–522.

Hollister, A.G., Corrier, D.E., Nisbet, D.J. and Delaoch, J.R. (1999) Effect of chicken derived cecal micro-organisms maintained in continuous culture on cecal colonization by *Salmonella typhimurium* in turkey poults. *Poultry Science* 78, 546–549.

Horsted, K., Hammershøj, M. and Hermansen, J.E. (2006) Short-term effects on productivity and egg quality in nutrient-restricted versus non-restricted organic layers with access to different forage crops. *Acta Agriculturae Scandinavica* Section A Animal Science 56, 42–54.

Horsted, K., Hermansen, J.E. and Ranvig, H. (2007) Crop content in nutrient-restricted versus non-restricted organic laying hens with access to different forage vegetations. *British Poultry Science* 48, 177–184.

Hossain, S.M. and Blair, R. (2007) Chitin utilisation by broilers and its effect on body composition and blood metabolites. *British Poultry Science* 48, 33–38.

Hovi, M., Sundrum, A. and Thamsborg, S.M. (2003) Animal health and welfare in organic livestock production in Europe: current state and future challenges. *Livestock Production Science* 80, 41–53.

Howlider, M.A.R. and Rose, S.P. (1987) Temperature and the growth of broilers. *World's Poultry Science Journal* 43, 228–237.

Hughes, B.O. (1984) The principles underlying choice feeding behaviour in fowls – with special reference to production experiments. *World's Poultry Science Journal* 40, 141–150.

Hughes, B.O. and Dun, P. (1983) A comparison of laying stock: housed intensively in cages and outside on range. *Research and Development Publication No. 18*. The West of Scotland Agricultural College, Auchincruive, Ayr, UK, 13 pp.

Hurling, R., Rodel, J.B. and Hunt, H.D. (1996) Fiber diameter and fish texture. *Journal of Texture Studies* 27, 679–685.

Inaoka, T., Okubo, G., Yokota, M. and Takemasa, M. (1999) Nutritive value of house fly larvae and pupae fed on chicken feces as food source for poultry. *Japanese Poultry Science* 36, 174–180.

Jelinek, C.F., Pohland, A.E. and Wood, G.E. (1989) Worldwide occurrence of mycotoxins in foods and feeds – an update. *Journal of the Association of Official Analytical Chemists* 72, 223–230.

Jensen, G.H. and Korschgen, L.J. (1947) Contents of crops, gizzards, and droppings of bobwhite quail force-fed known kinds and quantities of seeds. *Journal of Wildlife Management* 11, 37–43.

Jones, G.P.D. and Taylor, R.D. (2001) The incorporation of whole grain into pelleted broiler chicken diets: production and physiological responses. *British Poultry Science* 42, 477–483.

Karunajeewa, H. (1978) Free-choice feeding of poultry: a review. In: Farrell, D.J. (ed.) *Recent Advances in Animal Nutrition 1978*. University of New England, Armidale, Australia, pp. 57–70.

Kijlstra, A. and Eijck, I.A.J.M. (2006) Animal health in organic livestock production systems: a review. *NJAS – Wageningen Journal of Life Sciences* 54, 77–94.

Kjaer, J.B. and Sørensen, P. (2002) Feather pecking and cannibalism in free-range laying hens as affected by genotype, level of dietary methionine + cystine, light intensity during rearing and age at access to the range area. *Applied Animal Behaviour Science* 76, 21–39.

Kjaer, J.B., Su, G., Nielsen, B.L. and Sørensen, P. (2006) Foot pad dermatitis and hock burn in broiler chickens and degree of inheritance. *Poultry Science* 85, 1342–1348.

Klasing, K.C. (2005) Poultry nutrition: a comparative approach. *Journal of Applied Poultry Research* 14, 426–436.

Koene, P. (2001) Breeding and feeding for animal health and welfare in organic livestock systems – animal welfare and genetics in organic farming of layers: the example of cannibalism. In: *Proceedings of the fourth NAHWOA Workshop (Network for Animal Health and Welfare in Organic Agriculture)*. Wageningen, The Netherlands, pp. 62–85.

Kristensen, I. (1998) Organic egg, meat and plant production – bio-technical results from farms. In: Kristensen, T. (ed.) *Report of the Danish Institute of Agriculture Science* vol 1, pp. 95–169.

Lampkin, N. (1997) *Organic Poultry Production, Final report to MAFF*. Welsh Institute of Rural Studies, University of Wales, Aberystwth, UK, 84 pp.

Langhout, D.J. (1999) The role of the intestinal flora as affected by NSP in broilers. In: *Proceedings of the Twelfth European Symposium on Poultry Nutrition*. Veldhoven, The Netherlands, pp. 203–212.

Lawlor, P.G. and Lynch, P.B. (2001) Source of toxins, prevention and management of mycotoxicosis. *Irish Veterinary Journal* 54, 117–120.

Leeson, S. (1986) Nutritional considerations of poultry during heat stress. *World's Poultry Science Journal* 42, 69–81.

Leeson, S. and Summers, J.D. (1978) Voluntary food restriction by laying hens mediated through self-selection. *British Poultry Science* 19, 417–424.

Leeson, S. and Summers, J.D. (1979) Dietary self-selection by layers. *Poultry Science* 58, 646–651.

Lepkovsky, S., Chari-Bitron, A., Lemmon, R.M., Ostwald, R.C. and Dimick, M.K. (1960) Metabolic and anatomic adaptations in chickens 'trained' to eat their daily food in two hours. *Poultry Science* 39, 385–389.

Li, X., Liu, L., Li, K., Hao, K. and Xu, C. (2007) Effect of fructooligosaccharides and antibiotics on laying performance of chickens and cholesterol content of egg yolk. *British Poultry Science* 48, 185–189.

Makkar, H.P.S., Tran, G., Heuzé, V. and Ankers, P. (2014) State-of-the-art on use of insects as animal feed. *Animal Feed Science and Technology* 197, 1–33.

Mastika, I.M. and Cumming, R.B. (1985) Effect of nutrition and environmental variations on choice feeding of broilers. In: *Recent Advances in Animal Nutrition in Australia*. University of New England, New England, New South Wales, Australia, pp. 101–114.

Maurer, V., Hertzberg, H. and Hördegen, P. (2002) Status and control of parasitic diseases of livestock on organic farms in Switzerland. In: *Proceedings of the 14th IFOAM organic world congress*. EKO/Partalan kirjasto, 636.

Ndelekwute, E.K., Essien, E.B., Assam, E.D. and Ekanem, N.J. (2016) Potentials of earthworm and its by-products in animal agriculture and waste management – a review. *Bangladesh Journal of Animal Science* 45, 1-19.

Numi, E. and Rantala, M.W. (1973) New aspect of Salmonella infection in broiler production. *Nature* 241, 210–211.

Olver, M.D. and Malan, D.D. (2000) The effect of choice-feeding from 7 weeks of age on the production characteristics of laying hens. *South African Journal of Animal Science* 30, 110–114.

Osweiler, G.D. (1992) Mycotoxins. In: Leman, A.D., Straw, B.E., Mengeling, W.L., D'Allaire, S. and Taylor, D.J. (eds) *Diseases of Swine*, 7th edn. Iowa State University Press, Ames, Iowa, pp. 735–743.

Pasteiner, S. (1997) Coping with mycotoxin contaminated feedstuffs. *Feed International* May, 12–16.

Patterson, J.A., Orban, J.I., Sutton, A.L. and Richards, G.N. (1997) Selective enrichment of Bifidobacteria in the intestinal tract of broilers by thermally produced kestoses and effect on broiler performance. *Poultry Science* 68, 1351–1356.

Petkevicius, S., Knudsen, K.E., Nansen, P. and Murrell, K.D. (2001) The effect of dietary carbohydrates with different digestibility on the populations of Oesophagostomum dentatum in the intestinal tract of pigs. *Parasitology* 123, 315–324.

Ponte, P.I., Rosado, C.M., Crespo, J.P., Crespo, D.G., Mourão, J.L. *et al.* (2008) Pasture intake improves the performance and meat sensory attributes of free-range broilers. *Poultry Science* 87, 71–79.

Pousga, S., Boly, H. and Ogle, B. (2005) Choice feeding of poultry: a review. *Livestock Research for Rural Development* 17, Art. #45. Available at: http://www.lrrd.org/lrrd17/4/pous17045.htm (accessed 17 March 2018).

Rack, A.L., Lilly, K.G.S., Beaman, K.R., Gehring, C.K. and Moritz, J.S. (2009). The effect of genotype, choice feeding, and season on organically reared broilers fed diets devoid of synthetic methionine. *Journal of Applied Poultry Research* 18, 54–65.

Ravindran, V. and Blair, R. (1993) Feed resources for poultry production in Asia and the Pacific. III. Animal protein sources. *World's Poultry Science Journal* 49, 219–235.

Reinecke, A.J., Hayes, J.P. and Cilliers, S.C. (1991) Protein quality of three different species of earthworms. *South African Journal of Animal Science* 21, 99–103.

Rodenburg, T.B., van der Hulst-van Arkel, M.C. and Kwakkel, R.P. (2004) Campylobacter and Salmonella infections on organic broiler farms. *NJAS-Wageningen Journal of Life Sciences* 52, 101–108.

Rose, S.P. and Kyriazakis, I. (1991) Diet selection of pigs and poultry. In: *Proceedings of the Nutrition Society* 50, 87–98.

Roselli, M., Finamore, A., Britti, M.S., Bosi, P., Oswald, I. and Mengheri, E. (2005) Alternatives to in-feed antibiotics in pigs: evaluation of probiotics, zinc or organic acids as protective agents for the intestinal mucosa. A comparison of *in vitro* and *in vivo* results. *Animal Research* 54, 203–218.

Sales, J. (2014) Effects of access to pasture on performance, carcass composition, and meat quality in broilers: A meta-analysis. *Poultry Science* 93, 1523–1533.

Salmon, R.E. and Szabo, T.I. (1981) Dried bee meal as a feedstuff for growing turkeys. *Canadian Journal of Animal Science* 61, 965–968.

Santos, F.B.O., Sheldon, B.W., Santos, A.A. Jr, and Ferket, P.R. (2008). Influence of housing system, grain type, and particle size on salmonella colonization and shedding of broilers fed triticale or corn-soybean meal diets. *Poultry Science* 87, 405–420.

Shariatmadari, F. and Forbes, J.M. (1993) Growth and food intake responses to diets of different protein contents and a choice between diets containing two concentrations of protein in broiler and layer strain of chicken. *British Poultry Science* 34, 959–970.

Sirri, F., Castellini, C., Bianchi, M., Petracci, M., Meluzzi A. and Franchini, A. (2011) Effect of fast-, medium- and slow-growing strains on meat quality of chickens reared under the organic farming method. *Animal* 5, 312–319.

Smith, T.K. and Seddon, I.R. (1998) Synergism demonstrated between fusarium mycotoxins. *Feedstuffs* 22 June, 12–17.

Smulders, A.C.J.M., Veldman, A. and Enting, H. (2000) Effect of antimicrobial growth promoter in feeds with different levels of undigestible protein on broiler performance. In: *Proceedings of the 12th European Symposium on Poultry Nutrition*. WPSA Dutch Branch, Veldhoven, The Netherlands, 15–19 August, 1999.

Son, J.H. (2006) Effects of feeding earthworm meal on the egg quality and performance of laying hens. *Korean Journal of Poultry Science* 33, 41–47.

Sørensen, P. and Kjaer, J.B. (2000) Non-commercial hen breed tested in organic system. In: Hermansen, J.E., Lund, V. and Thuen, E. (eds) *Ecological Animal Husbandry in the Nordic Countries*, DARCOF Report vol 2, Tjele, Denmark, pp. 59–63.

Sosnicki, A.A. and Wilson, B.W. (1991) Pathology of turkey skeletal muscle: implications for the poultry industry. *Food Structure* 10, 317–326.

Steenfeldt, S., Engberg, R.M. and Kjaer, J.B. (2001) Feeding roughage to laying hens affects egg production, gastrointestinal parameters and mortality. In: *Proceedings of 13th European Symposium on Poultry Nutrition*. Blankeberge, Belgium, pp. 238–239.

Steenfeldt, S., Kjaer, J.B. and Engberg, R.M. (2007) Effect of feeding silages or carrots as supplements to laying hens on production performance, nutrient digestibility, gut structure, gut microflora and feather pecking behaviour. *British Poultry Science* 48, 454–468.

Sulonen, J., Kärenlampi, R., Holma, U. and Hänninen, M.L. (2007) Campylobacter in Finnish organic laying hens in autumn 2003 and spring 2004. *Poultry Science* 86, 1223–1228.

Taylor, R.D. and Jones, G.P.D. (2004) The influence of whole grain inclusion in pelleted broiler diets on proventricular dilatation and ascites mortality. *British Poultry Science* 45, 247–254.

Toghyani, M., Fosoul, S.S.A.S., Gheisari, A. and Tabeidian, S.A. (2014) Effect of choice feeding on footpad dermatitis and tonic immobility in broiler chickens. In: *Proceedings 25th Australian Poultry Science Symposium*. University of Sydney, 186.

USDA (1996) *Nationwide Broiler Chicken Microbiological Baseline Data Collection*. US Department of Agriculture, Food safety and inspection Service, Washington, DC.

Van Immerseel, F., Cauwerts, K., Devriese, L.A., Haesebrouck, F. and Ducatelle, R. (2002) Feed additives to control Salmonella in poultry. *World's Poultry Science Journal* 58, 501–513.

Van Kampen, M. (1977) Effects of feed restriction on heat production, body temperature and respiratory evaporation in the White Leghorn hen on a 'tropical' day. *TiJdrchrifr voor Diergeneeskunde* 102, 504–514.

Van Wagenberg, C.P.A., de Haas, Y., Hogeveen, H., van Krimpen, M.M., Meuwissen, M.P.M. *et al.* (2017) Animal Board Invited Review: Comparing conventional and organic livestock production systems on different aspects of sustainability. *Animal* 11(10), 1839–1851. Available at https://doi. org/10.1017/S175173111700115X (accessed 22 September 2017).

Waldroup, A.L., Skinner, J.T., Hierholzer, R.E. and Waldroup, P.W. (1993) An evaluation of fructooligo-saccharide in diets for broiler chickens and effects on Salmonellae contamination of carcasses. *Poultry Science* 72, 643–650.

Wang, D., Zhai, S.W., Zhang, C.X., Bai, Y.Y., An, S. and Xu, Y. (2005) Evaluation on nutritional value of field crickets as a poultry feedstuff. *Asian-Australasian Journal of Animal Sciences* 18, 667–670.

Washburn, K.W., Peavey, R. and Renwick, G.M. (1980) Relationship of strain variation and feed restriction to variation in blood pressure and response to heat stress. *Poultry Science* 59, 2586–2588.

Weese, J.S. (2002) Microbiologic evaluation of commercial probiotics. *Journal of the American Veterinary Medical Association* 220, 794–797.

Wilson, H.R., Wilcox, C.J., Voitle, R.A., Baird, C.D. and Dorminey, R.W. (1975) Characteristics of White Leghorn chickens selected for heat tolerance. *Poultry Science* 54, 126–130.

Zuidhof, M.J., Molnar, C.L., Morley, F.M., Wray, T.L., Robinson, F.E. *et al.* (2003) Nutritive value of house fly (*Musca domestica*) larvae as a feed supplement for turkey poults. *Animal Feed Science and Technology* 105, 225–230.

8

Conclusions and Recommendations for the Future

The organic poultry industry is small at present but is likely to expand in the future due to a strong demand from consumers for organic foods. Fortunately, poultry can be integrated more easily into many farming systems than other livestock can. Another attractive feature of organic poultry production is that its global warming potential is low. For instance, Herrero *et al.* (2013) showed that on a global basis pork produced 24 kg carbon per kilogram edible protein, and poultry only 3.7 kg carbon/kg protein, compared with around 58–1000 kg carbon/kg protein from ruminant meat.

It is hoped that the information presented in this volume will assist that expansion. Until recently poultry producers lacked advisory aids to assist them in developing successful organic systems.

Organic eggs and poultry meat sell at a premium over their conventional products, which helps to offset the higher costs of organic production. As pointed out earlier in this volume, the two main reasons for the higher costs of organic production are: (i) exposure of the birds to outdoor conditions, resulting in increased energy expenditure; and (ii) the increased cost of organic feedstuffs, which tend to be in scarce supply.

Supply of organic feedstuffs is particularly low in Europe, resulting in the aim of achieving 100% organic feed mixtures by 2018 in the EU to be shelved. A large

study, ICOPP (an acronym for 'Improved Contribution of local feed to support 100% Organic feed supply to Pigs and Poultry'), has examined the feasibility of that objective (Smith *et al.*, 2014). It was funded through the European CORE Organic II ERA-net programme to support organic research, and led by Aarhus University in Denmark with 15 partners across ten EU countries. It involved a range of feeding experiments with pigs (sows, piglets and finishers) and poultry (layers and broilers) that focused on concentrate feedstuffs, roughage and foraging from pasture land. Based on the compiled data the balance between feed supply and feed demand was calculated in terms of dry matter, energy, crude protein and essential amino acids (lysine, methionine and methionine + cystine). This analysis showed that for the countries involved in this project (ICOPP countries) there was a self-sufficiency rate for organic/conventional feed of 69%. Over 50% of the total demand for concentrate feed was for bovine animals, 16% was for pigs and 31% for poultry. The self-sufficiency rate for crude protein was 56%. Except for Lithuania, the organic crude protein demand exceeded availability, with an overall gap of approximately 135,000 t of crude protein existing within the ICOPP countries. The supply gap with essential amino acids was even higher than the supply gap with crude protein, being

© R. Blair 2018. *Nutrition and Feeding of Organic Poultry* (2nd edn)

just above 50% for lysine and about 40% for methionine + cystine.

It was therefore concluded as follows:

1. It seems quite unrealistic that the ICOPP countries will be able to cover the organic protein demand with their own efforts and increase production in the foreseeable future unless major shifts in production take place.

2. A large of amount of concentrate feed is fed to ruminants. It would be beneficial if part of the concentrate feed for ruminants (total around 1,000,000 t) could be used in feeding pigs and poultry.

3. In order to meet the essential amino acid requirements for the individual animal categories, the types of protein crops that can be produced organically in a country are relevant. There are different feeding possibilities, which were researched in the ICOPP and other research projects, but still there is a need for more innovative solutions.

Suggestions to address the shortage included improved forage utilization, the development of crops with improved nutrient content, use of earthworms and organically produced *Spirulina* algae as protein sources, processing plant feedstuffs such as sunflower seed to enrich the amino acid content, and use of novel feedstuffs such as insect larval meal and mussel meal. A great deal of research will be required to develop practical feeding systems based on these innovations.

Reviews of the efficiency of organic poultry production relative to conventional production have quantified the effects, as outlined in Chapter 7. The data show that egg-laying stock produce eggs at an efficiency of 87–99%, a feed/gain ratio of 106–128% and an energy use of 87–140%, relative to conventional layers (conventional = 100). The corresponding values for meat birds are: weight gain 76–84%, feed/gain ratio 140–153% and energy use 86–159%. Organic birds also require more land, 189–315% in the case of meat birds and 166–220% in the case of laying birds (relative to conventional birds at 100).

A satisfactory rate of gain and efficiency of feed conversion in poultry is important in terms of manure output. Poultry growing slowly require much more feed to reach market weight and during this period excrete more manure. As a result, there is an increased manure loading on the land to absorb all of the excreted nutrients, which may equate to an increase in stocking rate. The diets suggested in the text have been formulated to contain minimal excesses of minerals, and the routine use of phytase supplementation has been suggested to minimize the impact of excreted minerals on the environment. The acceptance of pure amino acids as permitted feed ingredients would assist in minimizing the excretion of nitrogen in manure, since it is obvious that at least some of the diets suggested (and in use) contain more crude protein than necessary. However, at present pure amino acids are not permitted in organic diets in most countries.

These data provide targets for research to improve the future efficiency of organic poultry production.

It is clear that the published database of organic feedstuffs composition needs strengthening. Results on nutrient composition from Europe indicate that organic feedstuffs are slightly lower in nutrient content than conventional feedstuffs, but results from North America suggest that both types of feedstuffs are more similar in nutrient content. A more extensive database would help to clarify this issue and would greatly assist organic poultry producers in formulating adequate feed mixtures.

In addition, research needs to be undertaken to establish the contribution of herbage, soil and associated organisms to the vitamin and mineral needs of organic poultry. The evidence at present on this issue is conflicting and some of the data are not of a standard suitable for acceptance by the scientific community. This research should also be extended to include a study of natural versus synthetic forms of vitamins. The provision that the vitamins used to supplement organic poultry diets should preferably be of natural origin conjures up an appealing image, but the practicality of such a proposal needs to be supported by research. The research should also include a study of the bioavailability and stability of natural sources of vitamins.

The scarcity of organic feedstuffs, particularly protein sources, in several regions has made derogations and permitted exceptions to the feeding standards necessary. This fact and the desire of organic producers to develop eventual self-sufficiency on farms should lead to an increase in the production of home-grown protein crops to utilize the nutrients left in pasture after being used by poultry. Also, as indicated above, new feedstuffs may have to be approved – such as insect larval meal or earthworm meal – to help alleviate the organic feedstuffs shortage. A related industry, which could be undertaken on organic farms, might be vermiculture as a means of providing an alternative source of protein. Processing and mixing equipment would then have to become more commonplace on farms, or on cooperative plants. Small-scale oil expeller plants for use with locally grown crops are a component of agricultural systems in several African and Asian countries (Panigrahi, 1995) and might be adapted for use on organic poultry units in more countries.

At present, the implementation of regulations and standards is the responsibility of local certifying agencies. The eventual adoption of agreed international regulations and standards will help to ensure uniformity of standards and promote trust in the consumer. The publication of detailed lists of approved feedstuffs by more countries, following the New Zealand example, will also assist the industry worldwide in producing diets that meet the required standards.

As has been pointed out in the text, organic foods will have to be produced efficiently so that they compete price-wise with conventional foods: there is evidence that the price that the consumer is willing to pay for organic foods is not unlimited (Siderer et al., 2005). This requires that the diets used in organic poultry production – the major cost in production – be formulated correctly to achieve a satisfactory rate of gain and efficiency of feed conversion in the market animals, and satisfactory meat and egg quality. The recent acceptance of synthetic vitamins

as permitted feed ingredients in organic poultry production is a welcome change in the regulations since it helps towards achieving these objectives.

One way in which organic egg producers could help to justify the additional cost of organic poultry produce might be to include feedstuffs such as flaxseed in organic poultry diets. Eggs from hens fed conventional diets containing flaxseed are being marketed as foods with an enhanced fatty acid profile beneficial in promoting cardiovascular health in humans (Van Elswyk, 1997). Such eggs from organic hens could be regarded as being doubly enhanced.

A welcome feature of a growing organic poultry industry is that it will stimulate interest in traditional breeds of poultry, many of which are in critically low numbers. An urgent need for research into the merits of these breeds for organic production is therefore highly desirable. In addition, the nutrient requirements of traditional, slow-growing poultry of the type favoured in organic production have not been defined adequately. Research in this area is also required. In the interim, the approach taken in this book has been to devise lower-energy diets containing a similar balance of nutrients to those in nutritional standards established for fast-growing hybrid poultry. This seems to be the most logical approach to be taken at present, but research needs to be undertaken to verify it and to define the nutritional needs of traditional breeds in sufficient detail. The effects of diet on poultry health need to be investigated in organic production, especially with the prohibition of antibiotics. The research to date indicates that probiotics and prebiotics are not as effective as antibiotics and more effective alternatives have to be discovered.

The genetically modified (GM) aspect of feed ingredients needs further consideration. The prohibition on feedstuffs derived by GM technology in organic production continues to be debatable but the ramifications of the prohibition need to be examined. For instance, most of the soybean crop grown in Brazil is GM-soy, causing feed supply problems for the organic industry in that country. As pointed out in the text, a strict interpretation of the GM prohibition may also ban the use of important vitamins in organic poultry diets and lead to deficiencies. In addition, the use of products such as protein concentrates from industrial starch production as sources of amino acids may be a source of GM products in organic diets. A study to determine the presence or absence of gene fragments derived from GM organisms in products such as protein concentrates is needed, to ensure that these products qualify as acceptable ingredients for organic production.

A related issue is the possible acceptability of pure amino acids derived from fermentation sources (lysine, tryptophan and threonine) in organic production. Several counties are pressing for their acceptability and their current banning is based in part on the premise that they may be derived from GM organisms. A review is needed to determine whether the ban is justified on scientific grounds, since amino acids derived using fermentation technology are analogous to vitamins derived in this manner and which are acceptable for use in organic diets. A review of this nature could help to establish whether the continued ban on amino acids derived from fermentation technology can be sustained on the basis of

the scientific evidence, or whether, given their potential importance as feed ingredients that would allow scarce sources of protein to be utilized more effectively and aid environmental sustainability by minimizing nitrogen excretion in manure, their banning should be overturned.

Another reason given for the current ban on pure amino acids such as methionine is that they are of synthetic origin, in keeping with the existing organic regulations that natural sources of nutrients are preferred or mandated over synthetic sources. That position on synthetic versus natural has been shown to be over-simplistic, as is evidenced by the need to approve synthetic vitamins in organic feeds in order not to impair animal welfare. In keeping with the approach suggested above with pure amino acids derived from fermentation sources, it is suggested that an analysis of the benefits/drawbacks of synthetic amino acids be conducted, using scientific data. This would help to determine whether the existing ban of synthetic amino acids is justified scientifically and ethically.

All of the above proposals indicate that experts in the area of poultry and animal nutrition should be more closely involved in the establishment of future standards for organic poultry production.

References

Herrero, M., Havlík, P., Valin, H., Notenbaert, A., Rufino, M.C. *et al.* (2013) Biomass use, production, feed efficiencies, and greenhouse gas emissions from global livestock systems. In: *Proceedings National Academy of Sciences*. Washington, DC, 110, 20878–20881.

Panigrahi, S. (1995) The potential for small-scale oilseed expelling in conjunction with poultry production in developing countries. *World's Poultry Science Journal* 51, 167–175.

Siderer, Y., Maquet, A. and Anklam, E. (2005) Need for research to support consumer confidence in the growing organic food market. *Trends in Food Science & Technology* 16, 332–342.

Smith, J., Gerrard, C. and Hermansen, J. (2014) Improved contribution of local feed to support 100% organic feed supply to pigs and poultry. *ICOPP Consortium 2014*. Available at: http://orgprints.org/28078 (accessed 21 July 2016).

Van Elswyk, M.E. (1997) Nutritional and physiological effects of flax seed in diets for laying fowl. *World's Poultry Science Journal* 53, 253–264.

Index

Note: bold page numbers indicate tables; italic page numbers indicate figures.

β-carotene 16, 30, 58, 93, 112, 114
β-glucan/β-glucanase 19, 61, 66, 73, 99, 115, **115**
β-xylanase *see* xylanase

AAFCO (Association of American Feed Control
    Officials) 73, 104
A, vitamin 16, 29, 30, 33–34, 58, 63, 112, 114, 128
    dietary requirements for **38**, **39**, **42**, **44**,
          45, **46**, **47**
Acamovic review 36
aflatoxin 86, 182, 198, 239, 240, 241, 245
Africa 62, 69, 88, 254
    organic standards in 9
African goose 213, 214
Agricultural Research Council (ARC) 36
Alabio duck 213
alfalfa *see* lucerne
alpha-galactosidase 98, 114, 115, **115**
amino acids (AA) 2, 4, 7, 10, 16, 21–23, **23**, **24**
    and bioavailability 22
    in cereals 58, 60, 64, 65, 66, 67, 69, 74
    digestibility of 51, **78**
    essential/semi-essential/non-essential 21–22
    in forages/roughages 105, 107
    in legume seeds 95, 97
    in milk products 109, 110, 111
    in oilseeds 76, 77, 80, 82, 86, 87, 88, 90, 93
    regulations/standards for 36, 55–56
    requirements for Bobwhite quail **48**
    requirements for broilers **42**, **179**, **180**
    requirements for ducks **45**
    requirements for geese **45**
    requirements for Japanese quail **47**

requirements for Leghorns **37**, **39**
requirements for ring-necked pheasants **46**
requirements for turkeys **43**
supply gap for 252–253
synthetic/fermentation-derived 55–56, 255
in tubers 103, 105
see also specific amino acids
amylase 15, 16, 78, 97, **115**, 116
anaemia 27–28
animal by-products 4, 5, 50
anti-caking agents 5, 7, 51, **54–55**
anti-nutritional factors (ANFs)
    in cereals 63, 69, 71, 102, 104, 107
    in legume seeds 95–96, 98, 99, 101
    in oilseeds 83, 86, 88, 90–91, 93
    see also β-glucan/β-glucanase
antibiotics 3, 242
antioxidants 6, 31, 51, **55**, 83, 111, 112, 236
    lupins as 100
    sesame meal as 89
    vitamin E as 28, 79
antitrypsins 19
ARC (Agricultural Research Council) 36
Argentina 8–9, 69, 82, 89
arginine 21, 65, 76, 86
    dietary requirements for **23**, **24**, **37**, **39**, **42**,
          43, **45**, **47**
Ascaridia galli 237, 238
ascites 207, 210, 241
ascorbic acid *see* C, vitamin
ash 20, 34, 108, 199, 235
    in cereals 61, 65, 66
    in oilseeds 84, 86, 88
    in potatoes 103, 104, 105

Asia 26, 84, 88, 111, 203, 208–209,
        213, 215, 234, 254
    nutrient requirement models in 35
    organic standards in 10–12
    see also specific countries
Aspergillus flavus 86, 240
Australia 32, 67–68, 69, 70, 72, 73, 75, 78, 84,
        87, 96, 100, 215, 228
    nutrient requirement models in 35
    organic standards in 9–10
Australorp 209
Avena sativa see oats
avian influenza 237
Aylesbury duck 213, 214

B complex vitamins 29, **30**, 32–33, 34, 58,
        93, 108, 109, 111
B$_{12}$, vitamin (cobalamin) 24, 27, 29, 32, 58, 114
    dietary requirements for **38**, **40**, **42**, **44**, **47**
bacteria *see* microorganisms
Bali duck 213
barley (*Hordeum* spp.) 18, 19, 23, 59–61, *59*
    hull-less 60–61
    and layer diets 60
    nutritional features of 60, **129**
    reduced phytate phosphorus in 58
    variability in **57**
    see also brewer's grains
Bedouin fowl 222–223
binders 5, 7, 51, **54–55**, 108, 198
    and mycotoxins 241
biotin 32, 64, 78, 80, 87, 95, 114
    dietary requirements for **38**, **40**, **42**, **44**, **47**
blood-clotting 25, 32, 91
Blue Foot chicken 209–210
bone formation 25, 26, 30–31
Brassica spp. 19, 27
    see also canola/rapeseed
B. campestris 85
B. juncea 84, 85
B. napus 85
B. oleracea 105, **131**
B. rapa 85
Brazil 9, 60–61, 89, 254–255
breeding stock 18, 75, 79, 174, **189**, 203–204,
        205, 212
    geese/ducks **45–46**, 182, **182**, **183**, **191**
    group-housing of 3
    pheasants **46**
    quail **47–48**, 215
    regulations for 11
    turkeys **43**, **44**, 71, **187**
    and vitamins 31, 32, 33
brewer's grains 61–62
brewer's yeast (*Saccharomyces*) 11, 33, **55**, 62,
        116, **173**, 244

Britain (UK) 35, 73, 81, 202, 240
    broiler feed formulas in **180**
    layer feed formulas in **177**
British Society of Animal Science 36
broad bean *see* faba bean
broilers 32, *178*
    amino acid requirements of 22
    and barley 60
    breeds for 207–210
    and canola 78, 79, 84
    and choice-feeding 227–228, **228**,
        229, 241
    and cottonseed meal 80, 81
    and DDGS 65–66
    and faba beans 96
    and fat digestion 19
    feed formulas for 178–179, **179**, **196**
    and fishmeal 111–112
    and food intake 18
    gossypol toxicity in 80
    and grass meal 106
    and groundnuts 86
    health problems in 237–238
    and linseed 82, 83
    and live-bird/carcass markets 209
    and lupins 101
    and meat quality *see* meat quality
    and molasses 108
    and mustard 84, 85
    nutrient requirements for 35, **42**, **189**
    and oats 67
    and peas 98
    productivity of 26, 83, **221**, **222**, **223**, 228,
        228, 234, 253
    protein/amino acid requirements
        of 23, **23**
    and seaweeds 108
    and sorghum 70
    and soybeans 92
    and sunflower meal 93–94
    and temperature factor 220–221, 223
    trace mineral requirements of 26
    and triticale 71, 227
    and wheat 73, 75
    see also specific breeds
brown-egg layers **37**, **38**, **39**, **40**, 204, 210,
        211, **220**
    and 'fishy eggs' 78, 79
buckwheat (*Fagopyrum* spp.) 18, 62–63, **130**

C, vitamin 29, 33
cabbage (*Brassica oleracea*) 105, **131**
caeca *15*, *16*, 17
cage-housing systems 204, 205, **206**
calcium (Ca) 24, 25–26, 30, **53**, 95, 97, 109,
        111, **113**, 175

in cereals 58, 64, 74
    dietary requirements for **24, 37, 39, 42, 43,
        45, 46, 47, 48**
    in forages/roughages 107, 108
    functions of 25
    in oilseeds 76, 77, 83, 86, 87, 88, 90, 93
calcium carbonate 56, 104
Campylobacter spp. 242, 244–245
C. jejuni 241, 244
Canada 32, 58, 61, 66, 67, 68, 70, 71, 73, 76, 81,
        82, 84, 87, 90, 98, 99, 108, 211
    layer feed formulas in **178**
    organic standards/regulations in 6, 8, 51,
        55–56, 114
    roaster feed formulas in **181**
    turkey feed formulas in **181**
cannibalism 178, 204, 205, 219, 222, 223,
        224, 237
canola/rapeseed (*Brassica* spp.) 27, 75, 76–80
    anti-nutritional factors in 78
    B. napus/*B. campestris* 76, 77
    digestibility of amino acids in 78–79, **78**
    full-fat 79–80
    GM 77, 89
    nutritional features of 77–78, **132**
    standards for 76–77
canola meal 23, 27, 32, **76**, 77, **133**
Capillaria spp. 237
carbohydrates 16, 82
    digestibility of 18–19
carotenoids 64, 80, 102, 103, 105
Caribbean countries 8
carrots 27, 233, **233**
Carthamus tictorius see safflower meal
cassava 27, 102–103, **134**
Cayuga duck 214
cereals/cereal by-products 5, 23, 25, **52**,
        56–75
    analysis of 59
    digestibility of 58
    GM 59, 254–255
    nutritional features of 58
    particle size/whole grains 59, 192
    processing 192
    variability in 56–58, **57**
    *see also specific cereals/cereal
        by-products*
certification 3, 6, 7–8, 9
CGF (corn gluten feed) 64–65
Chantecler 211
chicory (*Chicorium intybus*) 229–230, **230**,
        231–232, **232**, 242
Chile 9
China 8, 70, 76, 80, 82, 84, 85, 89, 92,
        108, 209, 214
    organic standards in 10–12
chitin 19, 234

chloride 24, 26, **113**
    dietary requirements for **24, 37, 39, 42, 43,
        45, 46, 47, 48**
choice-feeding systems 18, 225–229, *225,*
        226, 241
    advantages/disadvantages of 225, 226
    and learning in birds 228
    and 'nutritional wisdom' of birds 18, 225
    recommendations for 226–227
cholesterol 84, 102, 106, 206, 213, 214
choline 32, 78, 79, 85, 90, 93, 95, 108, 114
    dietary requirements for **38, 40, 42, 44, 45,
        46, 47, 48**
chondroitin sulfate 28
chyme 15, 16
Clostridium perfringens 241, 242
clover 27, 229–231, **230**, 232
coagulants 5, 7, 51, **54–55**
cobalamin *see* B$_{12}$, vitamin
cobalt 24, 27, **54**, **113**
coccidiosis 28, 67, 84, 219, 237, 241
Codex Alimentarius 3, 5, 6
coenzymes **30**, 33
Colinus virginianus **48**, 215
collagen 27, 33
colouring agents 6, 51, 65, 68, 106
Columbian Rock 210–211
competitive exclusion 242
complete feeds 174–175, 199, 225, 228–229
    regulations/standards for 11, 245
consumer attitudes 201–203
copper 24, 25, 26, 27, **54**, 77, 108, **113**
    dietary requirements for **37, 39, 42,
        43, 47**
Cornish 209
cottonseed (*Gossypium* spp.) 27, 75
cottonseed meal 19, 29, **76**, 80–81, **135**
Coturnix japonica **47**, 215
crop (in digestive system) 14, 15, 17
crop rotations 7
cyclopropenoid fatty acids (CPFAs) 19, 80
cystine 22, 31, **78**, 86, 88, 103, 105
    in cereals 65, 74
    dietary requirements for **23, 24, 37, 39, 42,
        43, 45, 46, 47, 48**

D, vitamin 25, 29, 30–31, 33–34, 112, 114, 240
    dietary requirements for **38, 39, 42, 44,
        45, 46, 47**
    in premixes 185
Danish Skalborg 205–206, **206**
DDGS (distiller's dried grains plus solubles)
        65–66, 70
Denmark 202, 205, 219, 229
developing countries 12, 36, 81, 85, 112, 226
diarrhoea 17, 237, 243, 244

diet formulations 128, 174–199, 224–229, 254
    for broilers 178–179, **179**
    and cold/hot weather 221
    complete feeds *see* complete feeds
    computer software for 190–191
    for ducks/geese 181–183, **182, 183, 191**
    formulation stage 186–191
    information required for 184
    for layers 176–178, **177, 178**
    and mix-mills 174, **176**, 196
    mixing 196
    and particle size 191–195
    pellets/pelleting 192, 196–198
    and premixes *see* premixes
    preparing/weighing/batching/blending
        191–195
    and purchased feed 174–175
    quality control of 199–200
    quality/variability issues with 185
    selecting ingredients for 184–185
    standards for 1, 2, 187–188
    storage of 196
    supplements for mixing with grains 175,
        175, 227
    for turkeys 179–181, **187**, 188, **188**
    and vitamins/trace minerals 185,
        187–188
    whole grains *see* whole-grain feeding
digestibility 18–20
    of carbohydrates 18–19
    of cereals 58
    of fats 19
    of minerals 20
    of proteins 19
digestion/digestive system 14–20, *15*
    chyme 15, 16
    crop 14, 15, 17
    gizzard/proventriculus *see* gizzard
    jejunum/ileum 16, 18, 22, 241, 242
    large intestine 16–17
    mouth 14
    saliva/salivary glands 14, 15, 18, 72
    small intestine 15–16
    time taken for 17
diseases 25, 30, 31, 227, 236–239
    coccidiosis 28, 67, 84, 219, 237, 241
    colibacillosis 238
    and diet 238–239
    enteritis 240, 241, 242
    foot-pad dermatitis 29, 227, 238
    prevention of 4, 236–237, 239
    resistance to 5, 7, 59, 67, 205, 208, 212
distiller's dried grains 65–66, **136**
    plus solubles (DDGS) 65–66, 70
Dorking, Silver Gray 206
Dromaius novaehollandiae **194**, 201, 215
drugs 3, 4, 50

duck eggs 213, 214
ducks 201, 240
    breeds/strains 213–214
    and canola 78, 79
    carcass quality/fatness of 181–182
    and choice-feeding 229
    feed formulas for 181–183, **182, 183**
    and lucerne 107
    nutrient requirements for **45–46, 191**
    and rice 69
    and seaweeds 108–109

E, vitamin 16, 28, 29, 31, 33–34, 58, 79, 93,
        112–113
    dietary requirements for **38, 39, 42,
        44, 46, 47**
earthworms/earthworm meal **230**, 231, 234,
        235–236, 254
EC Regulation 834/2007 7, 12, **52–55**
EC Regulation 2092/91 6, 7
EEC Regulation 1804/1999 6–7, 208
egg hatchability 27, 29, 32, 33, 106
egg production 203–207
    and body temperature 220–221
    and housing systems 219–222
    see also layers
egg quality 94, 206, **207**
    and cholesterol 84, 106, 206, 213, 214
    and consumer attitudes 202–203
    and omega-3 fatty acids 82, 83, 205
    and shell colour 78, 79, 201, 204,
        210, 211
    and vitamins 29, 30, 31
egg size 19, 27, 60, 86–87, 178
eggshell formation 25–26, 28, 31
Eimeria tenella/E. maxima 84, 239
Embden goose 213, 214
embryos 28, 29, 31, 78, 83, 240
emus (*Dromaius novaehollandiae*) **194**,
        201, 215
encephalomalacia 31
energy requirements 20–21, *21*, 23
    and cereals 58
    for egg production **41**
    for Leghorns **37**
    standards for 36
enteritis 240, 241, 242
Enterococcus 116, 244
environmental impact 3
enzymes 2, 3–4, 6, 14, 16, 19, 51, **54, 55**
    and copper 27
    standards/regulation for 114–116, **115**
    and trace minerals 28
ergot 69, 71
erucic acid 19, 76, 77
Escherichia coli 237, 238, 242

Europe/European Union (EU) 1, 71, 72, 76, 80, 84, 92, 97, 98, 208–209
  consumer attitudes on 202–203
  feed labels in 175
  and national certifying bodies 7–8
  nutrient requirement models in 35
  organic standards/regulations in 4–5, 6–8, 9, 51, **52–55**, **115**, 116, 219
  see also specific countries
excreta 16–17, 99
  and organic standards 8, 9
  sticky/wet droppings 19, 60, 67, 83, 99, 103, 104, 110
exudative diathesis 28, 31

faba bean (*Vicia faba*) 89, 95–96, **137**
Fagopyrum spp. 18, 62–63, **130**
FAO (UN Food and Agriculture Organization) 3, 5
fats 94–95, 100
  analysis of 35
  digestion of 16, 19
Favorelle 211
feather formation 22, 29, 32
feather pecking 204–205, 222, 224, 229, 237
feed additives/processing aids 2, 3, 51, **54**, 56
  see also enzymes
feed analysis 34–35, 59, 174, 175
feed formulas *see* diet formulations
feed labels 174–175
feed selection/intake 11, 17–18, **39–40**
  and amino acids 22, 23
  colour/visual appearance factor 17, 64, 226
  and forage 229–230
  and minerals 23, 27
  particle size factor *see* particle size
  and social interaction 18
  taste/smell factor 17, 226
  and temperature factor 221–222, 223
  and vitamins 30
feeding programmes 219–245
  choice-feeding *see* choice-feeding systems
  and forage *see* forage/roughage
  and genotype 222–224, **222**, **223**, **224**
  and housing system 4, 50, 219–222, **220**
fermentation by-products 33
fertilizers 56, 58, 66, 72
fibre 14, 17, 18, 20, 77, 80, 93, 97
  analysis 34–35, 175
  crude (CF) 34, **57**, 60
  total dietary (TDF) 61, 85
field bean *see* faba bean
fish 18, 51, **53**
fish oil 84, 112
fishmeal 4, 5, 32, 55, 111–112, **140**, **147**, **172**, 179, 228

'fishy eggs' 78, 79, 84, 112
flax *see* linseed
flock sizes 219, 223, 224
fodder 5, 7, 51
folic acid/folacin 32, 90
  dietary requirements for **38**, **40**, **42**, **44**, **47**
food quality 201–202
food safety 201, 203, 244–245
foot-pad dermatitis (FPD) 29, 227, 238
forage/roughage 5, 7, 14, 33, 51, **53**, 56, 105–109, 229–238
  and cold/hot weather 221, 222
  and feed intake 229–230
  and intake of insects/earthworms 234–236
  and meat quality 233–234
  and organic standards 4, 11, 50
  and removal of herbage 231–232, **232**
  see also pasture, access to
foraging behaviour 204, 205
France 35, 70, 92, 202–203, 209
free-range poultry 202, 204, 219

geese 78, 79, 106, 201, 214
  breeds 213, 214
  feed formulas for 183
  nutrient requirements for **45**
genotypes 222–224, **223**, **224**
Germany 70, 71, 100, 102, 238
gizzard 14, 15, 28, 67, 192, 226, 227, 241
  compaction 219, 223
  erosion 112
glucosinolates 27, 76, 77, 78, 79, 84, 85
gluten 70, 72
glycine **37**, **42**, **43**, **46**, **47**
GM crops 1, 3, 7, 254–255
goitre/goitrogenic substances 27, 78
goose eggs 214
Gossypium spp. 27, 75
gossypol 19, 80, 81
grain by-products 3, 4, 25
grass meal 68, **76**, 105–106, **139**
grass/clover pasture 229–231, **230**, 232
grasses (*Poaceae*) 33, 56, 208, 229–230, **230**
grazing management 4, 50
grit 15, 67, 225, 226, 230, **230**, 231, 241
groundnut/peanut (*Arachis hypogeae*) 27, 75, 85–87, 182
  anti-nutritional factors in 86
  nutritional features of 86, **153**
groundnut meal 32, **76**
gut microflora 14, 242, 244

haemagglutinins *see* lectins
Harco Black Sex-link 211

health/welfare 236–245
   and disease *see* diseases
   and feather pecking 204–205, 222, 224,
      229, 237
   helminth infestations 237, 243
   human *see* human health
   and intestinal microflora 14, 242, 244
   and mycotoxins *see* mycotoxins
   and parasites 4, 50, 109, 237, 238
   and prebiotics 242–243
   probiotics 2, 6, 51, 116, 243–244
   and whole-grain feeding 241–242
heat stress 77, 108, 207, 220–221, 228
   and genotypes 222–223
helminth infections 237, 243
Heterakis gallinarum 237
histidine 21, 76, 105
   dietary requirements for **24, 37, 39, 42, 43, 47**
hominy feed 64, **141**
Hordeum spp. *see* barley
hormones 3, 4, 50
horse bean *see* faba bean
housing systems 4, 50, 219–222
   and health/disease 37
human health 1, 31, 82, 83, 241, 244–245, 254

ICOPP (Improved Contribution of local feed to
      support 100% Organic feed supply to
      Pigs and Poultry) 252–253
IFN (International Feed Number) 56
IFOAM (International Federation of Organic
      Agriculture Movements) 4–5, 6, 8, 9, 11
ileum 16, 18, 22, 241, 242
immune system 7, 27, 28–29, 82, 207, 236
India 63, 69, 76, 80, 84, 85, 87, 88, 92, 203
   organic standards in 12
Indian Runner duck 182, 213
insect larvae/larval meal 56, **230**, 231, 234, 235,
      253, 254
insects 19, 230, **231**, 234–235
International Feed Number (IFN) 56
International Feed Vocabulary 56
iodine 24, 25, 26, 27, **54, 113**
   dietary requirements for **37, 39, 42, 43, 47, 48**
iron 24, 26, 27–28, **54**, 81, 87, 90, 95, 108, **113**
   dietary requirements for **37, 39, 42, 43, 47**
ISA strain **24**, 209, 223, **224**
ISO (International Organization of
      Standardization) 6
isoleucine 21, 65, 74, 90
   dietary requirements for **23, 24, 37, 39, 42,
      43, 45, 47**

Japan 9, 12, 84, 214, 234
junglefowl 203, 204, 234

K, vitamin 16, 26, 29, 32, **38, 40, 42, 44, 46, 47**
Kabir 210
Kaiya duck 213
Khaki Campbell duck 213
Korea 12

Label Rouge system 178, **179**, 209
labelling 6, 9, 10, 203
   of feeds 174–175
   of probiotics 244
lactobacilli 116, 241, 242, 243, 244
lactose 16, 109, 110, 236
Landaise goose 214
Latin America 70, 80, 92, 215
   organic standards in 8–9
   see also specific countries
layer fatigue 26
layers
   and barley 60
   of brown eggs *see* brown-egg layers
   cannibalism in 178
   and canola 78, 79
   and cassava 103
   and choice-feeding 226–227, **226**, 228
   and cottonseed meal 81
   and DDGS 66
   and faba beans 96
   feed formulas for 176–178, **177, 178, 195**
   feed supplements for **175**
   and food intake 18
   and forage 229–232, **230, 232**
   and grass meal 106
   and groundnuts 86
   and lucerne 107
   and lupins 101–102
   mortality rates in 237
   nutrient requirements for 35, **185, 186, 189**
   and oats 67, 68
   and peas 98
   poultry breeds for 203–207
   productivity of 68, 75, 86–87, 206, 220,
      220, 223, **224**, 231, 253
   protein/amino acid requirements of 23, **24**
   and soybeans 19, 91–92
   and sunflower seed/meal 93, 94
   and trace minerals 27
   and wheat 73
   see also specific breeds
lectins 91, 95–96, 97, 99
Leghorn, White 66–67, 75, 204, 206, 211,
      222–223
Leghorns 61, 79
   and faba beans 96
   and fat digestion 19
   nutrient requirements for **37–40**
legumes/legume seeds 25, 33, **52**, 95–102

lentil (*Lens culinaris*) 55, 99
leucine 21, 105
    dietary requirements for **24**, **37**, **39**, **42**, **43**, **45**, **47**
lighting, artificial 5, 11
lignin 35, 97
limestone, ground 25, 227
linamarin/linatine 83
linoleic acid **38**, **40**, **42**, **44**, **46**, **47**, **48**, 99
    in cereals 58, 60, 63, 64, 68
    and egg size 60, 71
    in oilseeds 79, 82, 83, 90
linolenic acid 77, 79, 82, 83, 84, 90, 95, 97, 99, 215
linseed/flax (*Linum usitatissimum*) 27, 33, 75, 81–84, 239, 254
    anti-nutritional factors in 83
    and 'fishy eggs' 84
    nutritional features of 82–83, **138**
    omega-3 fatty acid in 82, 83
linseed meal 18, **76**
lipase 16, 69
lipid 28, 32, 67, 68, 80, 93, 94, 214, 215, 243
live-bird market 209
liver *15*, 16, 27, 32, 78, 81
Lohmann 230, 238
lucerne/alfalfa (*Medicago sativa*) 14, 17, 25, 33, 106–107, **143–144**, 228
lucerne meal 19, 25, 32, 68, **76**, 106
lupin (*Lupinus* spp.) **76**, 99–102, **145**
lysine 19, 21, 55, 103, 105, 175
    in cereals 58, 63, 65, 66, 72, 74
    dietary requirements for **23**, **24**, **37**, **39**, **42**, **43**, **45**, **46**, **47**
    in legume seeds 95, 97
    in oilseeds 76, 77, **78**, 80, 83, 86, 88, 90, 93

magnesium (Mg) 24, 25, 26, **53**, 58, 83, 90, 108
    dietary requirements for **24**, **37**, **39**, **42**, **43**, **45**, **47**
maize (*Zea mays*) 23, 25, 33, 63–66, 192, 227, 228
    by-products 64–66
    distillers' grains 21
    and fungal toxins 64
    GM 59, 89
    nutritional features of 58, 63–64, **146**
    selenium deficient 26
    and yellow-pigmented yolk 58, 63
maize gluten meal 65, 68, **76**
malt sprouts/cleanings 61, 62
management practices 3, 4, 50, 204
manganese 24, 25, 26, 28, 32, **54**, 95, 100, 108, **113**
    dietary requirements for **37**, **39**, **42**, **43**, **46**, **47**
manure output 253
Maya duck 214

meat birds *see* broilers
meat meal 4, 50, 63, **184**, 235
meat quality 201, 202, 203, 207–208, **208**
    and disease 238
    and fatty acids 82, 83, 202, 208
    and fishy flavour 112
    and forage 233–234
metabolic rate 17, 27, 222, 223
methionine 21, 22, 31, 63, 103, 105, 175
    in cereals 63, 65, 66, 74
    contribution from organic feedstuffs of 51
    dietary requirements for **23**, **24**, **37**, **39**, **42**, **43**, **45**, **46**, **47**
    in feed formulas 178, 179
    in oilseeds 77, **78**, 83, 86, 88
    synthetic 55, 255
Mexico 8, 9, 69, 87, 88
microorganisms 6, 51, **54**, 116
    intestinal microflora 242, 244
milk/milk products 4, 5, 16, 33, 51, **53**, 109–111, 236
    dried 109, 111, **149**
    skimmed 109, **148**
    whey 109–110, **170–171**
milo *see* sorghum/milo
mineral supplements 56, 192
minerals 2, 5, 10, 11, 16, 23–26, **24**
    in cereals 58, 64
    digestibility of 20
    in fishmeal 111
    in forages/roughages 106, 107, 108
    in legumes 95, 97, 100
    in milk products 109–110
    in oilseeds 80, 83, 86, 87, 90
    requirements for Bobwhite quail **48**
    requirements for broilers **42**
    requirements for ducks **45**
    requirements for geese **45**
    requirements for Japanese quail **47**
    requirements for Leghorns **37**, **39**
    requirements for ring-necked pheasants **46**
    requirements for turkeys **43**
    sources of 112, **113**
    standards/regulations for 36, 50–51, 112
    trace *see* trace minerals
    see also specific minerals
mix mills 174, *176*, 196
molasses **53**, 103, 107–108, **150–151**
monounsaturated fatty acid (MUFA) 208, 214, 215
Muscovy duck (*Cairina moschata*) 213–214
muscular dystrophy 31
mustard 84–85
mycotoxins 64, 86, 182, 184, 239–241
    effects of 240
    and human health 245
    treatment for 240–241
    and yeasts 116

Netherlands 237, 245
New Hampshire 206, 210, 223
New Zealand 10, 51, **52–55**, 254
niacin/nicotinic acid 29, 32, 33, 95, 108, 114
    in cereals 58, 64
    dietary requirements for **38, 40, 42, 44, 45,
        46, 47, 48**
    and ducks/geese 181
    in oilseeds 78, 87
Nigeria 81, 108
NIRS (near-infrared reflectance spectroscopy) 35
nitrogen (N) 5, 20, 51, 91, 100, 103
    in excreta 8, 9, 22, 23, 90, 114, 176,
        253, 255
    non-protein (NPN) 34, 51, 234, 235
non-starch polysaccharides (NSP) 60, 66, 75, 80,
    102, 115, 233
North America 21, 26, 70, 72, 108, 204, 208
    egg-producing breeds/strains in 204, 206
    labelling in 174–175, 244
    nutrient requirements in 35
    oilseeds in 76, 77, 78
    organic standards in 8
    see also Canada; Mexico; United States
Norway **57**, 202
NRC (Nutrient Requirements of Poultry) 21, 22,
        35, 187–188, 190, 222–223, **222**
    criticisms of 36
nutrition 14–48
    derivation of standards on 36
    digestibility *see* digestibility
    and digestion *see* digestion/digestive
        system
    energy requirements 20–21, *21*
    and feed analysis 34–35
    and feed selection/intake *see* feed
        selection/intake
    five elements of 14
    mineral/trace mineral requirements 23–29, **24**
    protein/amino acids requirements 21–23,
        23, 24
    publication on requirements 35–36
    vitamins *see* vitamins
    water 33
nutritional wisdom 18, 225

oats (*Avena sativa*) 18, 19, 58, 66–68
    naked/hull-less (*A. nuda*) 67–68
    nutritional features of 66, **152**
    variability in **57**
oilseeds/oilseed by-products 25, **52**, 75–95, **76,**
    94–95
    extraction methods for 75–76, 77, 82, 87,
        89–90
    *see also specific oilseeds/oilseed
        by-products*

oleic acid 58, 77, 84, 90, 97, 100, 214, 215
oligosaccharides 19, 97, 100
omega-3 fatty acids 82, 83, 205, 208, 239
organic feed 3, 50–116
    additives/processing aids *see* feed
        additives/processing aids
    cost of 1, 20, 74, 225, 226
    criteria for 5, 50–51
    data gap for 51, 253
    and International Feed Vocabulary 56
    lists of approved ingredients 51–56, **52–55**
    restrictions of 3–4
    supply gap for 1, 252–253, 254
    variability in 56–58, **57**, 72, 128, 185
    see also diet formulations
organic poultry production, aims/principles
        of 3–6
    and Codex Alimentarius definition 3
    environmental benefits 252
    four stages of 3
    and standards *see* organic standards
organic produce 1–2
    brand image of 3
    consumer attitudes to 201–203
    increasing demand for 1, 252
    selling price of 1, 3, 202, 219, **220**, 252, 254
organic standards 4–12, 50
    African 9
    Asian 10–12
    Australasian 9–10
    Codex Alimentarius 3, 5, 6
    criteria for feedstuffs 5
    European 6–8
    IFOAM 4–5, 6
    international harmonization of 6, 12
    Latin American 8–9
    North American 8
Orpington, Buff 206
Oryza sativa see rice
ostrich (*Struthio camelus*) 183, **184**, 188, **192,
        193**, 201, 215
outdoor-based systems 3, 5, 7, 11, 219–222
overstocking 4, 50
oyster shell 25, 228, 229, 230

Pakistan 80, 81, 203
pantothenic acid 33, 58, 80, 95, 108, 114
    dietary requirements for **38, 40, 42, 44, 45,
        46, 47, 48**
parasites 4, 50, 109, 237, 238, 243
particle size 18, 59, 98–99, 192–195
pasture, access to 2, 4, 25, 56, *220*, 229, *229*
pea, field (*Pisum sativum*) **76**, 96–99, **154**
peanut *see* groundnut
Pekin duck 181, 213–214
pellet binders *see* binders

pellets/pelleting 192, 196–198
pentosans 19, 108, 198
perosis 28, 32, 33
pheasants **46**, 183, 201
phenylalanine 21, 22, 105
    dietary requirements for **37**, **39**, **42**, **43**, **47**
phosphorus (P) 20, 24, 25–26, 30, **53**, 107, 109,
       111, **113**, 175
    in cereals 58, 60, 64
    dietary requirements for **24**, **37**, **39**, **42**, **43**,
       45, **46**, **47**, **48**
    in legume seeds 95, 97
    in manure 114
    in oilseeds 76, 83, 86, 87, 90, 93
    phytate *see* phytate phosphorus
phytase 20, 64, 69, 90, 115, **115**, 244
phytate phosphorus 20, 25, 26
    in cereals 58, 64, 67, 69
    and oilseeds **76**, 77, 88, 89, 90
    and zinc 29
phyto-oestrogens 82
Pilgrim goose 213, 214
Plymouth Rock
    Barred 204, 206, 210
    Buff 206
    White 206, 211
Poland 61, 66, 69, 70, 71, 72, 77, 82, 97
pollards *see* wheat middlings
polyunsaturated fatty acids (PUFA) 77, 82,
       83, 87, 88, 90, 93, 97, 112, 202,
       208, 214, 215
potassium 24, 26, 58, 77, 86, 90, 103, 108
    dietary requirements for **37**, **39**, **42**, **43**, **47**
potato (*Solanum tuberosum*) 55, 103–105
    by-products 104–105
    cooked 18, 103, 104, **155**
potato protein concentrate (PPC) 103, 105, **156**
Poulet de Bresse 209
poultry breeds 203–216
    dual-purpose 210–211, *210*
    for egg production 203–207
    endangered, information sources on 216
    and feeding programmes 222–224, **222**,
       223, **224**
    for meat production 207–210
    ostriches/emus 215–216
    quail 214–215
    selection of appropriate 4, 5, 7
    traditional *see* traditional/heritage breeds
    turkeys 211–212, *212*
    waterfowl 212–214
PPC (potato protein concentrate) 103, 105, **156**
prebiotics 242–243
premixes 26, 28, **175**, 176, 185–186, 187, 195, 228
    admixing with cereal 195
    for chicken grower/layer diets **185**, **186**
    for turkey grower/layer diets **187**, **188**

preservatives 6, 51, **54**
probiotics 2, 6, 51, 116, 243–244
protein 21–23, 175
    in cereals 58, 60–61, 62, 66, 69, 70, 72
    contribution from organic feedstuffs of 51
    digestibility of 19, 95
    digestion of 15, 16, 17
    in fishmeal 111
    ideal (IP) 23
    in legume seeds 95, 96, 97, 99, 100
    and minerals 25
    in oilseeds 75–76, 77, 80, 82, 84–85, 86, 87,
       88, 90, 91
    requirements for Bobwhite quail **48**
    requirements for broilers **42**
    requirements for ducks **45**
    requirements for geese **45**
    requirements for Japanese quail **47**
    requirements for Leghorns **37**, **39**
    requirements for ring-necked pheasants **46**
    requirements for turkeys **43**
    selection/intake of 18
    supplements 29, 56, 75, 80–81, 83, 86, 88,
       96, 101, 111, 234
    in tubers 103
    see also amino acids
provitamins 5, 10, **30**, 50, **54**, 58
    see also β-carotene
PUFA *see* polyunsaturated fatty acids
pullets 66, 83, 204, 211, 227, 229, 237
    feed formulas for **177**, 181, **189**
pyridoxine 33, 80, 83, 114
    dietary requirements for **38**, **40**, **42**, **44**, **46**, **47**

quail 68, 201, 214–215, 230, 240
    Bobwhite (*Colinus virginianus*) **48**, 215
    Japanese (*Coturnix japonica*) **47**, 215
quality control 199–200

rapeseed *see* canola/rapeseed
rapeseed oils 19
ratites *see* emus; ostriches
record-keeping 3
recycling of wastes/by-products 3, 7
Redbro Cou Nu breed 209
Rhenish goose 214
Rhode Island Red 78, 204, 206, 210–211
riboflavin 26, 33, 58, 78, 87, 95, 106, 109, 114
    dietary requirements for **38**, **40**, **42**, **44**, **45**,
       46, **47**, **48**
rice (*Oryza sativa*) 18, 68–69, 89
rice bran 21, 68, 69, 89, **157**
rickets 25, 30
Robusta maculate 210
rodents 64, 184, 196, 245

Ross 207, **208**, 210
Rouen duck 213
roughage *see* forage/roughage
Russia 12, 70, 82, 84, 92
rye (*Secale cereale*) 18, 19, 58, 70, **158**
rye malt sprouts 62

Saccharomyces see brewer's yeast
safflower meal (*Carthamus tictorius*) 32, 76,
    87–88
    nutritional features of 87, **159**
saliva/salivary glands 14, 15, 18, 72
salmonella 110, 184, 241, 242, 245
salt 111, **113**, 175, **175**
    iodized **24**, 27
saponin 19, 97, 98, 99, 107
Scandinavia 32, 202
SDS (sudden death syndrome) 207, 210
seaweeds **53**, 108–109, **142**
Secale cereale see rye
selenium (Se) 24, 25, 26, 28, 31, **54**, 74, 77, 83,
    113, 175
    dietary requirements for **37**, **39**, **42**, **43**,
        46, **47**
serine **37**, **42**, **43**, **46**, **47**
sesame meal (*Sesamum indicum*) 29, **76**,
    88–89, **160**
Shaver Red Sex-link 211
shell colour 78, 79, 201, 204, 210, 211
silage 5, 7, 51, 56, 233, **233**
    processing aids for **55**
sinapine 78, 79
Sinapis alba 84
skin colour 60, 73, 201, 208, 209
sodium 24, 25, 26, **53**, 58, 86, 108, **113**, 175
    in cereals 58, 64
    dietary requirements for **24**, **37**, **39**, **42**, **43**,
        45, **46**, **47**, **48**
sodium selenite/selenate 28
soil 7, 24, 50
    trace elements in 26, 28
Solanum tuberosum see potato
solvent-extracted meal 80, 81, 82, 84, 85,
    86, 89
    restrictions on 4, 5, 50, 77
sorghum/milo (*Sorghum vulgare*) 19, 23, 58,
    69–70, **161**
South Africa 9, 215
South America *see* Latin America
soybean/soybean meal (*Glycine max*) 19, 23, 27,
    29, 32, 75, 76, **76**, 89, 89–92, 179
    anti-nutritional factors in 90–91
    extraction processes for 4, 89–90
    full-fat 91–92, 228
    GM 89
    nutritional features of 90, **162–163**

selenium deficient 26
    substitutes for 62
soy protein isolate **76**, 92
Spain 69, 100
Speckled Sussex 206
spelt (*Triticum aestivum* var. *spelta*) 70
Spirulina algae 253
spleen 27, 85, 236
stabilizing agents 5, 51, 68, 95, 109, 110, 185
starch 18, 19, **57**, 58, 59, 61, 63, 64–65, 99, 102, 103
steam-pelleting 19, 25, 75, 96, 101
stocking rates 4, 8, 10, 209, 253
stress 4, 17, 28, 236, 237
    see also heat stress
Struthio camelus 183, **184**, 188, **192**, **193**, 201, 215
sucrose 17, 97, 108, **150**, **151**
sudden death syndrome (SDS) 207, 210
sugarcane/sugarbeets *see* molasses
sulfur (S) 24, **24**, 26, **53**, 77, 93, 95, 97, 99
sunflower seeds/meal (*Helianthus annus*) 18,
    75, **76**, 92–94, *92*
    anti-nutritional factors in 93
    nutritional features of 93, **164**
    substitutes for 62
sustainability 3, 6, 70, 114, 174, 220, 237
    and aquatic feed sources 7, 55, 111
swede **165**
Sweden 61, 202, 206
Switzerland 204, 237

tannins 19, 69, 70, 78, 96, 97–98, 107
temperature factor
    cold weather 207, 211, 221
    and drinking water 34
    and protein/amino acid requirements 22
    see also heat stress
thiamin 29, 33, 58, 95, 114
    dietary requirements for **38**, **40**, **42**, **44**, **47**
Three Yellow (3Y) 209
threonine 21, **78**, 86, 90, 103, 105
    in cereals 58, 65, 66, 74
    dietary requirements for **23**, **24**, **37**, **39**, **42**,
        43, **47**
thyroid gland 27, 78, 85
tibial dyschondroplasia 27
Toulouse goose 214
trace minerals 5, 10, 11, 24, 26–29, 50–51, **54**,
    90, 93, 108
    premixes 26, 28, **175**, 185, **185**, **186**,
        187–188, **187**
    requirements for Bobwhite quail **48**
    requirements for broilers **42**
    requirements for ducks **45–46**
    requirements for geese **45**
    requirements for Japanese quail **47**
    requirements for Leghorns **37**, **39**

requirements for ring-necked pheasants **46**

requirements for turkeys **43**

sources of 112, **113**

standards/regulations for 50–51, **54**, 187–188

see also specific trace minerals

traditional/heritage breeds 201, 206, 254

dual-purpose 210, 211

ducks/geese 183, 213

local, lists of 203, 210

and NRC dietary standards 36, 287

turkeys 179, 212, *212*

triglycerides 83, 102, 107, 214

triticale (*Triticale hexaploide/tetraploide*) 58, 70–71, **166**, 227

'Bogo'/'Grado' cultivars 71

Triticum aestivum see wheat

T. aestivum var. *spelta* 70

trypsin inhibitors 19, 86, 90, 91, 95–96, 97, 107

tryptophan 3, 21, 29, 103

in cereals 58, 64, 72, 74

dietary requirements for **23**, **24**, **30**, **37**, **39**, **42**, **43**, **45**, **47**

in legume seeds 95, 97

in oilseeds 76, **78**, 80, 86, 88, 90

Tsaiya duck 213

tubers **52**, 102–105

Turkey X disease 240

turkeys 32, 201, 211–212, 235

breeds/strains of 212

and canola 79

and choice-feeding 228–229

and copper 27

and cottonseed meal 81

feed formulas for 179–181, **197**

and fishmeal 111

heritage breeds 179, 212, *212*

nutrient requirements for 23, **24**, **43–44**, 188, **189**

and oats 68

and peas 98

premixes for **187**, **188**

and temperature factor 223

and triticale 71

tyrosine 22, 105

dietary requirements for **24**, **37**, **39**, **42**, **43**, **47**

United States (USA) 69, 70, 80, 82, 84, 86, 87, 89, 90, 99, 202, 211, 215

egg-laying breeds in 206–207

food safety in 244

organic standards/regulations in 8, 9, 51, 112

selenium deficiency in 26

urine/uric acid 16–17, 22, 29

vaccination 237, 239

valine 21, 65, 74

dietary requirements for **23**, **24**, **37**, **39**, **42**, **43**, **45**, **47**

Vicia faba 89, 95–96, **137**

Vietnam 89, 229

vitamins 2, 5, 10, 11, 29–34, 107, 128

in cereals 58, 64

classification/characteristics of 29, **30**

digestion of 16, 17

and ducks/geese 181

and feed analysis 35

in legume seeds 95

in milk products 109, 111

in oilseeds 78, 80, 87, 88, 93

premixes **175**, 185, **185**, **186**, 187–188, **187**, **188**

requirements for Bobwhite quail **48**

requirements for broilers **42**

requirements for ducks **46**

requirements for geese **45**

requirements for Japanese quail **47**

requirements for Leghorns **38**, **39–40**

requirements for ring-necked pheasants **46**

requirements for turkeys **44**

responses to deficiency 33–34

sources of 112–114

standards/regulations for 36, 50–51, **54**, 187–188

supplements 27, 32, 56, 253–254

water-/fat-soluble 16, 17, 29, 30–33, **30**, 109, 112, 114

see also specific vitamins

water, drinking 33, 34, 110, 222, 242

waterfowl 5, 93, 212–214

see also ducks; geese

Weende System 34–35

welfare considerations 1, 2, 4, 7, 201, 202–203

wheat (*Triticum aestivum*) 18, 19, 23, 25, 32, 58, 70, 71–75, 228, 229–230

by-products 73–76

nutritional features of 72, **167**

variability in **57**, 72

wheat bran 21, 33, 73–74

substitutes for 62

wheat flour 73

wheat germ meal 74

wheat malt sprouts 62

wheat middlings (pollards) 21, 33, 74–75, **168**

wheat mill run 74

wheat red dog 74

wheat shorts 75, **169**

whey 109–110, **170–171**

WHO (World Health Organization) 3, 5

whole-grain feeding 15, 33, 59, 175, 225
    gradual introduction of 226, 227, 228
    and health 239, 241–242
World Trade Organization 6
Wyandotte, White 211

xanthophylls 65, 66, 73, 106
xylanase 69, 73, 99, 115, **115**

yeasts 28, 32, **55**, 116, 244
yolk colour 58, 63, 66, 68, 71, 73, 98, 103, 106, 229
yolk discoloration 80, 81

Zea mays see maize
zinc 24, 25, 26, 28–29, **54**, 77, 90, 108, **113**, 175
    dietary requirements for **37**, **39**, **42**, **43**, **46**, **47**
    and disease control 244

CABI – who we are and what we do

This book is published by **CABI**, an international not-for-profit organisation that improves people's lives worldwide by providing information and applying scientific expertise to solve problems in agriculture and the environment.

CABI is also a global publisher producing key scientific publications, including world renowned databases, as well as compendia, books, ebooks and full text electronic resources. We publish content in a wide range of subject areas including: agriculture and crop science / animal and veterinary sciences / ecology and conservation / environmental science / horticulture and plant sciences / human health, food science and nutrition / international development / leisure and tourism.

The profits from CABI's publishing activities enable us to work with farming communities around the world, supporting them as they battle with poor soil, invasive species and pests and diseases, to improve their livelihoods and help provide food for an ever growing population.

CABI is an international intergovernmental organisation, and we gratefully acknowledge the core financial support from our member countries (and lead agencies) including:

 UKaid
from the British people

Ministry of Agriculture
People's Republic of China

 Australian Government
Australian Centre for
International Agricultural Research

  Agriculture and
Agri-Food Canada

 Ministry of Foreign Affairs of the
Netherlands

 Schweizerische Eidgenossenschaft
Confédération suisse
Confederazione Svizzera
Confederaziun svizra
Swiss Agency for Development
and Cooperation SDC

Discover more

To read more about CABI's work, please visit: **www.cabi.org**

Browse our books at: **www.cabi.org/bookshop**,
or explore our online products at: **www.cabi.org/publishing-products**

Interested in writing for CABI? Find our author guidelines here:
www.cabi.org/publishing-products/information-for-authors/